Deep Learning-Based Machinery
Fault Diagnostics

Deep Learning-Based Machinery Fault Diagnostics

Editors

Hongtian Chen
Kai Zhong
Guangtao Ran
Chao Cheng

MDPI • Basel • Beijing • Wuhan • Barcelona • Belgrade • Manchester • Tokyo • Cluj • Tianjin

Editors

Hongtian Chen
Department of Chemical
and Materials Engineering,
University of Alberta,
Edmonton, AB T6G 1H9, Canada

Kai Zhong
Institutes of Physical Science
and Information Technology,
Anhui University,
Anhui 230601, China

Guangtao Ran
Department of Control
Science and Engineering,
Harbin Institute of Technology,
Harbin 150001, China

Chao Cheng
School of Computer Science
and Engineering, Changchun
University of Technology,
Changchun 130012, China

Editorial Office
MDPI
St. Alban-Anlage 66
4052 Basel, Switzerland

This is a reprint of articles from the Special Issue published online in the open access journal *Machines* (ISSN 2075-1702) (available at: https://www.mdpi.com/journal/machines/special_issues/dl_faul).

For citation purposes, cite each article independently as indicated on the article page online and as indicated below:

LastName, A.A.; LastName, B.B.; LastName, C.C. Article Title. *Journal Name* **Year**, *Volume Number*, Page Range.

ISBN 978-3-0365-5173-9 (Hbk)
ISBN 978-3-0365-5174-6 (PDF)

© 2022 by the authors. Articles in this book are Open Access and distributed under the Creative Commons Attribution (CC BY) license, which allows users to download, copy and build upon published articles, as long as the author and publisher are properly credited, which ensures maximum dissemination and a wider impact of our publications.

The book as a whole is distributed by MDPI under the terms and conditions of the Creative Commons license CC BY-NC-ND.

Contents

About the Editors vii

Hongtian Chen, Kai Zhong, Guangtao Ran and Chao Cheng
Deep Learning-Based Machinery Fault Diagnostics
Reprinted from: *Machines* **2022**, *10*, 690, doi:10.3390/machines10080690 1

Qunhong Tian, Tao Wang, Bing Liu and Guangtao Ran
Thruster Fault Diagnostics and Fault Tolerant Control for Autonomous Underwater Vehicle with Ocean Currents
Reprinted from: *Machines* **2022**, *10*, 582, doi:10.3390/machines10070582 5

Hong Zheng, Keyuan Zhu, Chao Cheng and Zhaowang Fu
Fault Detection for High-Speed Trains Using CCA and Just-in-Time Learning
Reprinted from: *Machines* **2022**, *10*, 526, doi:10.3390/machines10070526 23

Lu Qian, Qing Pan, Yaqiong Lv and Xingwei Zhao
Fault Detection of Bearing by Resnet Classifier with Model-Based Data Augmentation
Reprinted from: *Machines* **2022**, *10*, 521, doi:10.3390/machines10070521 37

Xiangyu Peng, Yalin Wang, Lin Guan and Yongfei Xue
A Local Density-Based Abnormal Case Removal Method for Industrial Operational Optimization under the CBR Framework
Reprinted from: *Machines* **2022**, *10*, 471, doi:10.3390/machines10060471 53

Shuyue Guan, Darong Huang, Shenghui Guo, Ling Zhao and Hongtian Chen
An Improved Fault Diagnosis Approach Using LSSVM for Complex Industrial Systems
Reprinted from: *Machines* **2022**, *10*, 443, doi:10.3390/machines10060443 67

Rongqiang Zhao and Xiong Hu
An Adaptive Fusion Convolutional Denoising Network and Its Application to the Fault Diagnosis of Shore Bridge Lift Gearbox
Reprinted from: *Machines* **2022**, *10*, 424, doi:10.3390/machines10060424 89

Zhongda Lu, Chunda Zhang, Fengxia Xu, Zifei Wang and Lijing Wang
Fault Detection for Interval Type-2 T-S Fuzzy Networked Systems via Event-Triggered Control
Reprinted from: *Machines* **2022**, *10*, 347, doi:10.3390/machines10050347 105

Zhigang Li, Zhijie Zhou, Jie Wang, Wei He and Xiangyi Zhou
Health Assessment of Complex System Based on Evidential Reasoning Rule with Transformation Matrix
Reprinted from: *Machines* **2022**, *10*, 250, doi:10.3390/machines10040250 123

Lijing Wang, Chunda Zhang, Juan Zhu and Fengxia Xu
Fault Diagnosis of Motor Vibration Signals by Fusion of Spatiotemporal Features
Reprinted from: *Machines* **2022**, *10*, 246, doi:10.3390/machines10040246 151

Jiarui Cui, Peining Wang, Xiangquan Li, Ruoyu Huang, Qing Li, Bin Cao and Hui Lu
Multipoint Feeding Strategy of Aluminum Reduction Cell Based on Distributed Subspace Predictive Control
Reprinted from: *Machines* **2022**, *10*, 220, doi:10.3390/machines10030220 169

Xiaoyu Cheng, Shanshan Liu, Wei He, Peng Zhang, Bing Xu, Yawen Xie and Jiayuan Song
A Model for Flywheel Fault Diagnosis Based on Fuzzy Fault Tree Analysis and Belief Rule Base
Reprinted from: *Machines* **2022**, *10*, 73, doi:10.3390/machines10020073 185

Pu Yang, Chenwan Wen, Huilin Geng and Peng Liu
Intelligent Fault Diagnosis Method for Blade Damage of Quad-Rotor UAV Based on Stacked Pruning Sparse Denoising Autoencoder and Convolutional Neural Network
Reprinted from: *Machines* **2021**, *9*, 360, doi:10.3390/machines9120360 209

Shubin Wang, Yukun Tian, Xiaogang Deng, Qianlei Cao, Lei Wang and Pengxiang Sun
Disturbance Detection of a Power Transmission System Based on the Enhanced Canonical Variate Analysis Method
Reprinted from: *Machines* **2021**, *9*, 272, doi:10.3390/machines9110272 229

Chen Xu and Yawen Mao
Auxiliary Model-Based Multi-Innovation Fractional Stochastic Gradient Algorithm forHammerstein Output-Error Systems
Reprinted from: *Machines* **2021**, *9*, 247, doi:10.3390/machines9110247 245

Ning Chen, Fuhai Hu, Jiayao Chen, Zhiwen Chen, Weihua Gui and Xu Li
A Process Monitoring Method Based on Dynamic Autoregressive Latent Variable Model and Its Application in the Sintering Process of Ternary Cathode Materials
Reprinted from: *Machines* **2021**, *9*, 229, doi:10.3390/machines9100229 261

About the Editors

Hongtian Chen

Hongtian Chen (Member, IEEE) received the B.S. and M.S. degrees in School of Electrical and Automation Engineering from Nanjing Normal University, China, in 2012 and 2015, respectively; and he received the Ph.D. degree in College of Automation Engineering from Nanjing University of Aeronautics and Astronautics, China, in 2019. He had ever been a Visiting Scholar at the Institute for Automatic Control and Complex Systems, University of Duisburg-Essen, Germany, in 2018. Now he is a Post-Doctoral Fellow with the Department of Chemical and Materials Engineering, University of Alberta, Canada. His research interests include process monitoring and fault diagnosis, data mining and analytics, machine learning, and quantum computation; and their applications in high-speed trains, new energy systems, and industrial processes.

Dr. Chen was a recipient of the Grand Prize of Innovation Award of Ministry of Industry and Information Technology of the People's Republic of China in 2019, the Excellent Ph.D. Thesis Award of Jiangsu Province in 2020, and the Excellent Doctoral Dissertation Award from Chinese Association of Automation (CAA) in 2020. He currently serves as Associate Editors and Guest Editors of a number of scholarly journals such as IEEE Transactions on Instrumentation and Measurement, IEEE Transactions on Neural Networks and Learning Systems and IEEE Transactions on Artificial Intelligence.

Kai Zhong

Kai Zhong (Professor) received the Ph.D. from Dalian University of Technology in 2020. Now he is an associate professor with the Institutes of Physical Science and Information Technology, Anhui University. His main research interests include process monitoring, fault diagnosis and deep neural networks.

Guangtao Ran

Guangtao Ran received the B.E. and M.E. degrees from Qiqihar University, Qiqihar, China, in 2016 and 2019, respectively. He is currently pursuing the Ph.D. degree with the Department of Control Science and Engineering, Harbin Institute of Technology, Harbin, China. Now he is also a joint training student with the Department of Electrical and Computer Engineering, University of Alberta, Edmonton, AB T6G 1H9, Canada.

His research interests include fuzzy control, reinforcement learning, networked control systems, multi-agent systems, and robust control.

Chao Cheng

Chao Cheng (Professor) received the M.Eng. and Ph.D. degrees from Jilin University, Changchun, China, in 2011 and 2014, respectively. He is currently a professor with the Changchun University of Technology, Changchun. He has been a Post-Doctoral Fellow in process control engineering with the Department of Automation, Tsinghua University, Beijing, China, since 2018. He has also been a Post-Doctoral Fellow with the National Engineering Laboratory, CRRC Changchun Railway Vehicles Co., Ltd., China, since 2018. His research interest includes dynamic system fault diagnosis and predictive maintenance, wireless sensor network, artificial intelligence, and data-driven method.

Editorial
Deep Learning-Based Machinery Fault Diagnostics

Hongtian Chen [1,*], Kai Zhong [2], Guangtao Ran [1,3] and Chao Cheng [4]

1 Department of Chemical and Materials Engineering, University of Alberta, Edmonton, AB T6G 1H9, Canada
2 Institutes of Physical Science and Information Technology, Anhui University, Hefei 230601, China
3 Department of Control Science and Engineering, Harbin Institute of Technology, Harbin 150001, China
4 School of Computer Science and Engineering, Changchun University of Technology, Changchun 130012, China
* Correspondence: chtbaylor@163.com

Citation: Chen, H.; Zhong, K.; Ran, G.; Cheng, C. Deep Learning-Based Machinery Fault Diagnostics. *Machines* 2022, *10*, 690. https://doi.org/10.3390/machines10080690

Received: 8 August 2022
Accepted: 8 August 2022
Published: 13 August 2022

Publisher's Note: MDPI stays neutral with regard to jurisdictional claims in published maps and institutional affiliations.

Copyright: © 2022 by the authors. Licensee MDPI, Basel, Switzerland. This article is an open access article distributed under the terms and conditions of the Creative Commons Attribution (CC BY) license (https://creativecommons.org/licenses/by/4.0/).

In recent years, deep learning has shown its unique potential and advantages in feature extraction and pattern recognition. The application of deep learning to fault diagnosis of complex machinery systems has begun its initial exploration stage. This Special Issue provides an international forum for professionals, academics, and researchers to present the latest developments from theoretical studies and computational algorithm development to applications of advanced deep learning-based machinery system fault diagnosis methods. The contents of these studies are briefly described as follows.

In [1], a possibilistic fuzzy C-means (PFCM) algorithm was proposed to realize the fault classification. Based on the results of fault diagnostics, a fuzzy control strategy was used to solve the fault tolerant control for AUV. Considering the uncertainty of ocean currents, a min-max robust optimization strategy was carried out to optimize the fuzzy controller, which was solved by a cooperative particle swarm optimization (CPSO) algorithm. Simulation and underwater experiments were used to verify the accuracy and feasibility of the proposed method in fault diagnostics and fault-tolerant control.

In [2], the authors proposed a fault detection (FD) model, named as CCA-JITL by using canonical correlation analysis (CCA) and just-in-time learning (JITL) to process scalar signals of high-speed train gears. After data pre-processing and normalization, CCA transformed covariance matrices of high-dimension historical data into low-dimension subspace and maximized correlations between the most important latent dimensions. Then, JITL components formulated the local FD model by utilizing the subsets of testing samples with larger Euclidean distances to training data. A case study demonstrated that a CCA-JITL FD model significantly outperformed traditional CCA models. The proposed approach can also be integrated with other dimension reduction FD models, such as the principal component analysis and partial least squares models.

In [3], the authors designed a Resnet-based classifier with the model-based data augmentation skill, which was applied for bearing fault detection. In particular, a dynamic model was first established to describe the bearing system by adjusting model parameters, such as speed, load, fault size, and the different fault types. Large amounts of data under various operation conditions can then be generated. The training dataset was constructed through the simulated data, which was then applied to train the Resnet classifier. Moreover, in order to reduce the gap between the simulation data and the real data, the envelop signals were used instead of the original signals in the training process. Finally, the effectiveness of the proposed method was demonstrated by the real bearing data. It was remarkable that the application of the proposed method can be further extended to other mechatronic systems with a deterministic dynamic model.

In [4], a local density-based abnormal case removal method was proposed to remove the abnormal cases so as to prevent performance deterioration in industrial operational optimization. More specifically, the reasons why classic case-based reasoning (CBR) would retrieve abnormal cases were analyzed from the perspective of case retrieval. Then, a local density-based abnormal case removal algorithm was designed based on the local outlier factor (LOF) and properly integrated into the traditional case retrieval step. Finally, the effectiveness and the superiority of the local density-based abnormal case removal method was tested on a numerical simulation and the cut-made process of cigarette production. The results showed that the proposed method improved the operational optimization performance of an industrial cut-made process by 23.5% compared with classic CBR and 13.3% compared with case-based fuzzy reasoning.

In [5], in order to improve the performance of fault diagnosis, the authors designed a novel approach by using particle swarm optimization (PSO) with wavelet mutation and least square support vector machine (LSSVM). The implementation process can be concluded as adhering to the following three steps. Firstly, the original signals were decomposed through an orthogonal wavelet packet decomposition. Secondly, the decomposed signals were reconstructed to obtain the fault features. Finally, the extracted features were used as the inputs of the fault diagnosis model. This joint optimization method not only solved the problem that PSO is easy to fall into a local optimum but also improved the classification performance of fault diagnosis effectively. Through experimental verification, the wavelet mutation particle swarm optimization and least square support vector machine (WMPSO-LSSVM) fault diagnosis model has a maximum fault recognition efficiency that was 12% higher than LSSVM and 9% higher than extreme learning machine (ELM). The error of the corresponding regression model under the WMPSO-LSSVM algorithm was 0.365 less than that of the traditional linear regression model.

In [6], traditional fault diagnosis methods were limited in the condition detection of shore bridge lifting gearboxes due to their limited ability to extract signal features and their sensitivity to noises. In order to solve this problem, an adaptive fusion convolutional denoising network (AF-CDN) was proposed in this paper. First, a novel 1D and 2D adaptive fused convolutional neural network structure was built. The fusion of both the 1D and 2D convolutional models can effectively improve the feature extraction capability of the network. Then, a gradient updating method based on the Kalman filter mechanism was designed. Finally, the effectiveness of the developed method was evaluated by using the benchmark datasets and the actual data collected for the shore bridge lift gearbox.

In [7], the authors investigated the event-triggered fault diagnosis (FD) problem. Firstly, an FD fuzzy filter was proposed by using IT2 T-S fuzzy theory to generate a residual signal. The evaluation functions were referenced to determine the occurrence of system faults. Secondly, under the event-triggered mechanism, a fault residual system (FRS) was established with parameter uncertainties, external disturbances and time delays, which can reduce signal transmission and communication pressures. Thirdly, the stability conditions of the faulty residual system were proposed by using the Lyapunov theory. For the energy bounded condition of external noise interference, the performance criterion was established by linear matrix inequalities. The matrix parameters of the target FD filter were obtained via a convex optimization method. Finally, the simulation examples were provided to illustrate the effectiveness and the practicalities of the proposed method.

In [8], the authors thought that the relationship between the indicator reference grades and pre-defined assessment result grades was regarded as a one-to-one correspondence. However, in engineering practice, this strict mapping relationship was difficult to meet. Therefore, a new evidential reasoning (ER) rule-based health assessment model for complex systems with a transformation matrix was adopted. First, on the basis of the rule-based transformation technique, expert knowledge was embedded on the transformation matrix to solve the inconsistent problems between the input and the outputs, which keeps the completeness and consistency of information transformation. Second, a complete health assessment model was established via the calculation and optimization of the model parameters. Finally, the effectiveness of the proposed model was validated in contrast with other methods.

In [9], the authors constructed a spatiotemporal feature fusion network (STNet) to enhance the influence of signal spatiotemporal features on the diagnostic performance during motor fault diagnosis. The network used dual-stream branching to extract the fault features of motor vibration signals via a convolutional neural network and gated recurrent unit (GRU) simultaneously. The features were also enhanced by using the attention mechanism. Then, the temporal and spatial features were fused and input into the SoftMax function for fault discrimination. After that, the fault diagnosis of motor vibration signals was completed. In addition, several sets of experimental evaluations were conducted to verify the effectiveness of the proposed method.

In [10], a data-driven distributed subspace predictive control feeding strategy was proposed. Firstly, the aluminum reduction cell was divided into multiple sub-systems that affect each other according to the position of the feeding port. Based on the subspace method, the prediction model of the whole cell was identified, and the prediction output expression of each sub-system was deduced by decomposition. Secondly, the feeding controller was designed for each aluminum reduction cell subsystem, and the input and output information can be exchanged between each controller through the network. Thirdly, with consideration of the influence of other subsystems, each subsystem solved the Nash-optimal control feeding quantity so that each subsystem realized distributed feeding. Finally, the simulation results showed that the proposed strategy can significantly improve the problem of the uniform distribution of alumina concentration.

In [11], a new belief rule base (BRB) model, called the FFBRB (fuzzy fault tree analysis and belief rule base) was given, which solved the problems existing in the BRB effectively. The FFBRB used the Bayesian network as a bridge, used the FFTA (fuzzy fault tree analysis) mechanism to build the BRB's expert knowledge, used ER (evidential reasoning) as its reasoning tool, and used P-CMA-ES (projection covariance matrix adaptation evolutionary strategies) as its optimization model algorithm. The feasibility and superiority of the proposed method were verified by an example of a flywheel friction torque fault tree.

In [12], the authors introduced a new intelligent fault diagnosis method based on stack pruning sparse denoising autoencoder and convolutional neural network (sPSDAE-CNN). Firstly, a one-dimensional sliding window was introduced for data enhancement. In addition, transforming one-dimensional time-domain data into a two-dimensional gray image can further improve the learning ability of models. At the same time, pruning operation was introduced to improve the training efficiency and accuracy of the network. Actual experiments showed that for the fault of unmanned aerial vehicle (UAV) blade damage, the sPSDAE-CNN model the authors used has better stability and reliable prediction accuracy than traditional convolutional neural networks. The experimental results showed that the sPSDAE-CNN model still has a good diagnostic accuracy rate in high-noise environment. In the case of a signal-to-noise ratio of −4, it still has an accuracy rate of 90%.

In [13], aiming at the characteristics of dynamic correlation, periodic oscillation, and weak disturbance symptom of power transmission system data, an enhanced canonical variate analysis (CVA) method, called SLCVAkNN was presented. In the proposed method, CVA was first used to extract the dynamic features by analyzing the data correlation and established a statistical model with two monitoring statistics. Then, in order to handle the periodic oscillation of power data, the two statistics were reconstructed in phase space, and the k-nearest neighbor (kNN) technique was applied to design the nearest neighbor distance as the enhanced monitoring indices. Further considering the detection difficulty of weak disturbances with the insignificant symptoms, statistical local analysis (SLA) was integrated to construct the primary and improved residual vectors of the CVA dynamic features. The verification results on the real industrial data showed that the SLCVAkNN method can detect the occurrence of power system disturbance more effectively than the traditional data-driven monitoring methods.

In [14], the authors proposed an auxiliary model-based multi-innovation fractional stochastic gradient method. The scalar innovation was extended to the innovation vector for increasing data based on the multi-innovation identification theory. By establishing appropriate auxiliary models, the unknown variables were estimated and the improvement in the performance of parameter estimation was achieved owing to the fractional-order calculus theory. Compared with the conventional multi-innovation stochastic gradient algorithm, the proposed method was validated to obtain better estimation accuracy through the simulation results.

In [15], a process monitoring method based on the dynamic autoregressive latent variable model was proposed in this paper. First, from the perspective of process data, a dynamic autoregressive latent variable model (DALM) with process variables as input and quality variables as output was constructed to adapt to the variable time lag characteristic. In addition, a fusion of Bayesian filtering, smoothing and expectation maximization algorithm was used to identify model parameters. Then, the process monitoring method based on DALM was constructed, in which the process data were filtered online to obtain the latent space distribution of the current state, and two statistics were constructed. Finally, by comparing with the existing methods, the feasibility and effectiveness of the proposed method were tested on the sintering process of ternary cathode materials. Detailed comparisons were given to show the superiority of the proposed method.

As guest editors of this Special Issue, we would like to thank all of the authors for their contributions. We wish that the readers can benefit from the above fifteen papers. We would like to thank *Machines* for giving us the opportunity to serve as the guest editor for the Special Issue. Finally, we would like to thank the reviewers for their excellent job on evaluating these papers.

Funding: This research received no external funding.

Conflicts of Interest: The authors declare no conflict of interest.

References

1. Tian, Q.; Wang, T.; Liu, B.; Ran, G. Thruster Fault Diagnostics and Fault Tolerant Control for Autonomous Underwater Vehicle with Ocean Currents. *Machines* **2022**, *10*, 582. [CrossRef]
2. Zheng, H.; Zhu, K.; Cheng, C.; Fu, Z. Fault Detection for High-Speed Trains Using CCA and Just-in-Time Learning. *Machines* **2022**, *10*, 526. [CrossRef]
3. Qian, L.; Pan, Q.; Lv, Y.; Zhao, X. Fault Detection of Bearing by Resnet Classifier with Model-Based Data Augmentation. *Machines* **2022**, *10*, 521. [CrossRef]
4. Peng, X.; Wang, Y.; Guan, L.; Xue, Y. A Local Density-Based Abnormal Case Removal Method for Industrial Operational Optimization under the CBR Framework. *Machines* **2022**, *10*, 471. [CrossRef]
5. Guan, S.; Huang, D.; Guo, S.; Zhao, L.; Chen, H. An Improved Fault Diagnosis Approach Using LSSVM for Complex Industrial Systems. *Machines* **2022**, *10*, 443. [CrossRef]
6. Zhao, R.; Hu, X. An Adaptive Fusion Convolutional Denoising Network and Its Application to the Fault Diagnosis of Shore Bridge Lift Gearbox. *Machines* **2022**, *10*, 424. [CrossRef]
7. Lu, Z.; Zhang, C.; Xu, F.; Wang, Z.; Wang, L. Fault Detection for Interval Type-2 TS Fuzzy Networked Systems via Event-Triggered Control. *Machines* **2022**, *10*, 347. [CrossRef]
8. Li, Z.; Zhou, Z.; Wang, J.; He, W.; Zhou, X. Health Assessment of Complex System Based on Evidential Reasoning Rule with Transformation Matrix. *Machines* **2022**, *10*, 250. [CrossRef]
9. Wang, L.; Zhang, C.; Zhu, J.; Xu, F. Fault Diagnosis of Motor Vibration Signals by Fusion of Spatiotemporal Features. *Machines* **2022**, *10*, 246. [CrossRef]
10. Cui, J.; Wang, P.; Li, X.; Huang, R.; Li, Q.; Cao, B.; Lu, H. Multipoint Feeding Strategy of Aluminum Reduction Cell Based on Distributed Subspace Predictive Control. *Machines* **2022**, *10*, 220. [CrossRef]
11. Cheng, X.; Liu, S.; He, W.; Zhang, P.; Xu, B.; Xie, Y.; Song, J. A Model for Flywheel Fault Diagnosis Based on Fuzzy Fault Tree Analysis and Belief Rule Base. *Machines* **2022**, *10*, 73. [CrossRef]
12. Yang, P.; Wen, C.; Geng, H.; Liu, P. Intelligent Fault Diagnosis Method for Blade Damage of Quad-rotor UAV Based on Stacked Pruning Sparse Denoising Autoencoder and Convolutional Neural Network. *Machines* **2021**, *9*, 360. [CrossRef]
13. Wang, S.; Tian, Y.; Deng, X.; Cao, Q.; Wang, L.; Sun, P. Disturbance Detection of a Power Transmission System Based on the Enhanced Canonical Variate Analysis Method. *Machines* **2021**, *9*, 272. [CrossRef]
14. Xu, C.; Mao, Y. Auxiliary Model-based Multi-innovation Fractional Stochastic Gradient Algorithm for Hammerstein Output-error Systems. *Machines* **2021**, *9*, 247. [CrossRef]
15. Chen, N.; Hu, F.; Chen, J.; Chen, Z.; Gui, W.; Li, X. A Process Monitoring Method Based on Dynamic Autoregressive Latent Variable Model and Its Application in the Sintering Process of Ternary Cathode Materials. *Machines* **2021**, *9*, 229. [CrossRef]

Article

Thruster Fault Diagnostics and Fault Tolerant Control for Autonomous Underwater Vehicle with Ocean Currents

Qunhong Tian [1,*], Tao Wang [1], Bing Liu [1] and Guangtao Ran [2,3]

1. College of Mechanical and Electronic Engineering, Shandong University of Science and Technology, Qingdao 266590, China; wangt@sdust.edu.cn (T.W.); metrc@sdust.edu.cn (B.L.)
2. Department of Control Science and Engineering, Harbin Institute of Technology, Harbin 150001, China; ranguangtao@hit.edu.cn
3. Department of Electrical and Computer Engineering, University of Alberta, Edmonton, AB T6G 1H9, Canada
* Correspondence: tianqunhong@sdust.edu.cn

Abstract: Autonomous underwater vehicle (AUV) is one of the most important exploration tools in the ocean underwater environment, whose movement is realized by the underwater thrusters, however, the thruster fault happens frequently in engineering practice. Ocean currents perturbations could produce noise for thruster fault diagnosis, in order to solve the thruster fault diagnostics, a possibilistic fuzzy C-means (PFCM) algorithm is proposed to realize the fault classification in this paper. On the basis of the results of fault diagnostics, a fuzzy control strategy is proposed to solve the fault tolerant control for AUV. Considering the uncertainty of ocean currents, it proposes a min-max robust optimization problem to optimize the fuzzy controller, which is solved by a cooperative particle swarm optimization (CPSO) algorithm. Simulation and underwater experiments are used to verify the accuracy and feasibility of the proposed method of thruster fault diagnostics and fault tolerant control.

Keywords: autonomous underwater vehicle; thruster fault diagnostics; fault tolerant control; robust optimization; ocean currents

1. Introduction

An autonomous underwater vehicle (AUV) is one of the most important exploration tools in the ocean underwater environment. As an important part of AUV, the thruster directly determines the efficiency and safety with strong working intensity for AUV, However, the thruster fault usually happens in engineering practice [1,2]. Therefore, how to make thruster fault diagnosis and fault tolerant control for AUV is the premise for completing underwater missions [3,4].

There have been many works applied to AUV fault diagnosis. A Gaussian particle filtering algorithm is presented to estimate the AUV failure model, the Bayes algorithm is used to realize the AUV thruster fault detection [5]. For solving the fault diagnosis of AUV actuators, a diagnostic network is proposed based on extreme learning and a wide convolutional neural network [6]. Through experimental data analysis, a feature calculation method is presented to solve the weak faults thruster faults, which provides accurate and concise information for fault severity identification [7]. A fault diagnosis method is presented based on deep learning and attention mechanism for AUV, a data attention mechanism is developed for realizing dynamic decorrelation, multi-layer perceptron is used for fault detection [8]. From training datasets gathered in previous AUV operations directly, the Bayesian nonparametric technique is used for modelling the vehicle's performance including faults, in the light of the Kullback–Leibler divergence measure, a nearest-neighbor classifier is used to accomplish the fault diagnosis [9]. In summary, the above studies have given some methods to solve the AUV fault diagnosis. However, ocean currents perturbations could produce noise for thruster fault diagnosis, the above methods are difficult to be used for AUV fault diagnosis with ocean currents in practice effectively. The above methods also do not consider how to control AUV to complete the underwater missions with minor faults.

Fault tolerant control is the technology to ensure the AUV for completing the underwater mission with faults [10,11]. In order to realize the fault tolerant control, it develops a model-parameter-free control strategy for AUV trajectory tracking, tracking controller is designed through the employment of sliding mode control technology without utilizing model parameters. However, the sliding mode control easily lead to the chattering of the AUV control system [12]. In order to solve the problem of thruster fault tolerant control for AUV, a fault tolerant control method is proposed in the light of the sliding mode theory, the adaptive law is developed for the proposed controller to mitigate the chattering phenomenon [13]. In order to further improve the performance of the fault tolerant control, some intelligent methods are investigated [4,14,15]. An iterative learning algorithm is proposed to process the propeller failure for AUV based on an extended state observer, a fuzzy logic controller is introduced to deal with the fuzzification of the parameters of a saturated proportional-derivative controller and extended state observer [14]. Combined with the backstepping method, a single critic network based on adaptive dynamic programming is used to deal with the AUV fault tolerant control. It designs an online policy iteration algorithm in light of the estimated system states [4]. To further conduct the effect of the ocean currents, the fault tolerant issue is transformed into an optimal control problem by the adaptive dynamic programming method, the neural-network estimator is developed to estimate ocean currents [15], however, it is difficult to establish the ocean current accurately in practice. In summary, although the above research has given some methods for fault tolerant control for AUV, they are difficult to be used in an environment with ocean currents.

Ocean currents perturbations could produce noise for thruster fault diagnosis. In this paper, in order to solve the problem of the thruster fault diagnostics and fault tolerant control for AUV with ocean currents, the possibilistic fuzzy C-means (PFCM) algorithm is proposed for realizing the thruster fault diagnostics effectively. Once the thruster fault is diagnosed, based on the fault diagnosis results, a fault tolerant control is presented by the fuzzy controller, to improve the performance of the fuzzy controller, a robust optimization problem is proposed by considering the uncertainty of ocean currents, which is solved by the proposed co-evolutionary (CPSO) algorithm, finally, it forms a mechanism of diagnostics and control strategy to accomplish the missions.

The rest of this paper is given as follows. Section 2 presents the AUV mathematical models; Section 3 gives the algorithm for AUV fault diagnostics and fault tolerant control; Section 4 discusses the effectiveness of the proposed method based on different scenarios; Section 5 concludes the paper.

2. Mathematical Models of AUV

In this section, the problem description is given for the AUV firstly, and then the AUV models are discussed.

2.1. Problem Description

AUV works in a complex marine environment, which is a complex dynamic system with strong nonlinearity. Due to the complexity and unpredictability of the marine environment, thrusters are easy to fail. However, when the thruster fails, the expected task of the AUV cannot be completed, or AUV may be destroyed directly, which will cause extremely serious losses and may pollute the environment.

Thruster fault diagnostics is the premise to solve the above problems, which include the type of motor fault, propeller enwinding by foreign matter, propeller blade damage, thruster idling, and so on. However, AUV is greatly affected by the external disturbance of ocean currents, the external interference and fault are difficult to be separated, which takes great difficulty for AUV fault diagnosis. Meanwhile, the ocean currents increase the difficulty of controlling the AUV to accomplish the missions with thruster fault. Figure 1 gives the design process of thruster fault diagnostics and faults tolerant control for AUV, the PFCM algorithm is proposed to realize the thruster fault diagnostics. Once the thruster fault is diagnosed, based on the fault diagnosis results, a fault tolerant control is presented by the fuzzy controller, in order to improve the performance of the fuzzy controller, a robust optimization problem is proposed by considering the uncertainty of ocean currents, which is solved by the proposed CPSO algorithm, finally, it forms a mechanism of fault diagnostics and tolerant control strategy to accomplish the missions.

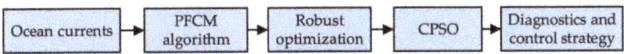

Figure 1. Design process of thruster fault diagnostics and fault tolerant control for AUV.

2.2. AUV KINEMATIC model

Figure 2 gives two coordinate systems for AUV to obtain the kinematic model, one is the earth-frame $\{O - X, Y, Z\}$, the other is the body-fixed frame $\{O_1 - X_1, Y_1, Z_1\}$. The AUV kinematic model can deal with the geometric aspects of motion, which is written in a general form as follows [16–19]:

$$\dot{\eta} = \begin{bmatrix} J_1 & 0_{3\times 3} \\ 0_{3\times 3} & J_2 \end{bmatrix} v \qquad (1)$$

$$\begin{bmatrix} \dot{x} & \dot{y} & \dot{z} & \dot{\phi} & \dot{\theta} & \dot{\psi} \end{bmatrix}^T = \begin{bmatrix} J_1 & 0_{3\times 3} \\ 0_{3\times 3} & J_2 \end{bmatrix} \begin{bmatrix} u & v & w & p & q & r \end{bmatrix}^T \qquad (2)$$

where the vector $\eta = \begin{bmatrix} x & y & z & \phi & \theta & \psi \end{bmatrix}^T$ denotes the position and orientation of AUV in the Earth-frame, x, y, z represent the position, ϕ, θ, ψ are the Euler angles of roll, pitch and yaw angles respectively; $v = \begin{bmatrix} u & v & w & p & q & r \end{bmatrix}$ denotes the translational and rotational velocities in the body-fixed frame, u, v, w are the surge, sway and heave components respectively, p, q, r are the roll, pitch and yaw rates respectively; J_1 and J_2 are the coordinate transformation matrixes, which are given as follows:

$$J_1 = \begin{bmatrix} \cos\theta\cos\psi & \sin\theta\sin\phi\cos\psi - \cos\phi\sin\psi & \sin\theta\sin\psi + \sin\theta\cos\phi\cos\psi \\ \cos\theta\sin\psi & \sin\theta\sin\phi\sin\psi + \cos\phi\cos\psi & \sin\theta\cos\phi\sin\psi - \sin\phi\cos\psi \\ -\sin\theta & \sin\phi\cos\theta & \cos\phi\cos\theta \end{bmatrix} \qquad (3)$$

$$J_2 = \begin{bmatrix} 1 & \sin\phi\tan\theta & \cos\phi\tan\theta \\ 0 & \cos\phi & -\sin\phi \\ 0 & \sin\phi/\cos\theta & \cos\phi/\cos\theta \end{bmatrix} \qquad (4)$$

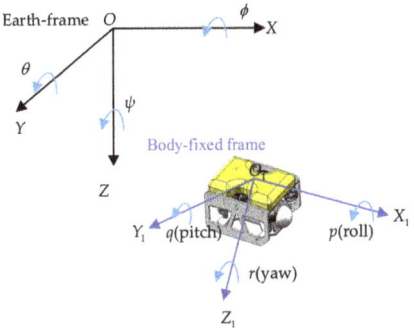

Figure 2. Coordinate systems for AUV.

2.3. AUV Dynamic Model

It can describe the general motion of AUV with six degrees of freedom dynamic equation as follows [17,18,20]:

$$M\dot{v} + C(v)V + D(v)v + g(\eta) = \tau \tag{5}$$

where $M \in R^{6\times 6}$ is the inertial matrix; v is the position and orientation vector; $C \in R^{6\times 6}$ is the matrix of Coriolis and Centripetal terms; $g \in R^{6\times 6}$ is the gravitational terms matrix; $D \in R^{6\times 6}$ is the damping matrix; τ is the control forces vector. Figure 3 shows the planform of the designed AUV, it assumes that the center of gravity is at the same point as the center of buoyancy for AUV, the translational motion and rotational motion are expressed by six equations as follows:

$$\begin{aligned}&m(\dot{u} - vr + wq - x_g(q^2 + r^2) + z_g(pr+q)) \\ &= X_{HS} + X_u|u| + X_{\dot{u}}u + X_{wq}wq + X_{qq}qq + X_{vr}vr + X_{rr}rr + X_{prop}\end{aligned} \tag{6}$$

$$\begin{aligned}&m(\dot{v} - wp + ur - z_g(qr - \dot{p}) + x_g(pq + \dot{r})) \\ &= Y_{HS} + Y_{v|v|}v|v| + Y_{r|r|}r|r| + Y_{\dot{v}}\dot{v} + Y_{\dot{r}}\dot{r} + Y_{ur}ur + Y_{wp}wp + Y_{pq}pq + Y_{uv}uv + Y_{uu\delta_r}u^2\delta_r\end{aligned} \tag{7}$$

$$\begin{aligned}&m(\dot{w} - uq + vp - z_g(q^2 + p^2) + x_g(rp + \dot{q})) \\ &= Z_{HS} + Z_{w|w|}w|w| + Z_{q|q|}q|q| + Z_{\dot{w}}\dot{w} + Z_{\dot{q}}\dot{q} + Z_{uq}uq + Z_{vp}vp + Z_{pr}pr + Z_{uw}uw + Z_{uu\delta_s}u^2\delta_s\end{aligned} \tag{8}$$

$$\begin{aligned}&I_{xx}\dot{p} + (I_{zz} - I_{yy})qr + m|-z_g(\dot{v} - wp + ur)| \\ &= K_{HS} + K_{p|p|}p|p| + k_{p|p|}p|p| + k_{\dot{p}}\dot{p} + k_{prop}\end{aligned} \tag{9}$$

$$\begin{aligned}&I_{yy}\dot{q} + (I_{xx} - I_{zz})pr + m|z_g(\dot{u} - vr + wq) - x_g(\dot{w} - uq + vp)| = \\ &M_{HS} + M_{w|w|}q|q| + M_{q|q|}q|q| + M_{\dot{w}}\dot{w} + M_{\dot{q}}\dot{q} + M_{uq}uq + M_{vp}vp + M_{rp}rp + M_{uw}uw + M_{uu\delta_s}u^2\delta_s\end{aligned} \tag{10}$$

$$\begin{aligned}&I_{zz}\dot{r} + (I_{yy} - I_{xx})qp + m|x_g(\dot{v} - wp + ur)| = \\ &N_{HS} + N_{v|v|}v|v| + N_{r|r|}r|r| + N_{\dot{v}}\dot{v} + N_{\dot{r}}\dot{r} + N_{ur}ur + N_{wp}wp + N_{pq}pq + N_{uv}uv + N_{uu\delta_r}u^2\delta_r\end{aligned} \tag{11}$$

$$F = \begin{bmatrix} 0 & 0 & \cos\beta & \cos\beta & -\cos\beta & -\cos\beta \\ 0 & 0 & \sin\beta & -\sin\beta & \sin\beta & -\sin\beta \\ 0 & 1 & 0 & 0 & 0 & 0 \\ b_v & b_v & \sin\beta \cdot c_h & \sin\beta \cdot c_h & -\sin\beta \cdot c_h & \sin\beta \cdot c_h \\ a_v & a_v & \cos\beta \cdot c_h & \cos\beta \cdot c_h & -\cos\beta \cdot c_h & -\cos\beta \cdot c_h \\ 0 & 0 & B_1 & B_2 & B_3 & B_4 \end{bmatrix} \begin{bmatrix} F_1 \\ F_2 \\ F_3 \\ F_4 \\ F_5 \\ F_6 \end{bmatrix} \tag{12}$$

where $B_1 = \cos\beta \cdot b_h + \sin\beta \cdot a_h$, $B_2 = -\cos\beta \cdot b_h - \sin\beta \cdot a_h$, $B_3 = -\cos\beta \cdot b_h - \sin\beta \cdot a_h$, $B_4 = \cos\beta \cdot b_h - \sin\beta \cdot a_h$. F_1 is the thrust of the left vertical thruster; F_2 is the thrust of the right vertical thruster. F_3 is the thrust of the left front horizontal thruster; F_4 is the thrust of the right front horizontal thruster; F_5 is the thrust of the left rear horizontal thruster thrust; F_6 is the thrust of the right rear horizontal thruster thrust. a_v is the distance between the center of vertical thruster and $X_1O_1Z_1$ plane; b_v is the distance between the center of vertical thruster and $Y_1O_1Z_1$ plane; a_h is the distance between the center of horizontal thruster and $X_1O_1Z_1$ plane; b_h is the distance between the center of horizontal thruster and $Y_1O_1Z_1$ plane; c_h is the distance between the center of horizontal thruster and $X_1O_1Y_1$ plane.

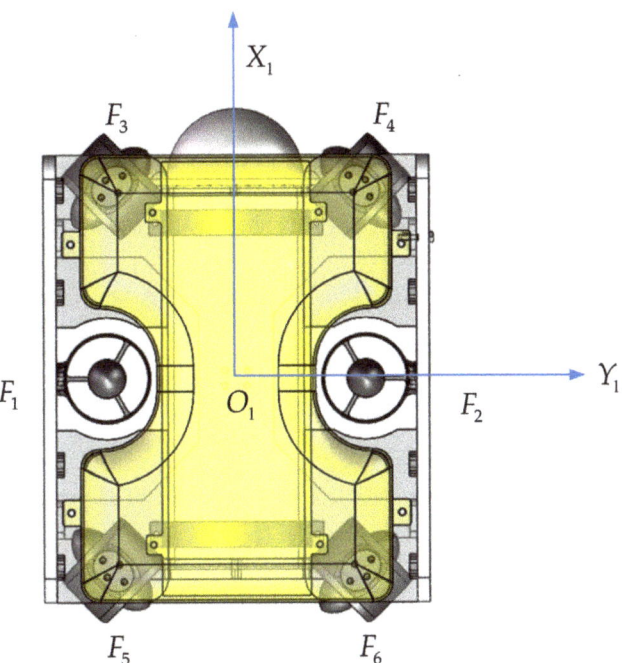

Figure 3. Planform of the designed AUV.

3. Thruster Fault Diagnostics and Fault Tolerant Control

Thruster fault diagnostics and fault tolerant control problem is discussed in this section. Based on the results of thruster fault diagnostics by PFCM algorithm, a fuzzy controller is proposed for AUV fault tolerant control, which is optimized with the proposed CPSO algorithm, the control performance is ensured with the robust optimization design.

3.1. Thruster Fault Diagnostics for AUV

AUV thruster is usually constituted by a motor, reducer, propeller, controller, and so on. The battery pack provides the drive energy, motor, and reducer as the actuator, the output voltage is controlled by the received upper computer instruction by the driver controller, further to control motor speed, motor and speed reducer drive screw rotation. Generally, the speed is proportional to the thrust. The propeller provides the thrust to realize the variable speed sailing of AUV, the drive controller uploads the real-time operation parameters of the propeller to the upper computer. The propeller protection cover is used to avoid the propeller damage caused by the impact of fish or other objects. In this paper, the type of the thruster fault is given as follows: motor fault, propeller enwinding by foreign matter, propeller blade damage, thruster got stuck, thruster idling. The voltage, current, and speed of the thruster are used to judge whether the thruster is faulty.

It is well known that it is difficult for fault diagnosis of nonlinear systems [21,22]. Moreover, ocean currents perturbations could produce noise and further increase the difficulty of thruster fault diagnosis. To solve the thruster diagnosis of AUV nonlinear system with ocean currents, it proposes the PFCM algorithm. PFCM algorithm is one popular clustering method, it is highly sensitive to noise and outliers, and the size of the clusters [23,24]. The algorithm is an unsupervised technique, the data is clustered based on similarities and dissimilarities, which are measured via distances of the cluster centers to the data points. The clustering results are described by introducing membership and probability partition matrixes.

PFCM algorithm is proposed with optimization as follows [25].

$$\min \quad J_M(U,T,V) = \sum_{i=1}^{n}\sum_{j=1}^{c}(au_{ij}^m + bt_{ij}^p)d_{ij}^2 + \sum_{i=1}^{n}\eta_i\sum_{j=1}^{c}(1-t_{ij})^p \qquad (13)$$

$$s.t. \quad \sum_{i=1}^{c} u_{ij} = 1$$

$$\eta_i = K\sum_{j=1}^{n} u_{ij}^m d_{ij}^2 / \sum_{j=1}^{n} u_{ij}^m \qquad (14)$$

where c is the number of clusters; $m > 1$ is the degree of fuzziness; n is the number of data points; a and b are the constants ($a > 0$, $b > 0$), which represent relative importance of fuzzy and possibilistic terms respectively, the larger value of b, the better the ability to resist noise points; $U = [u_{ij}]_{c \times n}$ is the membership degrees matrix ($0 \leq u_{ij}$); $T = [t_{ij}]_{c \times n}$ is the typicality matrix ($t_{ij} \leq 1$); $V = [v_{ij}]_{c \times n}$ is the cluster centers matrix; $p > 1$ is the possibilistic exponent, d_{ij} is the distance between the cluster center (v_i) and data point (x_i); η_i is the penalty factor, K is a constant.

The objective function in Equation (13) can be solved via an iterative procedure as follows:

$$u_{ij} = \frac{1}{\sum_{k=1}^{c}\left(\frac{d_{ij}}{d_{kj}}\right)^{2/m-1}} \qquad (15)$$

$$t_{ij} = \frac{1}{1 + \left(\frac{b}{\eta_i}d_{ij}^2\right)^{\frac{1}{p-1}}} \qquad (16)$$

$$v_i = \frac{\sum_{j=1}^{n}\left(au_{ij}^m + bt_{ij}^p\right)x_j}{\sum_{j=1}^{n}\left(au_{ij}^m + bt_{ij}^p\right)} \qquad (17)$$

according to the above Equations (15)–(17), it can obtain the optimal degree of membership and cluster center.

3.2. Fault Tolerant Control for AUV

Fuzzy theory can describe the uncertainty of the system, it has been used to solve the problem of fault diagnosis and control effectively [26–28]. Therefore, in this paper, fuzzy controller is proposed to solve the fault tolerant control problem for AUV path tracking, the fuzzy control includes the fuzzification, fuzzy inference and defuzzification. For the fuzzification operation, the Gaussian function is selected as the membership function of fuzzy variable; Table 1 gives the fuzzy control rule for fuzzy inference; centroid method is used to realize the defuzzification operation. The input parameters of the fuzzy controller are position error $e(t)$ and its derivative $\dot{e}(t)$, the output parameters of the fuzzy controller are angles and those derivatives. Figures 4 and 5 give the membership functions for position error and its derivative respectively. $\lambda_1, \lambda_2, \lambda_3, \lambda_4, \lambda_5, \lambda_6, \lambda_7$ are the mean of the normal distribution for Gaussian membership function of position error. It includes seven fuzzy states: $NB(\lambda_1), NM(\lambda_2), NS(\lambda_3), ZO(\lambda_4), PS(\lambda_5), PM(\lambda_6), PB(\lambda_7)$. $\beta_1, \beta_2, \beta_3, \beta_4, \beta_5, \beta_6, \beta_7$ are the mean of the normal distribution for Gaussian membership function of the derivative of position error. It includes seven fuzzy states: $NB(\beta_1), NM(\beta_2), NS(\beta_3), ZO(\beta_4), PS(\beta_5), PM(\beta_6), PB(\beta_7)$.

Table 1. Fuzzy control rules for AUV.

$e(t)/\dot{e}(t)$	NB	NM	NS	ZO	PS	PM	PB
NB	NB	NB	NM	NS	NS	ZO	PM
NM	NB	NM	NS	ZO	ZO	PS	PM
NS	NB	NM	NS	ZO	PS	PS	PM
ZO	NB	NM	NS	ZO	PS	PM	PB
PS	NM	NS	NS	ZO	PS	PM	PB
PM	NM	NS	ZO	ZO	PS	PM	PB
PB	NM	ZO	PS	PS	PM	PB	PB

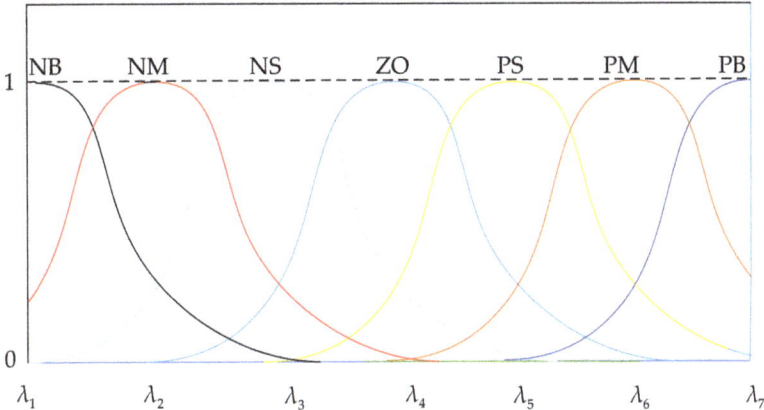

Figure 4. Membership function for position error.

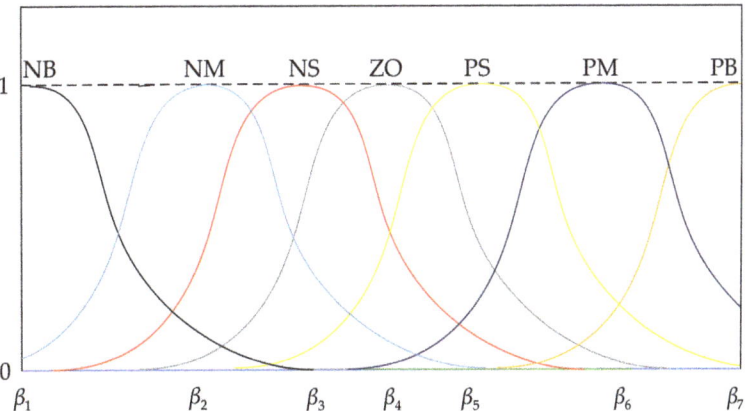

Figure 5. Membership function for the derivative of position error.

3.3. Robust Optimization for AUV

Tracking error between the desired and tracking path is the important performance index for path tracking results, which is replaced by the average of the total absolute of the position errors in this paper, it can be given as follows:

$$f_E = \frac{1}{T}\sum_{t=0}^{T}|e_t| \qquad (18)$$

where e_t is one position error at the point q_t of the tracking path; $|e_t|$ is the absolute value of e_t, f_E is the average of the total $|e_t|$, T is the total number of points of AUV tracking path.

It defines that the start and target points are (x_0, y_0, z_0) and (x_T, y_T, z_T) for the tracking path respectively, the tracking path can consist of a sequence of points $A = [(x_0, y_0, z_0), \cdots, (x_t, y_t, z_t), \cdots, (x_T, y_T, z_T)]$.

$$\begin{aligned} \min \quad & f_E(A) \\ \text{s.t.} \quad & x_{\min} \leq x_t \leq x_{\max} \\ & y_{\min} \leq y_t \leq y_{\max} \\ & z_{\min} \leq z_t \leq z_{\max} \\ & \dot{\eta} = \begin{bmatrix} J_1 & 0_{3\times 3} \\ 0_{3\times 3} & J_2 \end{bmatrix} V \end{aligned} \tag{19}$$

where x_m, y_m, z_m are the decision variables of the optimization problem for AUV path tracking; $(x_{\max}, y_{\max}, z_{\max})$ and $(x_{\min}, y_{\min}, z_{\min})$ are the maximum and minimum coordinate position points, respectively.

Considering the uncertain ocean currents, Equation (19) can be transformed into a robust optimization problem as follows:

$$\begin{aligned} \min_{(x_m,y_m,z_m)} \max_{(u_c,v_c,w_c)} & f(x_m, y_m, z_m, u_c, v_c, w_c) \\ \text{s.t.} \quad & x_{\min} \leq x_m \leq x_{\max} \\ & y_{\min} \leq y_m \leq y_{\max} \\ & z_{\min} \leq z_m \leq z_{\max} \\ & \dot{\eta} = \begin{bmatrix} J_1 & 0_{3\times 3} \\ 0_{3\times 3} & J_2 \end{bmatrix} V \\ & u_{\text{minc}} \leq u_c \leq u_{\text{maxc}} \\ & v_{\text{minc}} \leq v_c \leq v_{\text{maxc}} \\ & w_{\text{minc}} \leq w_c \leq w_{\text{maxc}} \end{aligned} \tag{20}$$

where $[u_{\text{minc}} \; v_{\text{minc}} \; w_{\text{minc}}]$ and $[u_{\text{maxc}} \; v_{\text{maxc}} \; w_{\text{maxc}}]$ are the minimum and maximum values of the components of ocean currents. Equation (20) is the robust optimization problem, which is also called "min-max" optimization problem for the AUV path tracking, whose goal is to find the robust solution for the tracking path with the best performance in all the worst ocean currents.

CPSO is proposed to solve the robust optimization problem (20), which can find a good solution to the "min-max" optimization problem for the AUV path tracking. The CPSO algorithm involves two populations P_1 and P_2, each population evolves independently and tied together via the fitness evaluation [29,30]. The first population P_1 is used to evolve the decision variables (x_m, y_m, z_m), the second population P_2 is used to evolve the ocean currents (u_c, v_c, w_c).

For the first population, the fitness function of decision variables is given by

$$G(x_m, y_m, z_m) = \max_{u_c,v_c,w_c \in P_2} f(x_m, y_m, z_m, u_c, v_c, w_c) \tag{21}$$

which is to be minimized.

For the second population, the fitness function of ocean currents is given by

$$H(u_c, v_c, w_c) = \min_{x_m, y_m, z_m \in P_1} f(x_m, y_m, z_m, u_c, v_c, w_c) \quad (22)$$

which is to be maximized.

3.4. Thruster Fault Diagnostics and Fault Tolerant Control Algorithm

For the CPSO algorithm, in the light of Equation (21), the global best value in P_1 is gotten as the solution. Based on Equation (22), the globally best values in P_2 are obtained as the scenarios for ocean currents. According to the above design principles, the optimal tracking path can be obtained.

Figure 6 gives the flowchart for the thruster fault diagnostics and fault tolerant control of AUV, the corresponding steps are given in detail in Algorithm 1.

Algorithm 1: Thruster fault diagnostics and fault tolerant control for AUV

1: Initializing the parameters m, p, U, T, V and so on for thruster fault diagnostics;
2: Calculating the penalty factor η_i based on Equation (14), updating U, T, V based on Equations (15)–(17) respectively;
3: If the iterations N_d are smaller than the given maximum number of times (N_{d_max}), obtaining the final U, T, V; else if go to Step 2.
4: Considering the effects of ocean currents, establishing the robust optimization model for AUV fault tolerant control systems.
5: Establishing the models of the evaluation functions (21) and (22) for P_1 and P_2 respectively.
6: Initializing the two populations randomly, evaluating each population co-evolutionarily by using (21) and (22), respectively.
7: Evolving the population P_1 based on (21); replacing the global best (g_{best}) and personal best (p_{best}) particle positions.
8: If the iterations (N_{i_1}) is smaller than the given maximum number of times (N_{m_1}), go to the next step, else if go to Step 7.
9: Evolving the population P_2 based on (22); replacing g_{best} and p_{best} particle positions.
10: If the iterations N_{i_2} are smaller than the given maximum number of times (N_{m_2}), go to the next step, else if, go to step 9.
11: If the iterations N_{i_3} are smaller than the given maximum number of times (N_{m_3}), obtaining the optimal parameters of the membership function, then getting the final tracking points, end the program; else if go to Step 6.

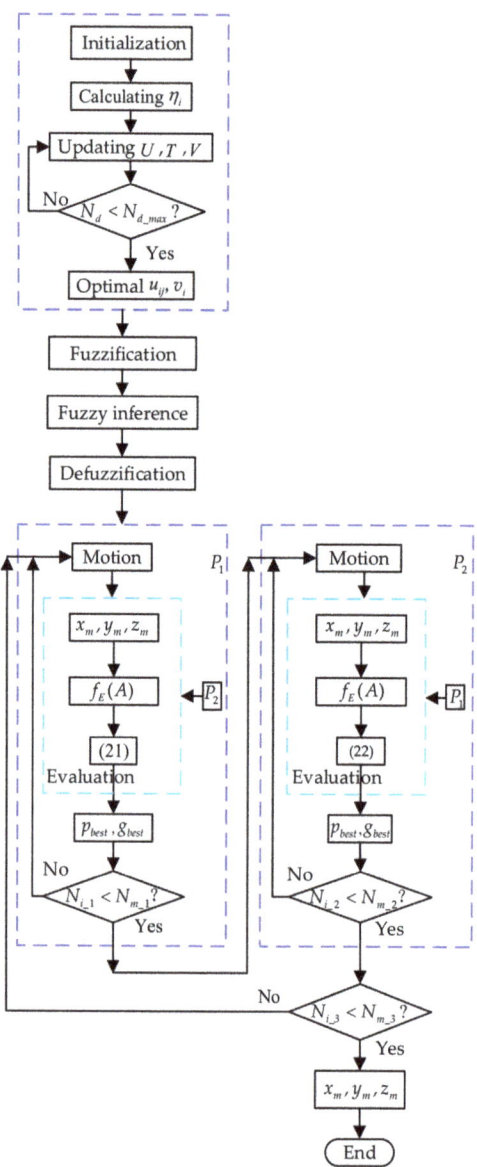

Figure 6. Flowchart for thruster fault diagnostics and fault tolerant control for AUV.

4. Simulation and Experiment Analysis

In this paper, different experiments are given to analyze the performance of the proposed thruster fault diagnostics and fault tolerant control method for AUV, Figure 7 shows the designed AUV, and Figure 8 gives the AUV thruster in practice. Based on the test data of the thruster fault, the results of the fault type are obtained for AUV thruster fault diagnostics. AUV is equipped with six underwater propellers, among which the propeller at the top of the AUV is used to control the sinking and floating of the AUV, and the other four propellers at the front and back are used to control the forward, backward, and steering of the AUV. The speed of AUV is set as 0.15 m/s, and the speed of uncertain ocean current is set as 0–0.08 m/s, whose direction is random. Sine path, circular path, rectangular path, and irregularity path are given to illustrate the tracking effect with thruster fault. The algorithm is coded in MATLAB R2019a and simulations are run on the PC with 2.00 GHz CPU/8 GB RAM.

Figure 7. The designed AUV.

Figure 8. The thruster of AUV.

In order to test the effectiveness of the proposed thruster fault diagnostics, a data set with 300 groups is obtained from the underwater experiment for our designed AUV. Each group data is composed of voltage, current, and speed of the thrusters, which can denote the characteristic of six types of thruster operation: motor fault, propeller enwinding by foreign matter, propeller blade damage, thruster got stuck, thruster idling, and normal operation. Therefore, it assumes that the number of the clustering centers is 6. Figure 9 shows the classification results by the proposed PFCM algorithm. The center of clustering of the thruster stuck state is $(11.98\ V, 0.58\ A, 0.75\ r/s)$, the center of clustering of the propeller enwinding state is $(12.2\ V, 0.41\ A, 12.07\ r/s)$, the center of clustering of the thruster normal operation is $(12.48\ V, 0.35\ A, 16.9\ r/s)$, the center of clustering of the propeller damage is $(12.78\ V, 0.27\ A, 21.01\ r/s)$, the center of clustering of the thruster idling state is $(13\ V, 0.17\ A, 25\ r/s)$, the center of clustering of the motor fault is $(13\ V, 0.05\ A, 0.77\ r/s)$. The signals of six different fault types are closely clustered around their respective clustering centers after classification by the PFCM algorithm. The proposed fault detection algorithm can accurately identify the fault types of AUV and effectively classify them.

Because the thrusters of AUV are often immersed in seawater, the probability of failure is significantly improved after the corrosion of seawater. AUV operation in the ocean may be enwound by marine plants or marine organisms, which affects the performance of the thrusters. The thruster is the main forward power of AUV, if the above phenomenon occurs, it affects the velocity, heading angle, and the safety of AUV, and even leads to the AUV being unrecoverable. Therefore, in this paper, after detecting the fault type of AUV based on the PFCM algorithm, the corresponding fault tolerant control is adopted according to the fault type and degree. By reducing the thrust of other thrusters, the AUV can continue to move to accomplish the missions. If the AUV loses all power, it can be stopped and floated up for recovery. Therefore, the fault tolerant control proposed in this paper only applies to the fault types of AUV power loss caused by AUV enwinding or propeller damage.

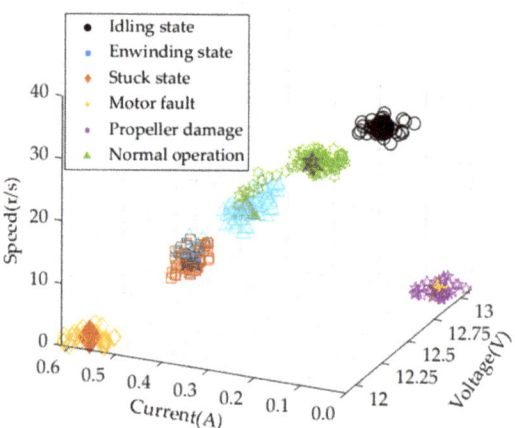

Figure 9. The classification results by the proposed PFCM algorithm.

When a thruster fault happens, it loses thrust and its torque balance is broken, which leads the change of heading angle and takes off its desired path. For example, if the left thruster (F_5) of the AUV fault happens and its thrust decreases, the thrust of its adjacent thruster (F_6) should be correspondingly reduced to balance the torque of the AUV. If the two adjacent thruster faults happen, the thrust of the thruster with a larger thrust is reduced accordingly to make its torque reach balance. If these two thrusters (F_5, F_6) have a large degree of fault, the thrusters (F_3 and F_4) are responsible for AUV regression on the opposite side, which can complete the follow-up tasks or realize turning back.

In the complex environment, there are many factors that affect the AUV operation, such as obstacles, ocean currents, and fish schools. Therefore, the AUV path is not a line, the curved path is an essential to the AUV path. This paper gives a sine curve path as follows:

$$\begin{cases} x(t_1) = t_1 \\ y(t_1) = 80\sin(t_1\pi/80) \\ z(t_1) = 0 \end{cases} \qquad (23)$$

where $t_1 \in [0:0.25:360]$, it assumes that the sine curve is constituted of 1440 points $[(0, 0, 0), (0.25, 0.78, 0), \cdots, (360, 0, 0)]$, the tracking start point is $(0, 0, 0)$. The initial position, angle, initial velocity, and expected velocity are set as $(x, y, z) = (0, 0, 0)$ m, $(\varphi, \theta, \phi) = (0°, 0°, 0°)$, $(u, v, w) = (0, 0, 0)$ [kn], $(u, v, w) = (0.2, 0, 0)$ [kn] for AUV. If the thruster is enwound with foreign matter. Based on the results of thruster fault diagnostics, Figure 10 gives the tracking results for the sine path by the proposed fault tolerant control algorithm, the tracking path length is 832.91 m. Figure 11 shows the position error for the corresponding sine path tracking, the average position errors is 0.27 m, the range of the tracking error is $[-1.44\text{ m}, 1.91\text{ m}]$, and the standard deviation of the tracking error is 0.16 m. The proposed algorithm can realize the sine path tracking with thruster enwinding effectively.

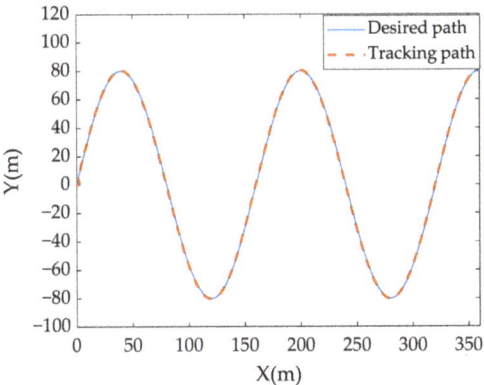

Figure 10. Tracking for sine path.

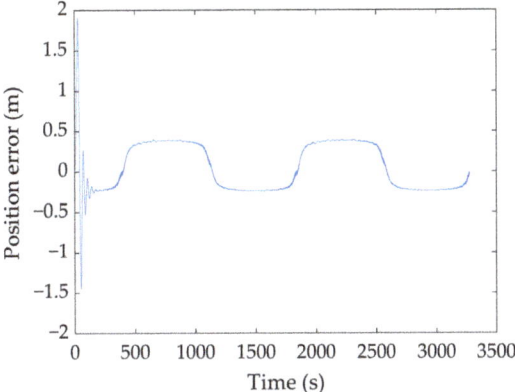

Figure 11. Position error for the sine path tracking.

In the practice, in order to complete some specific tasks, AUV needs to move around the detected object in a circle. Therefore, the circular curve is also one of the key paths of AUV path tracking. The circular path is described as follows:

$$\begin{cases} x(t_2) = 55 + 40\cos(t_2) \\ y(t_2) = 40\sin(t_2) \\ z(t_2) = 0 \end{cases} \quad (24)$$

where $t_2 \in [\pi : -\pi/400 : -\pi]$, it assumes that the curve is constituted of 800 points $[(15, 0, 0), (15.002, 0.31, 0), \cdots, (15, 0, 0)]$, the stat tracking point is $(0, 0, 0)$. The initial position, initial angle, initial velocity, and expected velocity are set as $(x, y, z) = (5, -5, 0)$ m, $(\varphi, \theta, \phi) = (0°, 0, 0)$, $(u, v, w) = (0, 0, 0)$ [kn]. $(u, v, w) = (0.2, 0, 0)$ [kn] for AUV respectively. If the thruster is enwound with foreign matter. Based on the results of thruster fault diagnostics, Figure 12 gives the path tracking results for the circle path. Figure 13 shows the position error for the circle path tracking. The tracking path length is 261.73 m, the average position errors is 0.59 m, the range is $[-10 \text{ m}, 0.71 \text{ m}]$ for AUV tracking position errors, the standard deviation of the tracking error is 0.15 m. The proposed algorithm can realize the circle path tracking with thruster enwinding effectively.

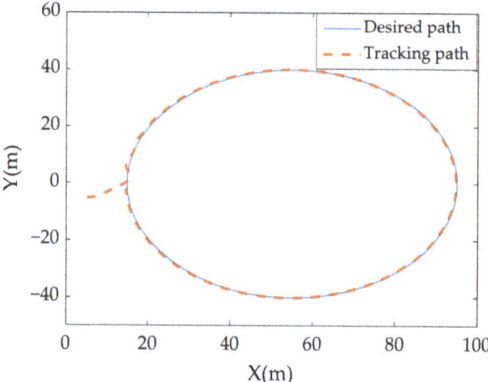

Figure 12. Tracking for circle path.

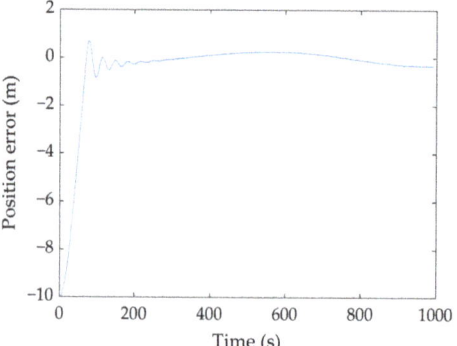

Figure 13. Position error for the circle path tracking.

In the real working environment, in order to successfully complete the missions, AUV needs to run along different paths, among which rectangular paths are common. Therefore, a rectangular path is given to simulate the actual path tracking, the basic path parameters are given for the rectangular path as follows:

$$\begin{cases} x = t_3 \\ y = t_3 + 10 & t_3 \in [10, 50] \\ y = -t_3 + 110 & t_3 \in [50, -30] \\ y = t_3 + 170 & t_3 \in [-30, -70] \\ y = -t_3 + 30 & t_3 \in [-70, 10] \end{cases} \quad (25)$$

It assumes that the path is constituted of the points $[(10, 20, 0), (1, 11, 0), \cdots, (10, 20, 0)]$. The start tracking point is $(10, 20, 0)$, The initial position, initial angle, initial velocity, and expected velocity are set as $(x, y, z) = (0, 0, 0)$ m, $(\varphi, \theta, \phi) = (0°, 0°, 0°)$, $(u, v, w) = (0, 0, 0)$ [kn], $(u, v, w) = (0.2, 0, 0)$ [kn]. If the thruster is enwound with foreign matter. Based on the results of thruster fault diagnostics, Figure 14 gives the tracking results for the rectangular path. Figure 15 shows the position error for the rectangular path tracking. The tracking path length is 255.06 m, the average position errors is 0.81 m, the range is $[-1.94$ m, 8.26 m$]$ for AUV tracking position errors, the standard deviation of the tracking error is 0.35 m. The proposed algorithm can realize the rectangular path tracking with thruster enwinding effectively.

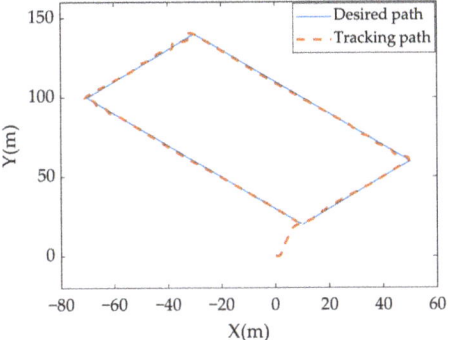

Figure 14. Tracking for the rectangular path.

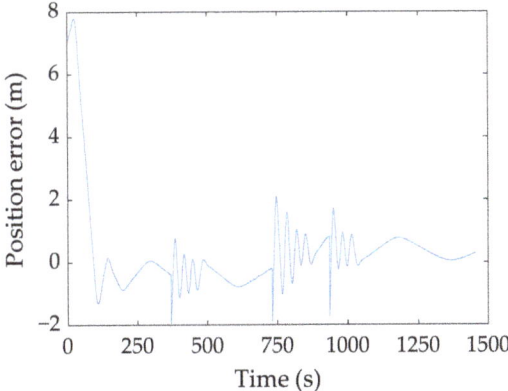

Figure 15. Position error for the rectangular path tracking.

In order to verify the path tracking effect of the controller proposed in the presence of obstacles, multiple circular obstacles are set in the environment. The tracking start point is $(x, y, z) = (0, 0, 0)$, the target point is $(x, y, z) = (100, 100, 0)$. The initial position, initial angle, initial velocity, and expected velocity are set as $(\varphi, \theta, \phi) = (0°, 0°, 0°)$, $(u, v, w) = (0, 0, 0)$ [kn]. $(u, v, w) = (0.2, 0, 0)$ [kn]. If the thruster is enwound with foreign matter, the ocean current is 0.07 [kn]. Based on the results of thruster fault diagnostics, it can obtain the fault tolerant control results by the proposed and existing traditional algorithms as shown in Figure 16. For the proposed algorithm, the tracking path length is 150.54 m, the average of the position errors is 0.78 m, and the standard deviation of the position errors is 0.18 m. For the traditional fuzzy control, the parameters of the membership function are optimized by the trial and error method. The tracking path length is 152.71 m, the average of the position errors is 0.92 m, the standard deviation of the position errors is 0.22 m. Table 2 gives the comparison results between the proposed and traditional algorithms. Figure 17 shows the position error for the function tracking. The ranges of the position errors are $[-1.65 \text{ m}, 7.53 \text{ m}]$ and $[-1.75 \text{ m}, 7.47 \text{ m}]$ for the proposed and existing algorithms respectively. Compared with the traditional algorithm, one can see that the tracking path length, average position error, and time are smaller by the proposed algorithm. The proposed algorithm can realize the path tracking in the environment with obstacles and ocean currents effectively.

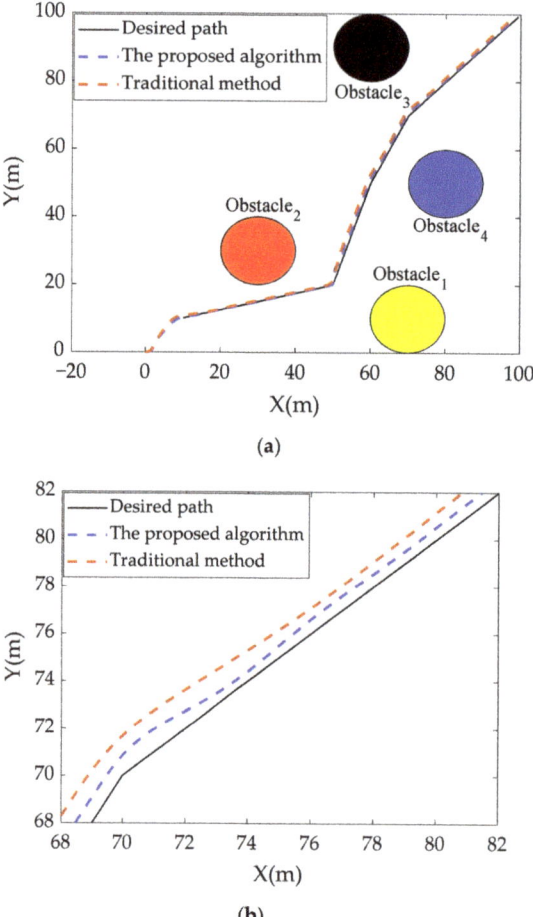

Figure 16. (a) Path tracking in the environment with obstacles. (b) Part of the enlarged view for path tracking curve.

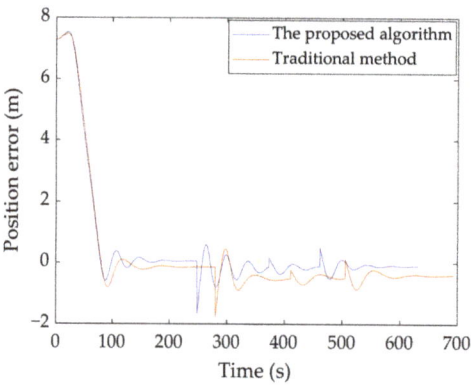

Figure 17. Position error for the path tracking in the environment with obstacles.

Table 2. Comparison results between the proposed and traditional algorithms.

Method	Path Length (m)	Mean Tracking Error (m)	Standard Deviation (m)
The proposed algorithm	150.54	0.78	0.18
Traditional method	152.71	0.92	0.22

5. Conclusions

Thruster is the driving mechanism for AUV movement, whose fault diagnostics and fault tolerant control are the premise to complete the underwater missions. In practice, ocean currents perturbations could produce noise for thruster fault diagnosis, in this paper, the PFCM algorithm is proposed to solve the problem of thruster fault diagnostics. It is not enough just to realize the thruster fault diagnostics, in order to successfully complete the missions with thruster fault, a fuzzy controller is presented. Considering the effect of ocean currents, the CPSO algorithm is developed to optimize the fuzzy controller, which guarantees the fault tolerant control performance. Based on the designed AUV, a date set is obtained to demonstrate the effectiveness of the thruster fault diagnostics. Different scenarios of path tracking are given to illustrate the performance of the proposed algorithm. Compared with the traditional fuzzy fault tolerant control, the tracking path length and tracking error are smaller by the proposed algorithm, which illustrates the proposed algorithm. In this paper, the proposed algorithm is difficult to be used for weak faults diagnosis of AUV thrusters. However, major faults are generally developed from weak faults. Therefore, in future work, we will try to solve the problem of accurate weak faults diagnosis of AUV thrusters in the presence of interference, which is one of the keys to preventing and reducing catastrophic accidents.

Author Contributions: Conceptualization, Q.T. and T.W.; methodology, Q.T.; software, T.W.; validation, Q.T., G.R. and B.L.; formal analysis, Q.T.; investigation, Q.T. and T.W.; resources, G.R.; data curation, T.W.; writing—original draft preparation, Q.T.; writing—review and editing, G.R.; visualization, T.W.; supervision, B.L.; project administration, B.L.; funding acquisition, B.L. All authors have read and agreed to the published version of the manuscript.

Funding: This research was funded by the China Postdoctoral Science Foundation (2022M710934), Postdoctoral Applied Research Project of Qingdao City, Project of Shandong Province Higher Educational Young Innovative Talent Introduction and Cultivation Team (Intelligent Transportation Team of Offshore Products).

Institutional Review Board Statement: Not applicable.

Informed Consent Statement: Not applicable.

Data Availability Statement: Not applicable.

Conflicts of Interest: The authors declare no conflict of interest.

References

1. Yin, B.; Zhang, M.; Lin, X.; Fang, J.; Su, S. A fault diagnosis approach for autonomous underwater vehicle thrusters using time-frequency entropy enhancement and boundary constraint–assisted relative gray relational grade. *Proc. Inst. Mech. Eng. Part I J. Syst. Control. Eng.* **2019**, *234*, 512–526. [CrossRef]
2. Yuan, J.; Wan, J.; Zhang, W.; Liu, H.; Zhang, H.; Syahputra, R. An Underwater Thruster Fault Diagnosis Simulator and Thrust Calculation Method Based on Fault Clustering. *J. Robot.* **2021**, *2021*, 6635494. [CrossRef]
3. Kadiyam, J.; Parashar, A.; Mohan, S.; Deshmukh, D. Actuator fault-tolerant control study of an underwater robot with four rotatable thrusters. *Ocean Eng.* **2020**, *197*, 106929. [CrossRef]
4. Che, G. Single critic network based fault-tolerant tracking control for underactuated AUV with actuator fault. *Ocean Eng.* **2022**, *254*, 111380. [CrossRef]
5. Sun, Y.; Ran, X.; Li, Y.; Zhang, G.; Zhang, Y. Thruster fault diagnosis method based on Gaussian particle filter for autonomous underwater vehicles. *Int. J. Nav. Archit. Ocean. Eng.* **2016**, *8*, 243–251. [CrossRef]
6. Jiang, Y.; Feng, C.; He, B.; Guo, J.; Wang, D.; Lv, P. Actuator fault diagnosis in autonomous underwater vehicle based on neural network. *Sens. Actuators A Phys.* **2021**, *324*, 112668. [CrossRef]

7. Yu, D.; Zhu, C.; Zhang, M.; Liu, X. Experimental Study on Multi-Domain Fault Features of AUV with Weak Thruster Fault. *Machines* **2022**, *10*, 236. [CrossRef]
8. Xia, S.; Zhou, X.; Shi, H.; Li, S.; Xu, C. A fault diagnosis method based on attention mechanism with application in Qianlong-2 autonomous underwater vehicle. *Ocean Eng.* **2021**, *233*, 109049. [CrossRef]
9. Raanan, B.Y.; Bellingham, J.; Zhang, Y.; Kemp, M.; Kieft, B.; Singh, H. Detection of unanticipated faults for autonomous underwater vehicles using online topic models. *J. Field Robot.* **2017**, *35*, 705–716. [CrossRef]
10. Liu, X.; Zhang, M.; Yao, F. Adaptive fault tolerant control and thruster fault reconstruction for autonomous underwater vehicle. *Ocean Eng.* **2018**, *155*, 10–23. [CrossRef]
11. Liu, F.; Tang, H.; Luo, J.; Bai, L.; Pu, H. Fault-tolerant control of active compensation toward actuator faults: An autonomous underwater vehicle example. *Appl. Ocean. Res.* **2021**, *110*, 102597. [CrossRef]
12. Zhu, C.; Huang, B.; Zhou, B.; Su, Y.; Zhang, E. Adaptive model-parameter-free fault-tolerant trajectory tracking control for autonomous underwater vehicles. *ISA Trans.* **2021**, *114*, 57–71. [CrossRef] [PubMed]
13. Lv, T.; Zhou, J.; Wang, Y.; Gong, W.; Zhang, M. Sliding mode based fault tolerant control for autonomous underwater vehicle. *Ocean Eng.* **2020**, *216*, 107855. [CrossRef]
14. Li, X.; Chao, H.; Wang, J.; Xu, Q.; Yang, K.; Mao, D. An Iterative Learning Extended-State Observer-Based Fuzzy Fault-Tolerant Control Approach for AUVs. *Mar. Technol. Soc. J.* **2021**, *55*, 33–46. [CrossRef]
15. Che, G.; Yu, Z. Neural-network estimators based fault-tolerant tracking control for AUV via ADP with rudders faults and ocean current disturbance. *Neurocomputing* **2020**, *411*, 442–454. [CrossRef]
16. Wang, L.; Liu, L.; Qi, J.; Peng, W. Improved Quantum Particle Swarm Optimization Algorithm for Offline Path Planning in AUVs. *IEEE Access* **2020**, *8*, 143397–143411. [CrossRef]
17. Cao, X.; Sun, H.; Jan, G.E. Multi-AUV cooperative target search and tracking in unknown underwater environment. *Ocean Eng.* **2018**, *150*, 1–11. [CrossRef]
18. Taheri, E.; Ferdowsi, M.H.; Danesh, M. Closed-loop randomized kinodynamic path planning for an autonomous underwater vehicle. *Appl. Ocean. Res.* **2019**, *83*, 48–64. [CrossRef]
19. Mahmoud Zadeh, S.; Yazdani, A.M.; Sammut, K.; Powers, D.M.W. Online path planning for AUV rendezvous in dynamic cluttered undersea environment using evolutionary algorithms. *Appl. Soft Comput.* **2018**, *70*, 929–945. [CrossRef]
20. Karkoub, M.; Wu, H.-M.; Hwang, C.-L. Nonlinear trajectory-tracking control of an autonomous underwater vehicle. *Ocean Eng.* **2017**, *145*, 188–198. [CrossRef]
21. Chen, H.; Chai, Z.; Dogru, O.; Jiang, B.; Huang, B. Data-Driven Designs of Fault Detection Systems via Neural Network-Aided Learning. *IEEE Trans. Neural Netw. Learn. Syst.* **2021**, *ahead of print*. [CrossRef]
22. Chen, H.; Li, L.; Shang, C.; Huang, B. Fault Detection for Nonlinear Dynamic Systems With Consideration of Modeling Errors: A Data-Driven Approach. *IEEE Trans. Cyber.* **2022**, 1–11. [CrossRef] [PubMed]
23. Chen, J.; Zhang, H.; Pi, D.; Kantardzic, M.; Yin, Q.; Liu, X. A Weight Possibilistic Fuzzy C-Means Clustering Algorithm. *Sci. Programm.* **2021**, *2021*, 9965813. [CrossRef]
24. Fazel Zarandi, M.H.; Sotodian, S.; Castillo, O. A New Validity Index for Fuzzy-Possibilistic C-Means Clustering. *Sci. Iran.* **2021**, *28*, 2277–2293. [CrossRef]
25. Askari, S. Fuzzy C-Means clustering algorithm for data with unequal cluster sizes and contaminated with noise and outliers: Review and development. *Expert Syst. Appl.* **2021**, *165*, 1–27. [CrossRef]
26. Ran, G.; Liu, J.; Li, C.; Lam, H.-K.; Li, D.; Chen, H. Fuzzy-Model-Based Asynchronous Fault Detection for Markov Jump Systems with Partially Unknown Transition Probabilities: An Adaptive Event-Triggered Approach. *IEEE Trans. Fuzzy Syst.* **2022**. [CrossRef]
27. Ran, G.; Li, C.; Rathinasamy, S.; Han, C.; Wang, B.; Liu, J. Adaptive Event-Triggered Asynchronous Control for Interval Type-2 Fuzzy Markov Jump Syst with Cyber-Attacks. *IEEE Trans. Control. Netw. Syst.* **2022**, *9*, 88–99. [CrossRef]
28. Ran, G.; Chen, H.; Li, C.; Ma, G.; Jiang, B. A Hybrid Design of Fault Detection for Nonlinear Syst Based on Dynamic Optimization. *IEEE Trans. Neural Netw Learn. Syst.* **2022**. [CrossRef]
29. Cramer, A.M.; Sudhoff, S.D.; Zivi, E.L. Evolutionary Algorithms for Minimax Problems in Robust Design. *IEEE Trans. Evol. Comput.* **2009**, *13*, 444–453. [CrossRef]
30. Tian, Q.; Zhao, D.; Li, Z.; Zhu, Q. A two-step co-evolutionary particle swarm optimization approach for CO_2 pipeline design with multiple uncertainties. *Carbon Manag.* **2018**, *9*, 333–346. [CrossRef]

Article

Fault Detection for High-Speed Trains Using CCA and Just-in-Time Learning

Hong Zheng [1], Keyuan Zhu [1], Chao Cheng [1,*] and Zhaowang Fu [2]

[1] School of Computer Science and Engineering, Changchun University of Technology Changchun, Changchun 130022, China; zhenghong@ccut.edu.cn (H.Z.); keyuanzhuccut@foxmail.com (K.Z.)
[2] 32184 Unit of PLA in China, Beijing 100000, China; fuzhaowang123@aliyun.com
* Correspondence: chengchao@ccut.edu.cn

Abstract: Online monitors of the running gears systems of high-speed trains play critical roles in ensuring operational safety and reliability. Status signals collected from high-speed train running gears are very complex regarding working environments, random noises and many other real-world constraints. This paper proposed fault detection (FD) models using canonical correlation analysis (CCA) and just-in-time learning (JITL) to process scalar signals of high-speed train gears, named as CCA-JITL. After data preprocessing and normalization, CCA transforms covariance matrices of high-dimension historical data into low-dimension subspaces and maximizes correlations between the most important latent dimensions. Then, JITL components formulate local FD models which utilize subsets of testing samples with larger Euclidean distances to training data. A case study introduced a novel system design of an online FD architecture and demonstrated that CCA-JITL FD models significantly outperformed traditional CCA models. The approach is applicable to other dimension reduction FD models such as PCA and PLS.

Keywords: canonical correlation analysis; just-in-time learning; fault detection; high-speed trains

Citation: Zheng, H.; Zhu, K.; Cheng, C.; Fu, Z. Fault Detection for High-Speed Trains Using CCA and Just-in-Time Learning. *Machines* **2022**, *10*, 526. https://doi.org/10.3390/machines10070526

Academic Editor: Yaguo Lei

Received: 19 May 2022
Accepted: 20 June 2022
Published: 28 June 2022

Publisher's Note: MDPI stays neutral with regard to jurisdictional claims in published maps and institutional affiliations.

Copyright: © 2022 by the authors. Licensee MDPI, Basel, Switzerland. This article is an open access article distributed under the terms and conditions of the Creative Commons Attribution (CC BY) license (https://creativecommons.org/licenses/by/4.0/).

1. Introduction

In the past twenty years, high-speed railway systems are gradually becoming one of the most popular transportation services because of their significant advantages in speed and energy efficiency [1–3]. The running gears are critical parts to ensure the safety of high-speed train operations. To precisely detect the real-time health status of running gears is very challengeable. In reality, sensor signals of running gears in high-speed trains have a very high degree of complexity, for instance, messy signals from bogie, bearing temperature, gear temperature, working environments and random noises. Moreover, there are only small-scale historical failure data available among large volumes of monitoring data streams. Incomplete training resources might easily raise detection errors.

With the rapid development of train sensor technology, data-driven FD methods have been well studied in the last century. Many multivariate statistical methods have been widely applied in the fault detection fields [4–6], for example, principal component analysis (PCA), partial least squares (PLS) and CCA. PCA was one of the earliest dimensionality reduction methods to process high-dimension signal data for FD purposes [7,8]. PCA projects high-dimension input data into low-rank subspaces while retaining the main information of the original data within a few top latent dimensions. Moreover, PCA FD models are derived from a large scale of normal status signals and generate fault alarm thresholds for incoming error signals. PLS and CCA are widely utilized to develop advanced FD models [6,9,10]. PLS decomposes the covariance matrices of two sets of variables into relational subspaces and residual subspaces. Then, the regression analysis to covariance structure estimates the multi-direction of one set of variables that explains the maximum multidimensional variance direction of another set of variables. CCA identifies linear combinations between two groups of variables to maximize the overall group correlation. Multi-set CCA resolved feature fusion of multiple groups of variables [11].

Chen and Ding [12] designed a general CCA-based FD infrastructure for non-Gaussian processes which aimed to boost the fault detection rate (FDR) under an acceptable false alarm rate (FAR). Peng and Ding [13] have proposed CCA-based distributed monitoring processes within partly-connected networks, which reduced communication costs and risks and avoided a significant drop in system performance. Chen and Chen [6] introduced a single-side CCA (SsCCA) model with promising FD performance using single-side neural networks. Chen and Li [14] had proposed a stacked approach, so called neural network-aided canonical variate analysis (SNNCVA), which showed satisfactory FD performance for nonlinear datasets. Garramiola and Poza [15] introduced a data-driven approach of fault diagnosis to build hybrid fusion models to detect, isolate and classify sensor faults. Kou and Qin [16] extended fault diagnosis methodology into tensor space to deal with multi-sensor data with high precision and convergence speed. Zhao and Yan [17] provided a comprehensive review which summarized state-of-the-art deep learning (DL) technologies applied on machine health monitoring (MHM). Niu and Xiong [18] proposed a novel fault Petri net fault detection and diagnosis (FDD) model to analyze signals of speed sensors of high-speed trains. Fu and Huang [19] proposed a fault diagnosis method based on the long-short-term memory (LSTM) recursive neural network (RNN) to reduce the steps of signal preprocessing and optimize prediction accuracy. Cheng and Guo [20] designed a real-time prediction framework for running state of running station based on multi-layer BRB and priority scheduling strategy. Guan and Huang [21] created a particle swarm optimization algorithm based on wavelet variation and a least squares support vector machine to avoid falling into local extremum problems. Sayyad and Kumar [22] introduced a survey to review service life prediction technologies of real-time health monitors of cutting tools from perspectives of modeling, systems, data sets and research trends. Capriglione and Carratu [23] proposed an FD method using a nonlinear autoregressive with Exogenous Inputs (NARX) neural network as a residual generator for online FD of travel sensors. Shabanian and Montazeri [24] proposed an online FD and diagnosis algorithm based on the neural fuzzy, and adaptive analytic method and neural network to track faults online.

JITL technologies involve collecting the most relevant samples as training data for online query and making predictions of local modeling running time [25–27]. Compared with similar samples in historical databases, the signal status of online query could be possibly acquired in real time. Robust JITL strategies to leverage the weights of high leakage points of signals such as outliers had been successfully applied to the FD tasks [28]. A simulation study showed that the combined JITL-PCA models outperformed PCA in the analyzing of nonlinear signals [26]. In addition, neural network methods and the stochastic hidden Markov model (HMM) were studied to improve FD performance of dynamic systems [29,30].

Motivated by the previous studies, we designed a novel CCA-JITL model to analyze real-time signals from running gears of high-speed trains. The model was built and testified using real-world datasets. The algorithm split the data input into two groups and verified the system performance by group comparison The evaluation demonstrated that the accuracy of FD detection was significantly improved. The algorithm detects the data in groups and verifies the two groups of results, and the proposed system infrastructure was also applicable to enhance PCA and PLS FD models.

The rest of this article is content as follows. Section 2 gives introduction the structure of running gears system, experiment design and datasets. Section 3 presents theoretical foundations of the proposed method. Section 4 presents evaluation results of a FD use case and discussion of the results. Finally, Section 5 summarizes this paper study.

2. Preliminaries

This section introduces the mechanical structure of a running gear system in a high-speed train. In this study, signal faults mainly caused by parts were selected as FD targets. Then, the research goals and problem statements of system design were discussed.

2.1. Introduction of a Running Gears System of a High-Speed Train

The running gear system is an important system that affects the smooth running of high-speed trains. It improves the traction performance of high-speed trains and has the functions of generating power, buffering and supporting. The running gears system of high-speed trains include many complicated parts such as the axle box, traction drive, detection sensor, and spring device. Any malfunction from these parts may cause the carriage to shake during running and result in unpredictable consequences.

This paper aimed to analyze the running gears model and establish a data-driven FD model. The running gears system has such a multi-level complex structure. Therefore, it is difficult to build an FD model. As shown in Figure 1, many temperature sensors are arranged inside the gears. The test points of temperature sensors, for example, A1–A4 for axle box bearing temperature measuring point, B1–B3 for motor temperature measuring point, and C1–C4 for gear box temperature are shown in Figure 1 [31].

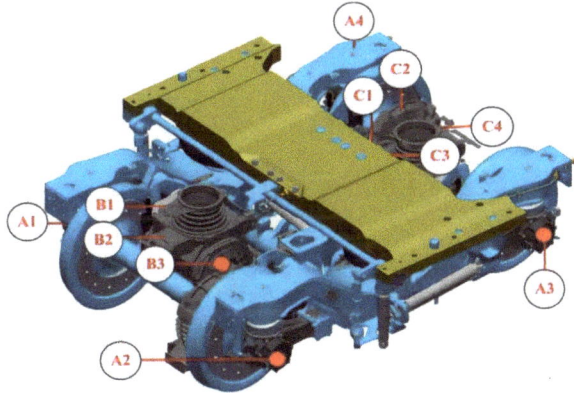

Figure 1. Structure of a running gears system.

2.2. Fault Description

The running gears system is equipped with many sensors to keep track of the actual status. The real data used in this study is based on data collected by a railway department in a specific year and then classified and processed to obtain the fault signals of gears. This paper uses the matrix to describe the data set for research purposes. This paper uses the matrix Z_w to describe the data set as followings

$$Z_w = [q_w(1), q_w(2), \cdots, q_w(8)] \quad (1)$$

where $Z_w \in R^{N \times m}$ with N samplings collected from m sensors. In this application $m = 8$, $N = 2000$. Furthermore, Z_w can be rewritten as

$$\begin{aligned} Z_w &= [\ X_x \quad Y_y\ ] \\ X_x(k) &= [q_w(1), q_w(2), q_w(3), q_w(4)] \\ Y_y(k) &= [q_w(5), q_w(6), q_w(7), q_w(8)] \end{aligned} \quad (2)$$

where $q_w(i)$ represents the data collected for the ith sensor. The data subset X_x is the input matrix, and Y is the output matrix. In the paper, we use types of faults as follows: (1)

Bogie 1 failure; (2) Bogie 2 failure; (3) Motor drive side bearing failure; (4) Non-drive side bearing failure; (5) Motor side big gear failure; (6) Wheel side pinion failure; (7) Wheel side motor big gear failure; (8) Motor side pinion failure. Moreover, in the process of data collection, the data collected from the same carriage in a train is selected. Without loss of generality, after splitting data into the two groups, we added fault data with labels to form experimental training data. Therefore, the fault data can be represented as

$$q_w(\eta) = \begin{bmatrix} x & f_{wt} \end{bmatrix}^T, \eta = 1, 2 \cdots, m \qquad (3)$$

Remark 1. *Divide the data into the two groups: (1) $q_w(1)$ to $q_w(4)$ in one group as input; (2) $q_w(5)$ to $q_w(8)$ in another group as output. We added fault data with labels to form experimental data.*

Remark 2. *In this study, all the FD models were constructed and compiled within the software environment of MATLAB, and all the experiments were executed and evaluated in a PC in CPU mode.*

2.3. Objective and Design Issues

The FD models for moving gear parts were often error-prone due to the scalability and complexity issues of signals. Our CCA-JITL FD model solved many challenges as below:

- Investigate effective data processing techniques, and maintain the original trend of the data.
- Design a series of statistical tests for model evaluation.
- Design a use case and apply the proposed method.

2.4. System Design

In order to solve the above problems, this paper proposed an FD method based on CCA and JITL. We mainly used the CCA component to preprocess and normalize data, transform high-dimensional data into low-dimensional variable covariance subspaces and maximize the correlation between the most important top latent dimensions. SVD was applied to decompose the covariance matrix of input and output into two separate singular subspaces and keep the original distribution trends of variable correlations. Subspace mapping procedures projects input and output matrices back to the singular subspaces only with the most important relations and generates two groups of variables, Px and Py, after dimension reduction which eliminates the noise, that is, the residual subspaces. Then, JITL was used to calculate the Euclidean similarity of the query sample and the training data, respectively, and selected a sample subset for online testing regarding distances between them. During the experiments, the data sets of Px and Py were equally divided as a training data set and a testing data set. Finally, the FD model formulated statistics to define thresholds of fault signals and performed to detect signal faults in the testing samples. The workflows of the model are shown in Figure 2.

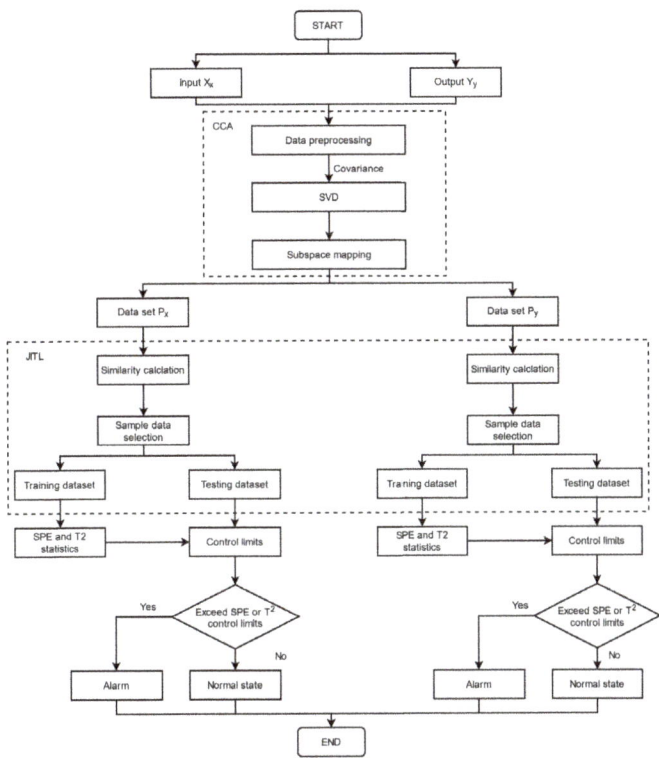

Figure 2. Flowchart of the proposed CCA-JITL FD model.

3. Methodology

In this section, combined with the characteristics of signals of the running gears, CCA-JITL FD model is introduced in details.

3.1. Canonical Correlation Analysis and Just-in-Time Learning Methods

CCA transforms covariance matrices of input and output datasets into two subspaces with the greatest correlation by computing the linear combination of the latent dimensions. The algorithm is adopted by using singular value decomposition (SVD), and it can preserve the original trend of the data [4,32]. The algorithm maximizes Pearson coherence between X_x and Y_y. Pearson correlation of data sets X_x and Y_y can be expressed as [4]

$$R(X_x, Y_y) = \max \frac{u^T S_{XY} v}{\sqrt{u^T S_{XX} u} \sqrt{v^T S_{YY} v}} \qquad (4)$$

where $S_{XY} = X_x^T Y_y$, $S_{XX} = X_x^T X_x$ and $S_{YY} = Y_y^T Y_y$. According to the data set X_x and Y_y given above, standardization is carried out, respectively. Calculation of matrix is [6]

$$W = S_{XX}^{-\frac{1}{2}} S_{XY} S_{YY}^{-\frac{1}{2}} = \frac{X_x^T Y_y}{\sqrt{X_x^T X_x} \sqrt{Y_y^T Y_y}} \qquad (5)$$

The matrix W is decomposed by singular value decomposition (SVD), and the W is decomposed as [4]

$$W = \Gamma D V^T \qquad (6)$$

where $\Gamma = (u_1, \cdots, u_t)$, $D = \begin{bmatrix} D_h & 0 \\ 0 & 0 \end{bmatrix}$, $V = (v_1, \cdots, v_n)$, and where h represents the number of top-ranking singular values and $\Sigma_h = \text{diag}(\rho_1, \cdots, \rho_h)$. Through the formula $\omega = \Gamma(:, 1:h)$, $\psi = V(:, 1:h)$, the related subspaces H_x and H_y are generated as [6]

$$H_x = S_{XX}^{-\frac{1}{2}} \omega = \frac{1}{\sqrt{X_x^T X_x}} \omega$$

$$H_y = S_{YY}^{-\frac{1}{2}} \psi = \frac{1}{\sqrt{Y_y^T Y_y}} \psi \qquad (7)$$

The latent space of X_x is divided into two subspaces, namely the related subspaces with Y_y and the unrelated subspaces with Y_y. Similarly, the latent space of Y_y is divided into two parts, namely the related subspaces with X_x and the unrelated subspaces with X_x. According to the above parameters, the original data inputs are mapped to the related latent spaces, H_x and H_y, and obtain two associated matrices, P_x and P_y. The correlation coefficient is $\rho 1$ between H_x and H_y if only considering the first canonical variate pair of CCA. The data matrix can be formulated as

$$P_x = u = H_x \left(X_x \left(H_x^T H_x \right)^{-1} H_x^T \right)^T$$

$$P_Y = v = H_y \left(Y_y \left(H_y^T H_y \right)^{-1} H_y^T \right)^T \qquad (8)$$

The following JITL algorithm is carried out on P_x and P_y, respectively, for data fitting. Different from the traditional global model approach, this JITL-based approach uses an online local model structure which could effectively track the current status of the algorithm.

JITL is to improve the prediction of the local FD model using similarity measures. After the most relevant normal data selected from the database, the distance measure, e.g., Euclidean distance $d(t'_s, t_c) = \|t'_s, t_c\|_2$, is employed to evaluate the similarity between t'_s and t_c. Here, t'_s is the data point of the training set, t_c is the data point of the test set; that is, a smaller value of distance implies a greater similarity between these two vectors [26]. The inverse of Euclidean distance is used to find the correlation between two vectors.

$$S_{i,k} = \frac{1}{\sqrt{e^{(\|t'_s, t_c\|_2)^2}}}, i = 1, 2, 3, \cdots, N \qquad (9)$$

where $S_{i,k}$ represents the magnitude of correlation.

Remark 3. *The JITL algorithm arranges $S_{i,k}$ values in descending order. The number of data to be selected is determined by calculating the accumulated contribution value of $S_{i,k}$ to the variance of the overall data, and the formula for the average value of $S_{i,k}$ is $\theta_i = \left(\sum_{i=1}^N S_{i,k} \right) / N$. The variance formula is $G = (S_{i,k} - \theta_i)^2$. The contribution parameter G is used to determine how many data points to be included in the testing sample data. For example, the algorithm picks 900 data points until the sum of G value reach 90%.*

Remark 4. *JITL selects testing data points which have lower correlations with training dataset. In the experiment, the system only takes the last 900 data points from the sorted testing dataset.*

3.2. Monitoring Statistics of FD Models

This section describes the test statistics used for FD. This article uses T^2 and SPE for FD. Firstly, a data matrix to be detected is given. Let the input and output matrices obtained after the JITL processing be $\mu_x = [\alpha_r(1), \cdots, \alpha_r(N)] \in R^{l \times N}$, $\mu_y = [\alpha_c(1), \cdots, \alpha_c(N)] \in R^{m \times N}$. According to the Formulas (5)–(7), the residual vector is obtained [6]

$$s = H_y^T \mu_x - M^T \mu_y \qquad (10)$$

where $M^T = D_h H_x^T$.

In the FD algorithm, statistics and their corresponding thresholds define the boundaries of system prediction. T^2 and SPE are the two most commonly used statistics in FD [33–36]. Taking the two data matrices P_x and P_y perform separately FD. The detection of the subspaces is the same as the routine detection process. Then, judging whether the input signals normally or not requires the following methods [4]

$$\begin{aligned} & SPE = s^T s \\ & SPE_x \leq J_{x,th} \quad \text{and} \quad SPE_y \leq J_{y,th} \Rightarrow \text{fault} - \text{free} \\ & SPE_x > J_{x,th} \quad \text{or} \quad SPE_y > J_{y,th} \Rightarrow \text{faulty} \end{aligned} \qquad (11)$$

where $J_{x,th}$ and $J_{y,th}$ are the thresholds for SPE_x and SPE_y, respectively. Then, judging whether the input signals normally or not need methods as following [4]

$$\begin{aligned} & T^2 = s^T \Lambda^{-1} s \\ & T_x^2 \leq T_{x,th} \quad \text{and} \quad T_y^2 \leq T_{y,th} \Rightarrow \text{fault} - \text{free} \\ & T_x^2 > T_{x,th} \quad \text{or} \quad T_y^2 > T_{y,th} \Rightarrow \text{faulty} \end{aligned} \qquad (12)$$

where s is residual matrix, $T_{x,th}$ and $T_{y,th}$ are the thresholds for T_x^2 and T_y^2, respectively.

3.3. Offline Training and Online Detection Algorithms

The procedures in Algorithm 1 are used for offline training. The steps in Algorithm 2 are used for online detection.

Algorithm 1 Offline training
1: Normalize the measurement data.
2: The data is divided into two data matrices via CCA model.
3: The JITL model is used to improve accuracy of data fitting.
4: Find the thresholds $J_{x,th}$ and $T_{x,th}$ associated with the data matrix P_x, and the thresholds $J_{y,th}$ and $T_{y,th}$ associated with the data matrix P_y.

Algorithm 2 Online detection
1: The collected fault data is normalized.
2: Find the two data matrices.
3: The JITL model is used to improve accuracy of data fitting.
4: Calculate SPE and T^2 via (11) and (12).
5: Determine whether a fault occurs comparing the test statistic with the thresholds.

3.4. System Evaluation Methodology

To measure the performance of the FD models, the most commonly used evaluation metrics are the false alarm rate (FAR) and fault detection rate (FDR). FAR uses the probability to quantify the occurrence of alarm when there is no fault. FDR uses the probability to quantify the occurrence of the alarm method in the case of actual failure.

According to the threshold calculated above, FARs and FDRs can be expressed as follows

$$FAR = \frac{M_j}{M_{th}} \times 100\% \qquad (13)$$

where M_j is the number of test statistics higher than the threshold in fault-free conditions, M_{th} is the total number of test statistics.

$$FDR = \frac{B_j}{B_{th}} \times 100\% \qquad (14)$$

where B_j is the number of test statistics higher than the threshold after injection of fault, and B_{th} is the total number of test statistics after fault injection.

Receiver Operating Characteristic (ROC) curves represent the performance of the model at different thresholds. The X axis of the curve is the false positive rate, and the Y axis is the true positive rate. The ideal is an inverted L-shaped curve [37]. The calculation formulas of the true positive rate (TPR) and false positive rate (FPR) are [38]

$$TPR = \frac{TP}{TP+FN}$$
$$FPR = \frac{FP}{TN+FP} \quad (15)$$

where TN is actually the number of samples classified into negative samples, FP is actually the number of samples classified into positive samples, FN is actually the number of samples classified into negative samples, TP is actually the number of samples classified into positive samples.

The Area Under a ROC Curve (AUC) is a comprehensive measure of sensitivity and specificity across all possible threshold ranges. It represents the probability that a classifier will rank randomly selected positive instances higher than randomly selected negative instances. The AUC ranges from 0 to 1. The closer AUC is to 1, the better FD performance [38]. The calculation formula is

$$AUC = \sum_{i=1}^{N} \frac{(TPR(i) + TPR(i+1))(FPR(i+1) - FPR(i))}{2} \quad (16)$$

4. Experimental Results and Discussion

High-speed train running gears systems are considered to verify the reliability of the proposed algorithms. When the data of the running gears is chosen, and in order to guarantee the consistency of the experiment input, signal data of the running parts was adopted from the same train and the same carriage. In order to guarantee the validity of the data, the monitoring data at the speed of 1000 r/min or above were utilized in the model. The paper uses real data of a running gears system with fault signals to simulate the settings of the experiments very close to the real situations.

4.1. Experimental Verification

Figure 3 shows two correlated subspaces of the input dataset. Figure 3a shows each input variable in the data set P_x. The charts of variables from top to bottom belong to bogie 1, bogie 2, motor-driven side bearing, and non-driven side bearing, respectively. Figure 3b shows each variable in the data set P_y. From top to bottom, the charts represent motor side big gear, wheel side pinion, wheel side motor big gear, and motor side pinion, respectively.

1. Fault Injection: Under the given speed 1000 r/min of high-speed trains, 1000 × 8 samples under health and fault conditions are collected from eight sensors as data sets. Fault data was injected from the 500th data points of the sample test dataset.
2. Fault Detection: Fault detection results of CCA-JITL are shown in Figure 4 where red dashed lines are thresholds and blue sold lines are test statistics.

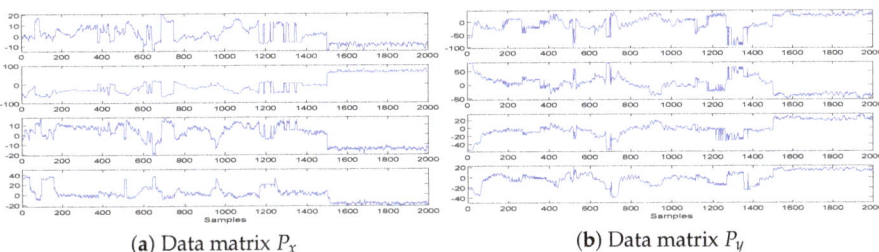

Figure 3. The input dataset. (**a**) Data matrix P_x; (**b**) Data matrix P_y.

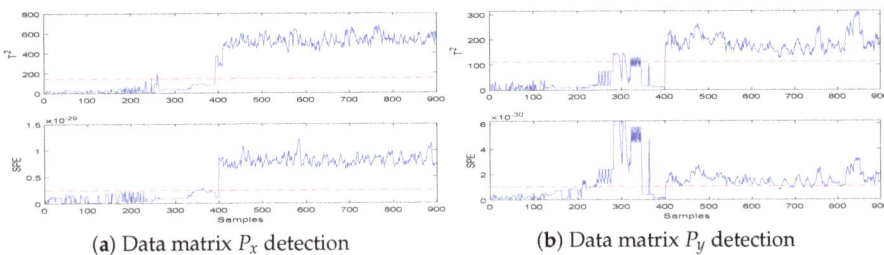

Figure 4. Experiment results of CCA-JITL FD model. (**a**) Data matrix P_x detection; (**b**) Data matrix P_y detection.

4.2. Discussions

In order to prove the reliability of the method in this paper, several points will be discussed: (1) the problem solved by this method; (2) the comparison analysis based on the FAR and FDR; (3) the feasibility of the proposed algorithms is testified.

CCA-JITL FD model was applied to detect fault signals of the running gears in two groups in which the results were compared to each other to improve detection accuracy. The method uses CCA to group data and a JITL algorithm to optimize selection of sample data points, so as to achieve better FD performance based on the data shown in Figure 4. Figure 4a depicts the CCA-JITL FD output based on data set P_x, and Figure 4b is the result of CCA-JITL FD based on data set P_y. The number of singular vectors, h values, decide the proximity of dimension reduction and affect FAR and FDR very much. We tuned parameters and concluded that when $h = 2$, CCA-JITL models achieved the best performance.

Figure 5 shows FD experiment results using only CCA. The system infrastructure of CCA-JITL was generalized to be utilized to other FD models using PCA and PLS. FD experiment results using SVD-based PLS and JITL are shown in Figure 6. FD experiment results using PCA and JITL are shown in Figure 7. FD experiment results using only PCA are shown in Figure 7.

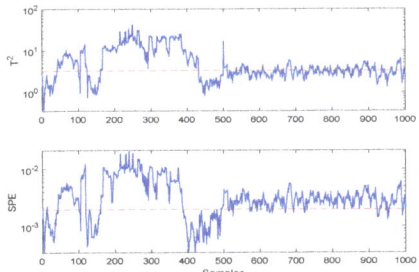

Figure 5. Experiment results of the FD model using CCA.

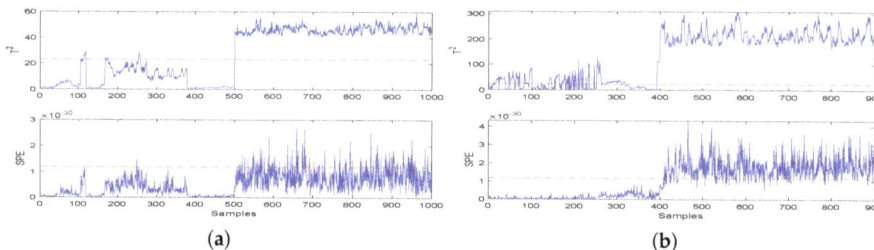

Figure 6. Experiment results of the FD model using PLS. (**a**) Online testing of the FD model using PLS; (**b**) Online testing of the FD model using PLS and JITL.

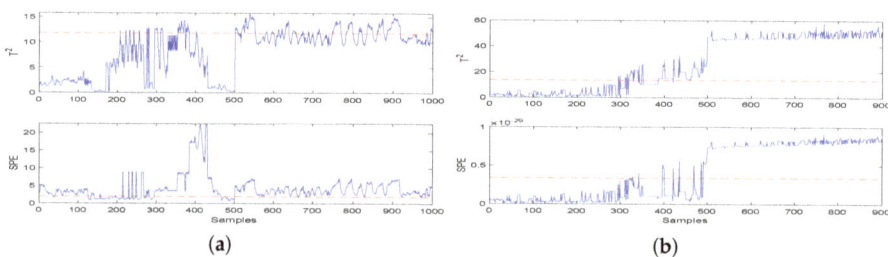

Figure 7. Experiment results of the FD model using PCA. (**a**) Online testing of the FD model using PCA; (**b**) Online testing of the FD model using PCA and JITL.

Based on the detection results shown in Figure 4, the detection of data set P_x after injection fault data is normal, and the detection results of data set P_y show short-term and transient fluctuations after fault injection. Then, the statistics fall back above the threshold. The detection results of the datasets P_x and P_y are compared with each other to verify the performance after injecting fault data, Figure 4a detects a fault, and Figure 4b shows short-term fluctuations. Then, the statistics fall back above the threshold. According to the comparison and verification of the FD results, it was proved that the fault detection at the 500th sample was accurate. Based on the detection results shown in Figure 5, it was observed that statistics were above the threshold before the injection failure time, so false positives have occurred. Moreover, after the injection fault, there is a fluctuation of the statistical value lower than the threshold value, and there is a situation of false negatives. Based on the detection results shown in Figure 6, in Figure 6a it was observed that statistics were above the threshold before the injection failure time, so false positives have occurred. Additionally, after the injection fault, there is a fluctuation of the statistical value lower than the threshold value, and there is a situation of false negatives. Based on the detection results shown in Figure 6b, detection after fault injection is normal, but the fluctuation of statistical value before injecting fault data was partly above the threshold, so a false positive situation had occurred.

Based on the detection results shown in Figure 7. Based on the detection results shown in Figure 7a. T^2 statistics showed a few statistical fluctuations higher than the threshold before the fault injection, so false positive situations have occurred. SPE statistics fluctuated a few times above the threshold before the injection of failure signals and kept below the threshold many times after the fault injection. There are serious false positives and omissions. In Figure 7b it was observed that the short-term or instantaneous fluctuations of T^2 scores were above the threshold before the fault injection time, so false positive situations had occurred. In the detection results of this method, SPE statistics fluctuation indicates that the SPE scores fluctuate higher than the thresholds before a fault signal is added and then if the SPE scores remain below the threshold, thus, false negative situations have occurred.

As shown in Figure 8, the receiver operating characteristic (ROC) curves of CCA-JITL, CCA, PLS-JITL, PLS, PCA-JITL and PCA are compared. The ROC curves of each method from top to bottom represent the ROC curves of the model when T^2 statistics and SPE statistics are used, respectively. Combined with the area under the curve (AUC) the score of each method shown in Table 1. It proved that the performance of CCA-JITL is better than other methods. The AUC values of CCA-JITL, PLS-JITL and PCA-JITL were mostly higher compared with those of PCA, CCA and PLS. The AUC scores of the models increases after adding JITL.

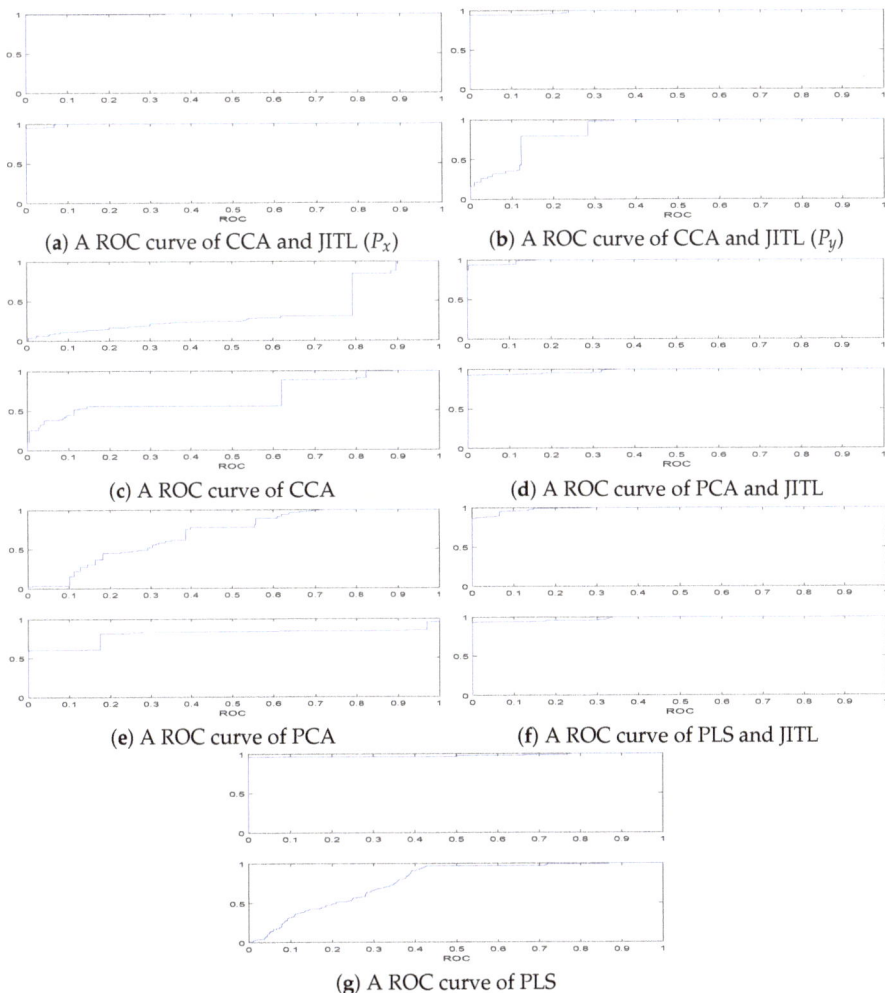

Figure 8. The ROC curves of FD models. (**a**) A ROC curve of CCA and JITL (P_x); (**b**) A ROC curve of CCA and JITL (P_y); (**c**) A ROC curve of CCA; (**d**) A ROC curve of PCA and JITL; (**e**) A ROC curve of PCA; (**f**) A ROC curve of PLS and JITL; (**g**) A ROC curve of PLS.

Comparisons of FAR and FDR measures on FD models using CCA-JITL, CCA, PLS-JITL, PLS, PCA-JITL and PCA are shown in Table 1. By comparing FAR and FDR among all algorithms, CCA-JITL worked best for online testing. FAR and FDR scores were calculated regarding T^2 and SPE statistics, respectively. The average scores of FAR and FDR were considered to be used in the result comparison. Compared with the CCA method, the average FAR score of CCA-JITL was reduced by 54.44% across both T^2 and SPE measures, and the average FDR score was increased by 34.05%. Compared with algorithms based on PCA-JITL or PLS-JITL, the FAR score of CCA-JITL is lower. The JITL component improved all the FDR scores of all the FD models. PLS, PCA, and CCA showed that the FDR score increased by 32.20%, 29.2%, and 34.05%, respectively, after using JITL. Since the variables of the real data used in this paper are not independent, CCA-JITL method are more favorable for FD of the data in the running gears. The feasibility of the proposed algorithms were testified by the above comparative experiments. JITL is also useful to shape the visual representation of data fitting so that the fault signals were displayed more distinguished.

Table 1. Online Testing Results of FD models.

Methods	FAR		FDR		AUC	
	T^2	SPE	T^2	SPE	T^2	SPE
PLS	8.81%	45.69%	100%	17.61%	0.9802	0.7729
PLS and JITL	44.75%	0%	100%	82%	0.8430	0.9261
PCA	14.75%	83.5%	41.6%	100%	0.5612	0.7708
PCA and JITL	28.26%	7.41%	100%	100%	0.9887	0.9798
CCA	69.8%	62.2%	38%	90%	0.3601	0.6775
CCA and JITL (P_x)	2.5%	6.5%	100%	100%	0.9961	0.9943
CCA and JITL (P_y)	7.75%	29.5%	100%	92.2%	0.9847	0.8778
CCA and JITL (average value)	5.125%	18%	100%	96.1%	0.9904	0.9361

The model was tested using a new 1000×8 data set as the independent testing set. According to Table 2, compared with the CCA model, the results of independent testing of the CCA-JITL FD model showed that the AUC score increased, FAR decreased by 33.2%, and FDR increased by 60.65%. It proved that this approach is generalizable and still had good performance when random new data was applied.

Table 2. Independent testing of FD models.

Methods	FAR		FDR		AUC	
	T^2	SPE	T^2	SPE	T^2	SPE
PLS	4.6%	0.6%	80%	78%	0.7851	0.7861
PLS and JITL	44.5%	0.5%	100%	66.8%	0.9384	0.8506
PCA	13%	66.8%	76.4%	79.6%	0.7370	0.3738
PCA and JITL	0%	2.75%	98.6%	98.2%	0.9975	0.9868
CCA	83.2%	54.2%	16.2%	62%	0.2158	0.7730
CCA and JITL (P_x)	13.25%	19%	100%	100%	0.9587	0.9874
CCA and JITL (P_y)	34.25%	67.5%	99%	100%	0.8081	0.8291
CCA and JITL (average value)	23.75%	43.25%	99.5%	100%	0.8834	0.9083

5. Conclusions and Future Studies

In this study, the proposed algorithms have demonstrated significant advantages on the fault detectability in the running gear systems. This paper presents an FD algorithm based on CCA and JITL. After data preprocessing and normalization, CCA transforms high-dimension historical input data matrices from the database into low-dimension subspaces to maximize correlations between the most important latent dimensions. Then, online input sample data is mapped to these subspaces with coordinates. Finally, JITL components measure Euclidean similarity between query samples and historical samples in subspaces and search subsets of query sample data points with largest distance to training data to build local fault detection models. The evaluation results of the case study showed CCA-JITL outperformed traditional CCA very much in terms of FAR and FDR. This approach was also applied to the FD models based on PCA and PLS and achieved better outcomes, which suggested our system infrastructure was transferable to PCA and PLS FD models.

In future, there are still many research directions that are worth further study. The evaluation results in Tables 1 and 2 suggested that PCA, PLS and CCA FD models have their unique strengths using different evaluation methods, and thus, the study of model fusion strategies will be promising. Moreover, only FD was investigated in this paper, without classifying and diagnosing positions and categories of faults. Different types of FDD machine learning models will be meaningful to detect specific failure points. Another possible direction for optimization is to change the fitting methods of JITL, such as clustering, and the derivation method of CCA, such as Kernel-based CCA, to enhance the performance of the systems. The third possibility is to improve the scope of the model, that is, how to apply the models to dynamic systems. Furthermore, the research investigation on how to support multi-sensor data acquisition will be very useful, for instance, the data acquisition system using FUSED deposition modeling [39]. Moreover, the method of using prior prediction to detect the remaining useful life is also an important research direction. These research topics will be considered in order to successfully implement and deliver real-world FDD applications for high-speed train running gear systems of high-speed trains.

Author Contributions: Conceptualization, H.Z.; supervision, C.C.; writing—original draft preparation, K.Z.; visualization, Z.F. All authors have read and agreed to the published version of the manuscript.

Funding: This research received no external funding.

Institutional Review Board Statement: Not applicable.

Informed Consent Statement: Not applicable.

Data Availability Statement: Not applicable.

Conflicts of Interest: The authors declare no conflict of interest.

References

1. Chen, H.; Jiang, B. A Review of Fault Detection and Diagnosis for the Traction System in High-Speed Trains. *IEEE Trans. Intell. Transp. Syst.* **2020**, *21*, 450–465. [CrossRef]
2. Gao, S.; Hou, Y.; Dong, H.; Stichel, S.; Ning, B. High-speed trains automatic operation with protection constraints: A resilient nonlinear gain-based feedback control approach. *IEEE/CAA J. Autom. Sin.* **2019**, *6*, 992–999. [CrossRef]
3. Chen, L.; Hu, X.; Tian, W.; Wang, H.; Cao, D.; Wang, F.Y. Parallel planning: A new motion planning framework for autonomous driving. *IEEE/CAA J. Autom. Sin.* **2019**, *6*, 236–246. [CrossRef]
4. Chen, H.; Jiang, B.; Ding, S.X.; Huang, B. Data-Driven Fault Diagnosis for Traction Systems in High-Speed Trains: A Survey, Challenges, and Perspectives. *IEEE Trans. Intell. Transp. Syst.* **2022**, *23*, 1700–1716. [CrossRef]
5. Ran, G.; Liu, J.; Li, C.; Lam, H.K.; Li, D.; Chen, H. Fuzzy-Model-Based Asynchronous Fault Detection for Markov Jump Systems with Partially Unknown Transition Probabilities: An Adaptive Event-Triggered Approach. *IEEE Trans. Fuzzy Syst.* **2022**, 1. [CrossRef]
6. Chen, H.; Chen, Z.; Chai, Z.; Jiang, B.; Huang, B. A Single-Side Neural Network-Aided Canonical Correlation Analysis With Applications to Fault Diagnosis. *IEEE Trans. Cybern.* **2021**, 1–13. [CrossRef]
7. Raveendran, R.; Kodamana, H.; Huang, B. Process monitoring using a generalized probabilistic linear latent variable model. *Automatica* **2018**, *96*, 73–83. [CrossRef]
8. Ge, Z.; Song, Z.; Ding, S.X.; Huang, B. Data Mining and Analytics in the Process Industry: The Role of Machine Learning. *IEEE Access* **2017**, *5*, 20590–20616. [CrossRef]
9. Li, X.; Yang, Y.; Pan, H.; Cheng, J.; Cheng, J. A novel deep stacking least squares support vector machine for rolling bearing fault diagnosis. *Comput. Ind.* **2019**, *110*, 36–47. [CrossRef]
10. Zhu, Q.; Qin, S.J. Supervised Diagnosis of Quality and Process Faults with Canonical Correlation Analysis. *Ind. Eng. Chem. Res.* **2019**, *58*, 11213–11223. [CrossRef]
11. Jiang, Q.; Yan, X. Learning Deep Correlated Representations for Nonlinear Process Monitoring. *IEEE Trans. Ind. Inform.* **2019**, *15*, 6200–6209. [CrossRef]
12. Chen, Z.; Ding, S.X.; Peng, T.; Yang, C.; Gui, W. Fault Detection for Non-Gaussian Processes Using Generalized Canonical Correlation Analysis and Randomized Algorithms. *IEEE Trans. Ind. Electron.* **2018**, *65*, 1559–1567. [CrossRef]

13. Peng, X.; Ding, S.X.; Du, W.; Zhong, W.; Qian, F. Distributed process monitoring based on canonical correlation analysis with partly-connected topology. *Control Eng. Pract.* **2020**, *101*, 104500. [CrossRef]
14. Chen, H.; Li, L.; Shang, C.; Huang, B. Fault Detection for Nonlinear Dynamic Systems With Consideration of Modeling Errors: A Data-Driven Approach. *IEEE Trans. Cybern.* **2022**, 1–11. [CrossRef]
15. Garramiola, F.; Poza, J.; Madina, P.; del Olmo, J.; Ugalde, G. A Hybrid Sensor Fault Diagnosis for Maintenance in Railway Traction Drives. *Sensors* **2020**, *20*, 962. [CrossRef]
16. Kou, L.; Qin, Y.; Zhao, X.; Chen, X. A Multi-Dimension End-to-End CNN Model for Rotating Devices Fault Diagnosis on High-Speed Train Bogie. *IEEE Trans. Veh. Technol.* **2020**, *69*, 2513–2524. [CrossRef]
17. Zhao, R.; Yan, R.; Chen, Z.; Mao, K.; Wang, P.; Gao, R.X. Deep learning and its applications to machine health monitoring. *Mech. Syst. Signal Process.* **2019**, *115*, 213–237. [CrossRef]
18. Niu, G.; Xiong, L.; Qin, X.; Pecht, M. Fault detection isolation and diagnosis of multi-axle speed sensors for high-speed trains. *Mech. Syst. Signal Process.* **2019**, *131*, 183–198. [CrossRef]
19. Fu, Y.; Huang, D.; Qin, N.; Liang, K.; Yang, Y. High-Speed Railway Bogie Fault Diagnosis Using LSTM Neural Network. In Proceedings of the 2018 37th Chinese Control Conference (CCC), Wuhan, China, 25–27 July 2018; pp. 5848–5852. [CrossRef]
20. Cheng, C.; Guo, Y.; Wang, J.h.; Chen, H.; Shao, J. A Unified BRB-based Framework for Real-Time Health Status Prediction in High-Speed Trains. *IEEE Trans. Veh. Technol.* **2022**, 1. [CrossRef]
21. Guan, S.; Huang, D.; Guo, S.; Zhao, L.; Chen, H. An Improved Fault Diagnosis Approach Using LSSVM for Complex Industrial Systems. *Machines* **2022**, *10*, 443. [CrossRef]
22. Sayyad, S.; Kumar, S.; Bongale, A.; Kamat, P.; Patil, S.; Kotecha, K. Data-Driven Remaining Useful Life Estimation for Milling Process: Sensors, Algorithms, Datasets, and Future Directions. *IEEE Access* **2021**, *9*, 110255–110286. [CrossRef]
23. Capriglione, D.; Carratù, M.; Pietrosanto, A.; Sommella, P. Online Fault Detection of Rear Stroke Suspension Sensor in Motorcycle. *IEEE Trans. Instrum. Meas.* **2019**, *68*, 1362–1372. [CrossRef]
24. Shabanian, M.; Montazeri, M. A neuro-fuzzy online fault detection and diagnosis algorithm for nonlinear and dynamic systems. *Int. J. Control. Autom. Syst.* **2011**, *9*, 665–670. [CrossRef]
25. Yan, B.; Yu, F.; Huang, B. Generalization and comparative studies of similarity measures for Just-in-Time modeling. *IFAC-PapersOnLine* **2019**, *52*, 760–765. [CrossRef]
26. Cheng, C.; Chiu, M.S. Nonlinear process monitoring using JITL-PCA. *Chemom. Intell. Lab. Syst.* **2005**, *76*, 1–13. [CrossRef]
27. Bittanti, S.; Picci, G. *Identification, Adaptation, Learning: The Science of Learning Models from Data*; Springer Science & Business Media: Berlin/Heidelberg, Germany, 1996; Volume 153.
28. Yu, H.; Yin, S.; Luo, H. Robust Just-in-time Learning Approach and Its Application on Fault Detection. *IFAC-PapersOnLine* **2017**, *50*, 15277–15282. [CrossRef]
29. Chen, H.; Chai, Z.; Jiang, B.; Huang, B. Data-Driven Fault Detection for Dynamic Systems with Performance Degradation: A Unified Transfer Learning Framework. *IEEE Trans. Instrum. Meas.* **2021**, *70*, 1–12. [CrossRef]
30. Ran, G.; Li, C.; Lam, H.K.; Li, D.; Han, C. Event-Based Dissipative Control of Interval Type-2 Fuzzy Markov Jump Systems Under Sensor Saturation and Actuator Nonlinearity. *IEEE Trans. Fuzzy Syst.* **2022**, *30*, 714–727. [CrossRef]
31. Cheng, C.; Qiao, X.; Luo, H.; Teng, W.; Gao, M.; Zhang, B.; Yin, X. A Semi-Quantitative Information Based Fault Diagnosis Method for the Running Gears System of High-Speed Trains. *IEEE Access* **2019**, *7*, 38168–38178. [CrossRef]
32. Cheng, C.; Wang, J.; Teng, W.; Gao, M.; Zhang, B.; Yin, X.; Luo, H. Health Status Prediction Based on Belief Rule Base for High-Speed Train Running Gear System. *IEEE Access* **2019**, *7*, 4145–4159. [CrossRef]
33. Chen, H.; Chai, Z.; Dogru, O.; Jiang, B.; Huang, B. Data-Driven Designs of Fault Detection Systems via Neural Network-Aided Learning. *IEEE Trans. Neural Netw. Learn. Syst.* **2021**, 1–12. [CrossRef]
34. Kresta, J.V.; Macgregor, J.F.; Marlin, T.E. Multivariate statistical monitoring of process operating performance. *Can. J. Chem. Eng.* **1991**, *69*, 35–47. [CrossRef]
35. Wise, B.M.; Gallagher, N.B. The process chemometrics approach to process monitoring and fault detection. *J. Process. Control* **1996**, *6*, 329–348. [CrossRef]
36. Chiang, L.H.; Russell, E.L.; Braatz, R.D. *Fault Detection and Diagnosis in Industrial Systems*; Springer Science & Business Media: Berlin/Heidelberg, Germany, 2000.
37. Kumar, D.; Ding, X.; Du, W.; Cerpa, A. Building sensor fault detection and diagnostic system. In Proceedings of the 8th ACM International Conference on Systems for Energy-Efficient Buildings, Cities, and Transportation, Coimbra, Portugal, 17–18 November 2021; pp. 357–360.
38. Li, G.; Zheng, Y.; Liu, J.; Zhou, Z.; Xu, C.; Fang, X.; Yao, Q. An improved stacking ensemble learning-based sensor fault detection method for building energy systems using fault-discrimination information. *J. Build. Eng.* **2021**, *43*, 102812. [CrossRef]
39. Kumar, S.; Kolekar, T.; Patil, S.; Bongale, A.; Kotecha, K.; Zaguia, A.; Prakash, C. A Low-Cost Multi-Sensor Data Acquisition System for Fault Detection in Fused Deposition Modelling. *Sensors* **2022**, *22*, 517. [CrossRef]

Article

Fault Detection of Bearing by Resnet Classifier with Model-Based Data Augmentation

Lu Qian [1], Qing Pan [2], Yaqiong Lv [1] and Xingwei Zhao [3],*

[1] School of Transportation and Logistics Engineering, Wuhan University of Technology, Wuhan 430070, China; qianlu@whut.edu.cn (L.Q.); y.q.lv@whut.edu.cn (Y.L.)
[2] State Key Laboratory of High Performance Complex Manufacturing, School of Mechanical and Electrical Engineering, Central South University, Changsha 410083, China; panqing0905@csu.edu.cn
[3] State Key Laboratory of Digital Manufacturing Equipment and Technology, Huazhong University of Science and Technology, Wuhan 430074, China
* Correspondence: zhaoxingwei@hust.edu.cn

Abstract: It is always an important and challenging issue to achieve an effective fault diagnosis in rotating machinery in industries. In recent years, deep learning proved to be a high-accuracy and reliable method for data-based fault detection. However, the training of deep learning algorithms requires a large number of real data, which is generally expensive and time-consuming. To cope with this, we proposed a Resnet classifier with model-based data augmentation, which is applied for bearing fault detection. To this end, a dynamic model was first established to describe the bearing system by adjusting model parameters, such as speed, load, fault size, and the different fault types. Large amounts of data under various operation conditions can then be generated. The training dataset was constructed by the simulated data, which was then applied to train the Resnet classifier. In addition, in order to reduce the gap between the simulation data and the real data, the envelop signals were used instead of the original signals in the training process. Finally, the effectiveness of the proposed method was demonstrated by the real bearing experimental data. It is remarkable that the application of the proposed method can be further extended to other mechatronic systems with a deterministic dynamic model.

Keywords: bearing fault detection; deep residual network; data augmentation

Citation: Qian, L.; Pan, Q.; Lv, Y.; Zhao, X. Fault Detection of Bearing by Resnet Classifier with Model-Based Data Augmentation. *Machines* **2022**, *10*, 521. https://doi.org/10.3390/machines10070521

Academic Editors: Hongtian Chen, Kai Zhong, Guangtao Ran and Chao Cheng

Received: 28 May 2022
Accepted: 24 June 2022
Published: 27 June 2022

Publisher's Note: MDPI stays neutral with regard to jurisdictional claims in published maps and institutional affiliations.

Copyright: © 2022 by the authors. Licensee MDPI, Basel, Switzerland. This article is an open access article distributed under the terms and conditions of the Creative Commons Attribution (CC BY) license (https://creativecommons.org/licenses/by/4.0/).

1. Introduction

As an indispensable element of rotating machinery, the rolling bearing plays an effective and crucial role in real industries, whose operation status profoundly influences the performance of rotating machinery equipment. If faults occur in critical bearings, it may cause costly downtime and catastrophic accidents. Therefore, having an effective and accurate fault diagnosis of bearings is critical to improving the reliability and safety of rotating machinery equipment.

During the past decades, the fault detection and diagnosis of roller bearings have been receiving increasing attention and have been a research hotspot. Due to the distinctive characteristics of the vibration signals produced by a faulty bearing, such as its periodicity and sensitivity to faults, great efforts have been made to develop a bearing fault diagnosis based on vibration-based methods. Model-based methods are devoted to revealing the fault generation mechanism and finding the fault-related information according to the map from inputs to responses [1,2]. Meanwhile, a few model-based methods have also been applied for degradation data analysis and the remaining useful life estimation and prediction [3,4]. In addition, numerous signal processing methods have been used to reduce the noise and extract and highlight the fault-related features in vibration signals to achieve an accurate fault diagnosis [5,6]. These methods can be classed into three categories on the basis of the fundamentals of signal processing methods. The first is time-domain analysis [7], such as the peak value, standard deviation and kurtosis, and so on. Frequency domain analysis, typified by fast Fourier frequency transform (FFT) [8], is the second category. The third kind is time-frequency domain analysis, including short-time Fourier transform (STFT) [9], wavelet transform [10,11] and empirical mode decomposition (EMD) [12], and so forth. However, most of the available traditional signal-based methods presently require human intervention and sufficient expert knowledge on the diagnosis of an object and signal processing, which limits their industrial application to mechanical equipment fault diagnosis. In this regard, alternative methods should be developed for a bearing fault diagnosis.

To overcome the limitations of demands of prior expertise based on the signal-based methods and achieve higher performance, machine learning techniques have already been widely applied in mechanical fault diagnoses [13]. Based on the machine learning techniques, fault diagnosis is regarded as a classification problem. In the traditional machine learning methods, representative features are first extracted from the raw signals, based on which pattern of recognition technology is applied to classify the health conditions of the equipment, for instance, support vector machines (SVM) [14], clustering algorithms [15] and artificial neural networks (ANN) [16,17] and so on. Shi et al. [18] applied linear discriminant analysis and gray wolf optimizer to improve the SVM algorithm and enhance the performance of fault classification. Zhang et al. [19] applied the BP neural network algorithm, which was based on the transfer component analysis, to detect the bearing fault states. In spite of the success achieved by these methods of fault diagnosis in the past years, it is still a challenge to ensure fault diagnosis accuracy with highly complex nonlinear signals. Due to the high performance in dealing with nonstationary signals, the deep learning method has recently been developed for feature extraction and pattern recognition [20]. Lei et al. [21] presented a framework for intelligent fault diagnosis, where a two-layer neural network with sparse filtering was constructed to learn the features from raw mechanical signals directly. Additionally, based on these learned features, the mechanical faults were identified by the classifier. Kolar et al. [22] propose a multi-channel deep convolutional neural network configuration for a rotary-machinery state classification. Janssens et al. [23] proposed a feature learning model for the bearings condition monitoring, based on convolutional neural networks, which removed the need for expert knowledge related to feature extraction compared with the classical statistical feature analysis. Mao et al. [24] proposed a multiple-fault diagnosis method that was based on deep output kernel learning, in which the depth features were extracted adaptively by an auto encoder neural network and thus, by means of solving the objective function constructed by the output kernel regularizer, the fault classifier was constructed. Due to the powerful capacity for classification and excellent convergence behaviors, deep learning methods can learn the deep features of different data and distinguish them automatically. However, deep learning methods require a large number of datasets to achieve a high accuracy of classification [25]. The industrial applications are limited by the requirement of in-service data under a wide range of operating conditions, which is generally an expensive and time-consuming practice to carry out dozens of experiments, especially for the key components in large machinery and equipment.

To deal with this issue, researchers have started to focus on the data augmentation method to extend the amount of available data with limited in-service data. Data augmentation is first applied in the field of two-dimensional images, and then the available images are transformed into new images by various means [26,27]. To solve the problem of the paucity of data, some approaches were developed based on data augmentation to deal with one-dimensional signals. To achieve the engineering prognostics, Kim et al. [28] proposed a run-to-fail (RTF) data augmentation method based on the dynamic time warping (DTW) technique, where a neural network was trained for the remaining useful life prediction of the current system by using the other system's RTF data. A semi-supervised learning (SSL) approach, based on data augmentation and metric learning, was proposed by Yu et al. [29]. Seven data augmentation strategies were applied to expand the feature space with limited labeled data. However, the data augmentation was realized by transforming the available signals into new signals in these studies, which led to limited distribution and feature space of the dataset. To overcome this barrier, simulation-driven machine learning methods were studied to create the training data, including a variety of operating conditions, which can be combined with available in-service fault data for the fault diagnosis. A data simulation by resampling (DSR) method was proposed by Hu et al. [30] to generate various working conditions of data for fault diagnoses. Lu et al. [31] proposed a vibration-based classification approach using model-based data augmentation for light-weight robotic-drilling-condition identification, where a dynamic model for a robotic drilling system was built to generate signals for the training data augmentation. Sobie et al. [32] generated training data by using information gained from high-resolution roller bearing dynamics simulations. Then, the machine learning algorithms were trained with the simulated data to classify the bearing faults. However, the roller bearing dynamics in this study are considered as a linear system, in which the race defect is modeled with a prescribed force, and the interaction between each element caused by faults is neglected. There exist certain differences with the actual situation.

Motivated by the aforementioned studies, we developed a fault detection approach based on data augmentation for roller bearing in this paper, which integrated a model-based method and deep learning method. To be specific, a dynamic model of a roller bearing was first established to reflect the correspondence between the bearing states and vibration signals. Then, the data augmentation was achieved by the simulated signals generated by the dynamic model, and based on this, a fault classifier was trained by a deep learning algorithm. Moreover, the envelop signals were used instead of the original signals in the training process to reduce the gap between the simulated data and the real data. Finally, the operation states of the roller bearings could be identified by the trained fault classifier by inputting the vibration signals to be classified.

The remainder of the paper is organized as follows: Section 2 introduces the framework of the proposed method in detail, including the model-based data augmentation and deep residual network classification. The experimental study will be presented and discussed in Section 3. Finally, some conclusions are given in Section 4.

2. Methodology of Data Augmentation

Model-based methods and data-driven methods have demonstrated the effectiveness and performance of fault diagnosis of machines [1,33,34]. Model-based methods show advantages in providing the map from inputs to responses and revealing the fault generation mechanism. However, it is less effective in dealing with data at a low signal-to-noise ratio (SNR). By contrast, intelligent fault diagnosis methods can achieve reliable diagnostic results with complex signals. However, massive datasets are required to ensure classification accuracy, which brings about a high cost of data collection and training. To detect the bearing state with less real data, a fault detection method for bearings based on data augmentation is proposed in this paper, which integrates the model-based methods and data-driven methods. To this end, based on the physics knowledge and failure mechanism of the bearings, a dynamic model was constructed to generate the vibration signals of the bearing to alleviate the problem of data acquisition. Then, the generated dataset was used to realize the data augmentation, and the deep learning algorithm was applied to train the fault classifier. Moreover, the envelop signals were used instead of the raw signals in the training process to reduce the gap between the simulated data and the real data. Finally, a reliable fault classifier insensitive to noise signals was obtained. The operation states of the rolling bearings can be delivered by the fault classifier by inputting the vibration signals to be classified. The framework of the proposed method is shown in Figure 1.

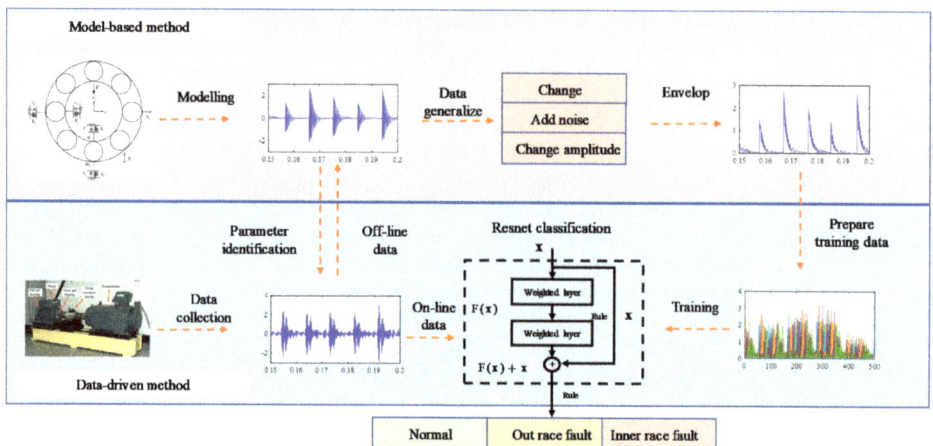

Figure 1. Framework of the fault detection by Resnet classifier with model-based data augmentation.

2.1. Model-Based Data Augmentation

To analyze the structural vibration characteristics of the rolling element bearing, the contact between the outer race and other components can be considered as a spring-mass system, in which the outer race is fixed on a pedestal, and the inner race is fixed with the shaft. The sensor is placed on the pedestal with an outer race to detect high-frequency natural vibrations of the bearing. Thus, to provide the vibration response signals of the rolling bearings containing different working states, a vibration model with four degrees of freedom (DOFs) was constructed by considering the movements in the horizontal and vertical directions of the inner race and outer race, as shown in Figure 2.

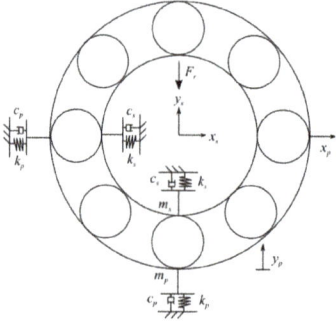

Figure 2. Vibration model of rolling element bearing.

Considering the vibration of the outer race and inner race in a vertical direction, the dynamic equation of the bearing system can be described as:

$$\begin{aligned} m_s\ddot{x}_s + c_s\dot{x}_s + k_s x_s + F_x &= 0 \\ m_s\ddot{y}_s + c_s\dot{y}_s + k_s y_s + F_y &= F_r \\ m_p\ddot{x}_p + c_p\dot{x}_p + k_p x_p &= F_x \\ m_p\ddot{y}_p + c_p\dot{y}_p + k_p y_p &= F_y \end{aligned} \quad (1)$$

where x_s and x_p denote the displacement of the inner race and outer race in the x direction, y_s and y_p represent the displacement of two raceways in the y direction, accordingly. F_x and F_y are the elastic contact force between the raceways and the rolling elements in the x

and y direction, and F_r is the radial load generally produced by the weight of the shaft and the rotor. Other parameters in Equation (1) can refer to the given nomenclature table.

According to the Hertz contact theory, the contact force between the raceways and the rolling elements can be given as

$$f = k_b \delta^n \tag{2}$$

where k_b represents the load-deflection factor which depends on the contact geometry and the elastic contacts of the material. δ is the overall contact deformation of the rolling elements, which is composed of the contact deformation of each rolling element. The exponent $n = 1.5$ for ball bearings and $n = 1.1$ for roller bearings.

When the bearings operate, part of the raceway will be in the load zone, and the other part of the raceway will be in the non-load zone, which is shown in Figure 3. The contact deformation of each rolling element is determined by the angular position of the rolling element, the relative displacement between the inner and outer races, and the bearing clearance. The calculation of the contact deformation of the jth rolling element can be given as

$$\delta_j = (x_s - x_p)\cos\phi_j + (y_s - y_p)\sin\phi_j - c, j = 1, 2, \ldots, n_b \tag{3}$$

where n_b denotes the number of the rolling elements. According to the elasto-hydrodynamic lubrication (EHL) theory, the clearance value c is set as negative, owing to the effect of oil EHL film [28]. The angular positions of the jth rolling element ϕ_j can be described as

$$\phi_j = \frac{2\pi(j-1)}{n_b} + \omega_c dt + \phi_0$$
$$\omega_c = (1 - \frac{D_b}{D_p})\frac{\omega_s}{2} \tag{4}$$

where ω_c is the angular velocity of the bearing cage, ϕ_0 denotes the initial angular position of the bearing cage, D_b is ball diameter and D_p is the pitch circle diameter of the bearing, ω_s is the angular velocity of the shaft.

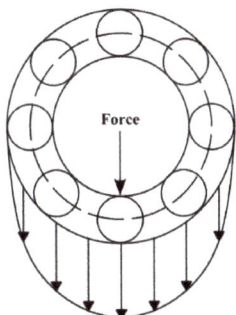

Figure 3. Load distribution of roller bearing.

According to Equations (2) and (3), summing up the contact forces of the n_b rolling elements, the overall nonlinear elastic contact forces of the bearing in the x and y directions can be calculated as

$$F_x = k_b \sum_{j=1}^{n_b} \gamma_j((x_s - x_p)\cos\phi_j + (y_s - y_p)\sin\phi_j - c)^{1.5}\cos\phi_j$$
$$F_y = k_b \sum_{j=1}^{n_b} \gamma_j((x_s - x_p)\cos\phi_j + (y_s - y_p)\sin\phi_j - c)^{1.5}\sin phi_j \tag{5}$$

where γ_j is a switch function which depends on the positive and negative values of the contact deformation δ_j, described as

$$\gamma_j = \begin{cases} 1 & if\ \delta_j > 0 \\ 0 & otherwise \end{cases} \tag{6}$$

It is noticed that the vibration model of the bearings presented above does not take different fault types into consideration. To simulate the vibration of the localized faults on the different components of bearings, the effects of the localized faults will be considered in the vibration model. The dynamic equation of the bearing system with faults can still be given by Equation (1). The main difference lies in the expression of the contact deformation of the rolling elements.

If a bearing operates in the health state at a steady speed, all forces in the bearing are in quasi-equilibrium. Once a localized fault occurs in the inner and outer races or the rolling elements, a certain deformation will be suddenly released when the fault contacts other components. As a result, a rapid change will take place in the elastic deformation of the components, and the force equilibrium state will be disturbed. Considering the new variations in the model with localized faults, the contact deformation of the jth rolling element is rewritten as

$$\delta_j = (x_s - x_p)\cos\phi_j + (y_s - y_p)\sin\phi_j - c - \overbrace{\beta_j c_d}^{\text{fault part}} \tag{7}$$

where c_d denotes the fault depth. β_j is a switch function to describe whether there is a contact loss due to the fault depth, which is closely related to the angular position of the faults. In addition, different fault types bring about different expressions of switch functions β_j. In what follows, the expressions of the switch functions for the different localized faults will be discussed in terms of the outer race fault, inner race fault, and the fault in the rolling elements.

a. Outer race fault

When the outer race exists as the local fault, such as a spall, the switch function β_j is expressed as

$$\beta_j = \begin{cases} 1 & if\ \phi_d < \phi_j < \phi_d + \Delta\phi_d \\ 0 & otherwise \end{cases} \tag{8}$$

In this case, the spall is fixed in the outer race located from the defined angular position ϕ_d to $\phi_j + \Delta\phi$. Here, phi_d is a constant value, and $\Delta\phi$ is related to the fault length.

b. Inner race fault

In the case of the inner race fault, the local fault rotates with the inner race and the shaft. The switch function β_j is given as

$$\beta_j = \begin{cases} 1 & if\ \omega_s t + \phi_{d0} < \phi_j < \omega_s t + \phi_{d0} + \Delta\phi_d \\ 0 & otherwise \end{cases} \tag{9}$$

In this case, the angular position of the fault ϕ_d will change with the speed of the shaft. Here, $\phi_d = \omega_s t + \phi_{d0}$, where ω_s denotes the angular velocity of the shaft and ϕ_{d0} is the initial angular position of the fault.

c. Fault in rolling elements

It is more complicated when a local fault occurs in a rolling element. The fault will rotate with the rolling element. The angular position of the fault is described as

$$\begin{aligned}\phi_s &= \omega_r t + \phi_{d0} \\ \omega_r &= \tfrac{\omega_s}{2}\tfrac{D_p}{D_b}\left(1 - \left(\tfrac{D_p}{D_b}\cos\alpha\right)^2\right)\end{aligned} \tag{10}$$

where ω_r is the angular velocity of the rolling element, and α is the contact angle.

When there exists a fault in the rolling element k, the fault will make contact with both the inner and outer races. The switch values and the fault periods will differ for both races due to the difference in the raceway curvature between the inner and outer races. Therefore, the switch function β_j is defined as

$$\beta_j = \begin{cases} 0, j \neq k \\ 1, \text{ if } 0 < \phi_s < \Delta\phi_{do}, j = k \\ \frac{c_{dr}+c_{di}}{c_{dr}-c_{do}}, \text{ if } \pi < \phi_s < \pi + \Delta\phi_{di}, j = k \\ 0, \text{ otherwise}, j = k \end{cases} \quad (11)$$

with

$$c_{dr} = \frac{D_b}{2} - \sqrt{\frac{D_b^2}{2} - x^2}, c_{di} = r_i - \sqrt{r_i^2 - x^2}, c_{do} = r_o - \sqrt{r_o^2 - x^2}$$

$$r_i = \frac{D_p - D_b}{2}, r_o = \frac{D_p + D_b}{2}$$

$$\Delta\phi_{do} = \frac{2x}{r_o}, \Delta\phi_{di} = \frac{2x}{r_i}$$

where x is the half of the spall width. For more details, please refer to [35].

2.2. Deep Residual Network for Fault Detection

Deep residual network (Resnet) is a deep learning method with extremely deep architecture, which shows outstanding performance on accuracy and convergence. It introduces the shortcut connection module into the framework to learn the residual, which avoids the degradation problem of deep networks. The high-level representative features can be better extracted by propagating the data information directly throughout the network [36,37].

A residual learning unit is shown in Figure 4, which can be expressed as:

$$\begin{aligned} \mathbf{y}_l &= h(\mathbf{x}_l) + F(\mathbf{x}_l, \mathbf{W}_l) \\ \mathbf{x}_{l+1} &= f(\mathbf{y}_l) \end{aligned} \quad (12)$$

where $\mathbf{X}_l$ and $\mathbf{X}_{l+1}$ denote the input and output vectors of the lth residual unit, which generally includes multi-layers. F is the residual function, which represents the learned residual, while $h(\mathbf{X}_l) = \mathbf{X}_l$ denotes the identity mapping, and $f(\mathbf{y}_l)$ is the activation function. Based on Equation (12), the learning features we obtained from the shallow layer l to the deep layer L are described as

$$\mathbf{x}_L = \mathbf{x}_l + \sum_{i=1}^{L-1} F(\mathbf{x}_i, \mathbf{W}_i) \quad (13)$$

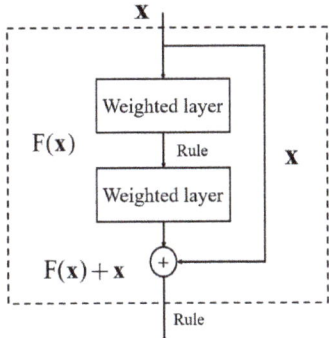

Figure 4. A residual learning unit.

With regard to backpropagation, assuming the loss function is E, the gradient of the reverse process can be obtained according to the chain rule of backpropagation.

$$\frac{\partial E}{\partial x_l} = \frac{\partial E}{\partial x_L} \cdot \frac{\partial x_L}{\partial x_l} = \frac{\partial E}{\partial x_L} \cdot (1 + \frac{\partial}{\partial x_l} \sum_{i=1}^{L-1} F(x_i, W_i)) \qquad (14)$$

where $\frac{\partial E}{\partial x_L}$ denotes the gradient of the loss function to L, the 1 in parentheses indicates that the shortcut mechanism can propagate the gradient lossless, and the other residual gradient needs to pass through the layer with weight; the gradient is not passed directly. The residual gradient is not all −1 coincidentally, and even if it is small, the presence of 1 will not result in the gradient disappearing. The advantage of the Resnet neural network is that it can be used to train complex networks and ensure high classification accuracy.

3. Experimental Verification

In this section, the open dataset from the Bearing Data Center of Case Western Reserve University is used to verify our method. In order to reduce the process of real data collection, model-based data augmentation is used to construct the training dataset. To reduce the parameters to be identified, Equation (1) can be written as:

$$\ddot{y}_s + \frac{c_s}{m_s}\dot{y}_s + \frac{k_s}{m_s}y_s = \frac{F_r - F_y}{m_s},$$
$$\ddot{y}_p + \frac{c_p}{m_p}\dot{y}_p + \frac{k_p}{m_p}y_p = \frac{F_y}{m_p}. \qquad (15)$$

An error index is used to evaluate the distance between the simulation results and the measured experimental results. To consider the influence of the wave shift, this index is defined in the frequency domain as

$$e_{inx} = |||FFT(\ddot{y}_{p,sim})| - |FFT(\ddot{y}_{p,real})||| / |||FFT(\ddot{y}_{p,real})||| \qquad (16)$$

where $|FFT(\ )|$ is the amplitude of the frequency.

Then, the system parameters are obtained by solving the optimization problem,

$$\operatorname*{argmin}_{P} e_{inx}$$
$$P \in \{c_s/m_s, c_p/m_p, k_s/m_s, k_p/m_p\} \qquad (17)$$

By comparing the simulation data with the experimental data, the parameters of the bearing model can be obtained, as shown in Table 1.

Table 1. Value of parameters in the dynamic equation.

Parameters	Description
c_s/m_s	1000 m/s
c_p/m_p	100 m/s
k_s/m_s	7×10^6 N/m
k_p/m_p	15×10^5 N/m

To better simulate the real situation, a disturbance signal is added in the Equation (1),

$$m_s \ddot{y}_s + c_s \dot{y}_s + k_s y_s + F_y + F_{ext} = F_r \qquad (18)$$

with

$$F_{ext} = A\sin(\omega t)$$

where A is the amplitude of the disturbance force, while ω is the frequency of the disturbance force. In the fault-free case, the information about the disturbance force can be extracted. Figure 5a shows the experiment in the fault-free case. A vibration with 30 Hz can be measured, which is considered the disturbance force.

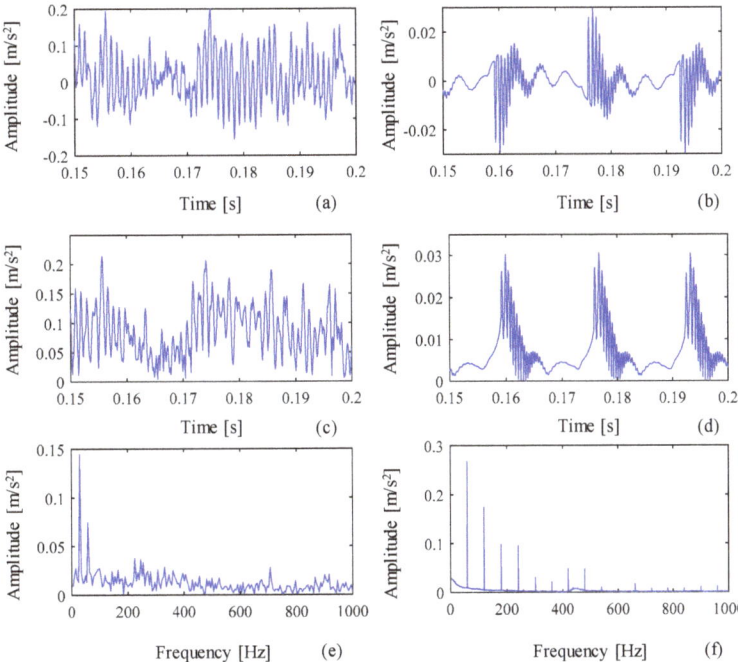

Figure 5. Real and simulation data in fault-free cases: (**a,c,e**) are the original data, the envelop data, and the frequency spectrum of the real envelop data; (**b,d,f**) are the original data, the envelop data, and the frequency spectrum of the simulation envelop data.

For the purpose of the data augmentation, we used the envelop of the signals instead of the original signals. The reason is that the envelop signals contain less noise. Additionally, the information of the eigenfrequency was not taken into consideration in the envelop signal. Thus, by using envelop signals, we did not need a sufficiently exact model, i.e., the parameter k_s and k_p could deviate to the real value to some extent. Figure 5 gives the real and simulation data in the fault-free case, and (a,c,e) are the original data, the envelop data, and the frequency spectrum of the real envelop data, and (b,d,f) are the original data, the envelop data, and the frequency spectrum of the simulation envelop data, respectively. Figure 6 shows the real and simulation data in the outer race fault case. The task of fault detection is to distinguish the normal case, outer race fault, and inner race fault. Therefore, a Resnet deep neural network was used to design the classifier. The deep neural network shows a powerful ability for classification, but requires mass data for training. Therefore, we used the dynamic model to generate the dataset to assist the training process.

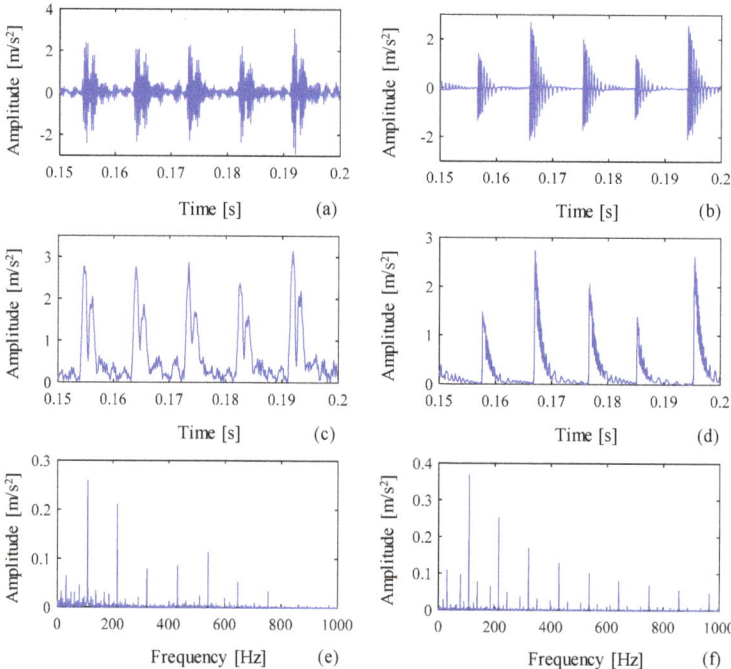

Figure 6. Real and simulation data with outer race fault: (**a,c,e**) are the original data, the envelop data, and the frequency spectrum of the real envelop data; (**b,d,f**) are the original data, the envelop data, and the frequency spectrum of the simulation envelop data.

In parameter identification, a group of experiment data is required. After parameter identification, the dynamic model can generate data under different conditions. For example, parameter identification is carried out when the rotation speed is 1797 rpm with a 12 kg load. The model can generate vibration data at different speeds and different loads. Figure 7 shows the generated data with the outer race fault, where (a,c,e) are the original data at different speeds, different pre-loads, and with noise, and (b,d,f) are the envelop data at different speeds, different pre-loads, and with noise. Similarly, the vibration data in the normal case and the inner race fault case can also be generated. Figure 8 shows the simulated vibration data in the inner race fault case. Only a few data are required (such as data with the outer race fault), and the model can generate rich data in different situations.

The Resnet classifier is used for the fault detection of the bearing faults. Table 2 shows the parameters of the Resnet. Three training datasets are constructed for the verification of the proposed method. The first one is the original data. Then, the real envelop dataset and the simulated envelop dataset are used to train the Resnet classifier. The testing dataset includes the normal, outer race fault, and inner race fault cases.

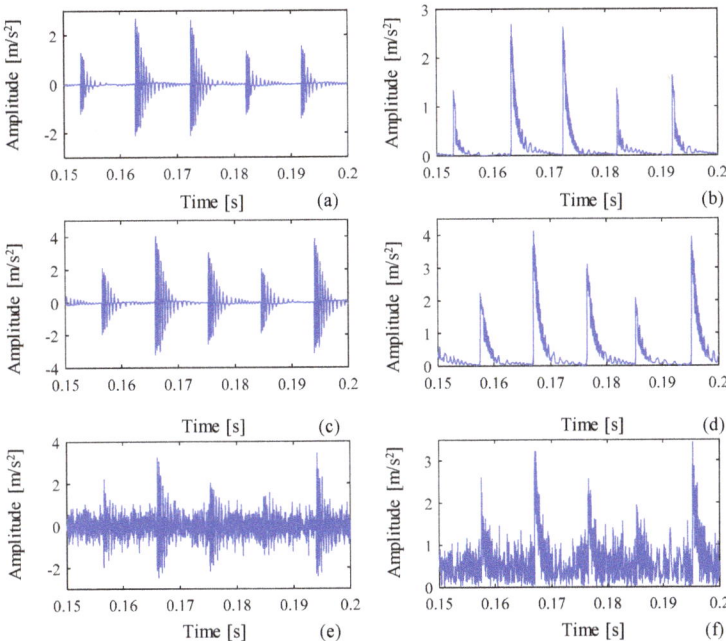

Figure 7. The generated data with the outer race fault: (**a,c,e**) the original data at different speeds, different pre-loads, and with noise; (**b,d,f**) are the envelop data at different speeds, different pre-loads, and with noise.

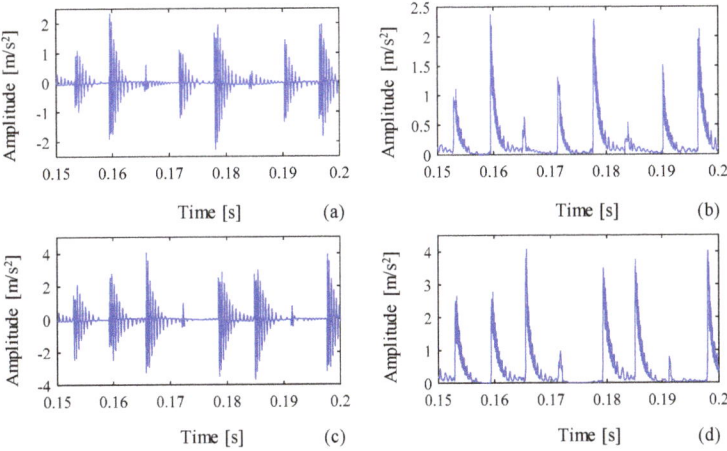

Figure 8. The generated data with inner race fault: (**a,b**) are the original data and the envelop data when the rotation speed is 1797 rpm; (**c,d**) are the original data and the envelop data when the rotation speed is 1730 rpm.

Table 2. Structure parameters of Resnet.

Description	Value		
The resolution of input signals	1×2000		
	Layer name	Output size	Layer
	Conv1	1×1000	$3 \times 3, 8$
	Conv2	1×500	$\begin{bmatrix} 3 \times 3, 16 \\ 3 \times 3, 16 \end{bmatrix} \times 3$
The size of the network	Conv3	1×250	$\begin{bmatrix} 3 \times 3, 32 \\ 3 \times 3, 32 \end{bmatrix} \times 4$
	Conv4	1×125	$\begin{bmatrix} 3 \times 3, 64 \\ 3 \times 3, 64 \end{bmatrix} \times 6$
	Conv5	1×64	$\begin{bmatrix} 3 \times 3, 128 \\ 3 \times 3, 128 \end{bmatrix} \times 3$
	GAP	1×2	128
Activation function	Sigmoid		

In general, if smaller differences exist between the simulation data and the real data, the classified results will be more accurate. However, the gap between the simulation and the real data will always exist. This is the reason why we use the envelop data instead of the original data. Figure 9 shows the comparison results between the simulation and real data. The error index is used to evaluate the performance of the simulation results for data augmentation. The error index for the original data is 0.9214, while the error index for the envelop data is 0.4622. The gap between the real and simulation results of the original data is much larger than that of the envelop data. This is the reason why we use the envelop data as the training dataset. training cost, which has a large application prospect.

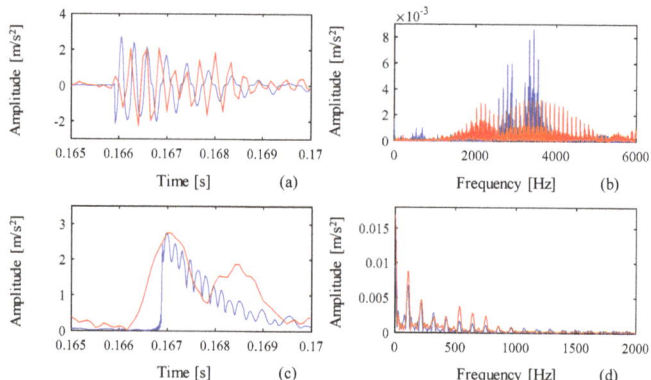

Figure 9. Comparison of experimental and simulation data: (**a,c**) are the data in the time domain; (**b,d**) are the data in the frequency domain, where the blue line is the experiment data, and the red line is the simulation data.

Figure 10 shows the distribution of the probability of each dataset after training. The training dataset contains 500 groups of data, while the testing dataset contains 150 groups of data. If the real data (the original data and the envelop data) are used for training, the classified accuracy can reach 100%. The reason for this is that the difference is great for the signal in the three cases. However, the collection of the data in different operation situations is expensive work. Figure 10c shows the classification results of the Resnet classifier, which is trained by pure simulation data. The classified accuracy is still 100%, but the possibility is lower than that by using the real data. The reason for this is that the simulation data is not the same as the real data. A gap between the simulation and real data therefore results in a low possibility. By using the envelop data, we can reduce the gap and achieve accepted classification results. The proposed method, based on the Resnet classifier with model-based data augmentation, can overcome the high costs of the classifier training cost, which has a large application prospect.

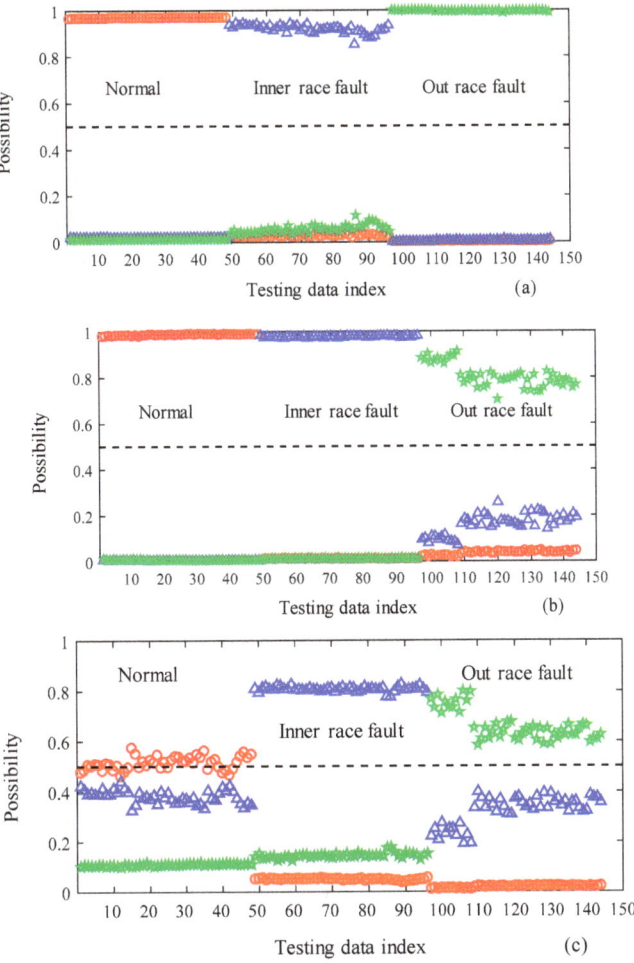

Figure 10. Classification results of the Resnet classifier: (**a**) the training and testing datasets are the real original data; (**b**) the training and testing datasets are the real envelop data; (**c**) the training dataset is the simulation data while the testing dataset is the real data.

4. Conclusions

In this paper, a bearing fault detection method, based on a Resnet classifier with model-based data augmentation, is proposed. For our purpose, a four-DOFs dynamic model is constructed to describe the bearing system. The dynamic model was identified by comparing the simulation and experimental results. Then, a large number of data under different conditions could then be generated, based on which the training dataset was constructed, and the Resnet classifier was trained for the bearing state classification. Furthermore, to reduce the gap between the simulation data and the real data, the envelop signals were applied in the training process rather than the original signals. The proposed method was testified by the real data from the Bearing Data Center of Case Western Reserve University. The trained Resnet classifier was able to identify the bearing states with 100% accuracy. The framework of the proposed method, based on data augmentation, which combines the theoretical model with the deep learning method, can be further used in other fields which have the deterministic model.

Author Contributions: For research articles with several authors, Conceptualization, L.Q. and X.Z.; methodology, L.Q. and Q.P.; software, Y.L.; validation, L.Q., Y.L. and X.Z.; formal analysis, Q.P.; investigation, L.Q.; resources, Y.L.; data curation, L.Q.; writing—original draft preparation, L.Q.; writing—review and editing, Y.L.; visualization, Q.P.; supervision, Y.L.; project administration, X.Z.; funding acquisition, X.Z. All authors have read and agreed to the published version of the manuscript.

Funding: This research was funded by [National Natural Science Foundation of China under Grant] grant number [51905184, 72101194], [Open Research Fund of State Key Laboratory of High Performance Complex Manufacturing, Central South University] grant number [Kfkt2020-12], [Humanities and Social Science Foundation of Ministry of Education of China] grant number [20YJC630096].

Informed Consent Statement: Not applicable.

Data Availability Statement: Not applicable.

Conflicts of Interest: The authors declare no conflict of interest.

Nomenclature

x_s, y_s	shaft/inner race DOF
x_p, y_p	pedestal/outer race DOF
m_s	mass of shaft/inner race
m_p	mass of pedestal/outer race
c_s	damping of shaft/inner race
c_p	damping of pedestal/outer race
k_s	stiffness of shaft/inner race
D_b	ball diameter
D_p	pitch circle diameter
ω_s	angular velocity of the shaft
ω_c	angular velocity of the cage
ω_r	angular velocity of the rolling element
ϕ_j	angular position of the rolling elements
δ	overall contact deformation
c	clearance value
ϕ_0	initial angular position of cage
c_d	fault depth
ϕ_d	angular position of the fault
ϕ_{d0}	initial angular position of the fault

References

1. Qian, L. *Observer-Based Fault Detection and Estimation of Rolling Element Bearing Systems*; Shaker Verlag: Herzogenrath, Germany, 2019.
2. Gao, Z.; Cecati, C.; Ding, S.X. A survey of fault diagnosis and fault-tolerant techniques—Part I: Fault diagnosis with model-based and signal-based approaches. *IEEE Trans. Ind. Electron.* **2015**, *62*, 3757–3767. [CrossRef]
3. Mebarki, N.; Benmoussa, S.; Djeziri, M.; Mouss, L.H. New Approach for Failure Prognosis Using a Bond Graph, Gaussian Mixture Model and Similarity Techniques. *Processes* **2022**, *10*, 435. [CrossRef]
4. Zhang, Z.; Si, X.; Hu, C.; Lei, Y. Degradation data analysis and remaining useful life estimation: A review on Wiener-process-based methods. *Eur. J. Oper. Res.* **2018**, *271*, 775–796. [CrossRef]
5. Lv, Y.; Zhao, W.; Zhao, Z.; Li, W.; Ng, K.K. Vibration signal-based early fault prognosis: Status quo and applications. *Adv. Eng. Inform.* **2022**, *52*, 101609. [CrossRef]
6. Pająk, M.; Muślewski, Ł.; Landowski, B.; Kałaczyński, T.; Kluczyk, M.; Kolar, D. Identification of Reliability States of a Ship Engine of the Type Sulzer 6AL20/24. *SAE Int. J. Engines* **2021**, *15*, 527–542. [CrossRef]
7. Dyer, D.; Stewart, R. Detection of rolling element bearing damage by statistical vibration analysis. *J. Mech. Des.* **1978**, *100*, 229–235. [CrossRef]

8. Choy, K.; Gunter, E.; Allaire, P. Fast fourier transform analysis of rotor-bearing systems. Topics in Fluid Film Bearing and Rotor Bearing System Design and Optimization. 1978. Available online: https://www.academia.edu/download/86674071/fast_fourier_transform_analysis_of_rotot1980v2_linked.pdf (accessed on 27 May 2022).
9. Too, J.; Abdullah, A.R.; Mohd Saad, N.; Mohd Ali, N. Feature selection based on binary tree growth algorithm for the classification of myoelectric signals. *Machines* **2018**, *6*, 65. [CrossRef]
10. Zhou, Y.; Wang, J.; Wang, Z. Bearing Faulty Prediction Method Based on Federated Transfer Learning and Knowledge Distillation. *Machines* **2022**, *10*, 376. [CrossRef]
11. Sun, Q.; Tang, Y. Singularity analysis using continuous wavelet transform for bearing fault diagnosis. *Mech. Syst. Signal Process.* **2002**, *16*, 1025–1041. [CrossRef]
12. Lei, Y.; Lin, J.; He, Z.; Zuo, M.J. A review on empirical mode decomposition in fault diagnosis of rotating machinery. *Mech. Syst. Signal Process.* **2013**, *351*, 108–126. [CrossRef]
13. Chen, H.; Chai, Z.; Dogru, O.; Jiang, B.; Huang, B. Data-driven designs of fault detection systems via neural network-aided learning. *IEEE Trans. Neural Netw. Learn. Syst.* **2021**, 1–12. [CrossRef]
14. Yuan, H.; Wu, N.; Chen, X.; Wang, Y. Fault diagnosis of rolling bearing based on shift invariant sparse feature and optimized support vector machine. *Machines* **2021**, *9*, 98. [CrossRef]
15. Yiakopoulos, C.T.; Gryllias, K.C.; Antoniadis, I.A. Rolling element bearing fault detection in industrial environments based on a K-means clustering approach. *Expert Syst. Appl.* **2011**, *38*, 2888–2911. [CrossRef]
16. Samanta, B.; Al-Balushi, K.; Al-Araimi, S. Artificial neural networks and support vector machines with genetic algorithm for bearing fault detection. *Eng. Appl. Artificial Intell.* **2003**, *16*, 657–665. [CrossRef]
17. Lv, Y.; Zhou, Q.; Li, Y.; Li, W. A predictive maintenance system for multi-granularity faults based on AdaBelief-BP neural network and fuzzy decision making. *Adv. Eng. Inform.* **2021**, *49*, 101318. [CrossRef]
18. Shi, Q.; Zhang, H. Fault diagnosis of an autonomous vehicle with an improved SVM algorithm subject to unbalanced datasets. *IEEE Trans. Ind. Electron.* **2020**, *68*, 6248–6256. [CrossRef]
19. Zhang, N.; Li, Y.; Yang, X.; Zhang, J. Bearing Fault Diagnosis Based on BP Neural Network and Transfer Learning. *J. Phys. Conf. Ser.* **2021**, *1881*, 022084. [CrossRef]
20. Ding, H.; Gao, R.X.; Isaksson, A.J.; Landers, R.G.; Parisini, T.; Yuan, Y. State of AI-based monitoring in smart manufacturing and introduction to focused section. *IEEE/ASME Trans. Mechatron.* **2020**, *25*, 2143–2154. [CrossRef]
21. Lei, Y.; Jia, F.; Lin, J.; Xing, S.; Ding, S.X. An intelligent fault diagnosis method using unsupervised feature learning towards mechanical big data. *IEEE Trans. Ind. Electron.* **2016**, *63*, 3137–3147. [CrossRef]
22. Kolar, D.; Lisjak, D.; Pająk, M.; Pavković, D. Fault diagnosis of rotary machines using deep convolutional neural network with wide three axis vibration signal input. *Sensors* **2020**, *20*, 4017. [CrossRef]
23. Janssens, O.; Slavkovikj, V.; Vervisch, B.; Stockman, K.; Loccufier, M.; Verstockt, S.; Van de Walle, R.; Van Hoecke, S. Convolutional neural network based fault detection for rotating machinery. *J. Sound Vib.* **2016**, *377*, 331–345. [CrossRef]
24. Mao, W.; Feng, W.; Liang, X. A novel deep output kernel learning method for bearing fault structural diagnosis. *Mech. Syst. Process.* **2019**, *117*, 293–318. [CrossRef]
25. Zhang, R.; Tao, H.; Wu, L.; Guan, Y. Transfer learning with neural networks for bearing fault diagnosis in changing working conditions. *IEEE Access* **2017**, *5*, 14347–14357. [CrossRef]
26. LeCun, Y.; Bottou, L.; Bengio, Y.; Haffner, P. Gradient-based learning applied to document recognition. *Proc. IEEE* **1998**, *86*, 2278–2324. [CrossRef]
27. Shijie, J.; Ping, W.; Peiyi, J.; Siping, H. Research on data augmentation for image classification based on convolution neural networks. In Proceedings of the 2017 Chinese Automation Congress CAC, Jinan, China, 2–22 October 2017; IEEE: Piscataway, NJ, USA; p. 4165.
28. Kim, S.; Kim, N.H.; Choi, J.H. Prediction of remaining useful life by data augmentation technique based on dynamic time warping. *Mech. Syst. Signal Process.* **2020**, *136*, 106486. [CrossRef]
29. Yu, K.; Lin, T.R.; Ma, H.; Li, X. A multi-stage semi-supervised learning approach for intelligent fault diagnosis of rolling bearing using data augmentation and metric learning. *Mech. Syst. Signal Process.* **2021**, *146*, 107043. [CrossRef]
30. Hu, T.; Tang, T.; Chen, M. Data simulation by resampling—A practical data augmentation algorithm for periodical signal analysis-based fault diagnosis. *IEEE Access* **2019**, *7*, 125133–125145. [CrossRef]
31. Lu, H.; Zhao, X.; Tao, B.; Ding, H. A state-classification approach for light-weight robotic drilling using model-based data augmentation and multi-level deep learning. *Mech. Syst. Signal Process.* **2022**, *167*, 108480. [CrossRef]
32. Sobie, C.; Freitas, C.; Nicolai, M. Simulation-driven machine learning: Bearing fault classification. *Mech. Syst. Signal Process.* **2018**, *99*, 403–419. [CrossRef]
33. Chen, H.; Jiang, B.; Ding, S.X.; Huang, B. Data-driven fault diagnosis for traction systems in high-speed trains: A survey, challenges, and perspectives. *IEEE Trans. Intell. Transp. Syst.* **2020**, *23*, 1700–1716. [CrossRef]
34. Chen, H.; Jiang, B. A review of fault detection and diagnosis for the traction system in high-speed trains. *IEEE Trans. Intell. Transp. Syst.* **2019**, *21*, 450–465. [CrossRef]

35. Sawalhi, N.; Randall, R.B. Simulating gear and bearing interactions in the presence of faults: Part I. The combined gear bearing dynamic model and the simulation of localised bearing faults. *Mech. Syst. Signal Process.* **2008**, *22*, 1924–1951. [CrossRef]
36. He, K.; Zhang, X.; Ren, S.; Sun, J. Deep residual learning for image recognition. In Proceedings of the IEEE Conference on Computer Vision and Pattern Recognition, Las Vegas, NV, USA, 27–30 June 2016; pp. 770–778.
37. Zhang, W.; Li, X.; Ding, Q. Deep residual learning-based fault diagnosis method for rotating machinery. *ISA Trans.* **2019**, *95*, 295–305. [CrossRef] [PubMed]

Article

A Local Density-Based Abnormal Case Removal Method for Industrial Operational Optimization under the CBR Framework

Xiangyu Peng [1], Yalin Wang [1], Lin Guan [1] and Yongfei Xue [2,*]

1 School of Automation, Central South University, Changsha 410083, China; 164601005@csu.edu.cn (X.P.); ylwang@csu.edu.cn (Y.W.); gl970305@csu.edu.cn (L.G.)
2 School of Computer and Information Engineering, Central South University of Forestry & Technology, Changsha 410004, China
* Correspondence: xueyongfei@csuft.edu.cn

Abstract: Operational optimization is essential in modern industry and unsuitable operations will deteriorate the performance of industrial processes. Since measuring error and multiple working conditions are inevitable in practice, it is necessary to reduce their negative impacts on operational optimization under the case-based reasoning (CBR) framework. In this paper, a local density-based abnormal case removal method is proposed to remove the abnormal cases in a case retrieval step, so as to prevent performance deterioration in industrial operational optimization. More specifically, the reasons as to why classic CBR would retrieve abnormal cases are analyzed from the perspective of case retrieval in industry. Then, a local density-based abnormal case removal algorithm is designed based on the Local Outlier Factor (LOF), and properly integrated into the traditional case retrieval step. Finally, the effectiveness and the superiority of the local density-based abnormal case removal method was tested by a numerical simulation and an industrial case study of the cut-made process of cigarette production. The results show that the proposed method improved the operational optimization performance of an industrial cut-made process by 23.5% compared with classic CBR, and by 13.3% compared with case-based fuzzy reasoning.

Keywords: data-driven; operational optimization; case-based reasoning; local outlier factor; abnormal case removal

1. Introduction

Frequent changes in operating conditions require the operating settings to change accordingly and appropriately, and unsuitable settings will bring about performance deterioration and disqualified products [1]. Therefore, operational optimization plays an essential role in industrial production since it ensures process safety and enhances economic benefit [2–4]. Generally, there are two kinds of operational optimization methods: model-based methods and data-based methods. In particular, the model-based methods firstly build a process model with some basic operational laws, such as material conservation and energy conservation, and then construct a constrained optimization problem with the pre-established process model [5,6]. On this basis, global optimal solutions are obtained with some optimization algorithms, such as sequential quadratic programming (SQP) [7], the genetic algorithm (GA) [8], and particle swarm optimization (PSO) [9]. Although model-based methods have been successfully applied to many fields, their shortages are inevitable when the industrial process is extremely complex. In fact, it is difficult to build an accurate model if the process is featured by a large scale, long procedure, and changeable environments [10]. Moreover, it is challenging to select an appropriate optimization algorithm to balance the efficiency and the accuracy of a certain operational optimization problems [11].

In response to the drawbacks of model-based methods, data-based methods–which are free from prior knowledge on process mechanisms [12]–have attracted much attention in both the academic and industrial community [13]. For example, Wang et al. designed an adaptive moving window convolutional neural network to extract useful information from the process time-series data, based on which the optimal decision is made according to the expected operational indices [14]. Ding et al. integrated the reinforcement learning strategy with Case-Based Reasoning (CBR) so that the optimal operational indices for a large mineral processing plant can be easily found [15]. Overall, data-based methods benefit from various kinds of sensors installed in modern industry, and they can make optimal decisions using plentiful historical data and operational experience.

Among the data-based methods, CBR does not rely on any process mechanism knowledge, so it is suitable for operational optimization problems where it is difficult to establish accurate process models. In detail, CBR solves the operational optimization problem by referring to previous operating experience, and it has been successfully applied to many processes. For example, Li et al. developed a principal component regression-based case reuse method under the CBR framework [16]. To be specific, the developed method could learn valuable experience from historical production data and finally obtain the global optimal operating settings for a coking flue gas denitration process. Ding et al. integrated a multi-objective evolutionary algorithm into the classic CBR, and the modified CBR was then employed to optimize some operating indexes of the largest hematite ore processing plant in western China [17]. Basically, since CBR could work out the optimal operating settings for certain conditions with some successful cases (also named historical optimal cases or case base), requirements of safety and stability are automatically satisfied for the acquired settings [18]. This is another advantage of CBR when it is employed to solve operational optimization problems in industry.

Conventionally, CBR includes the following steps: (1) Case retrieval; (2) Case reuse; (3) Case revision; and (4) Case retention [19]. Among them, case retrieval is one of the most important steps and its task is to retrieve the most useful cases from the pre-established case base to solve the target problem [20,21]. Currently, the majority of case retrieval is based on similarity [22], which is typically measured by various kinds of distances, such as the Euclidean distance, the Mahalanobis distance, the cosine angle distance, etc. [23]. However, similarity fails to consider the significance among different dimensions. Therefore, reference [24] employs the weighted Mahalanobis distance to measure the similarity, and reference [25] designed a new similarity measurement that combined the Euclidean distance and the cosine angle distance. To improve the accuracy of case retrieval facing nonlinearity, Li et al. introduced a new similarity index that can transfer traditional distance-based similarity into their corresponding Gaussian forms by Gaussian transformation [26]. In terms of industrial operational optimization, the Euclidean distance or the weighted Euclidean distance is adopted to calculate the similarity between two cases in most previous studies. Usually, the weights are allocated based on experience, and the allocation requires prior knowledge about the studied process. Moreover, the accuracy of case retrieval would be decreased if the process data include measuring error. Therefore, Zhang et al. utilized fuzzy logic to select the most suitable cases from a case base, and then obtained the global optimal solution for the target problem in an oil refinery [18].

Although plenty of works have improved the accuracy of case retrieval, it is still difficult to guarantee the quality of retrieved cases when applied to complex industrial processes when only using distance-based similarity. Firstly, measuring error is unavoidable in historical data [27], so it is hard to build the case base accurately. Secondly, industrial processes often run in many working conditions [28], so it is difficult to ensure the distance-based case retrieval would only retrieve cases from the same working conditions as the target problem. In this paper, these wrongly retrieved cases are named as abnormal cases because they are not helpful for the target problem. Furthermore, applying the operational settings of abnormal cases to the target problem is hazardous and may result in performance deterioration and disqualified products, or even stall the production of subsequent processes. Therefore, a local density-based abnormal case removal method is proposed in this paper to remove the abnormal cases in the case retrieval step, and finally to improve the performance of CBR for industrial operational optimization. The main contributions of this paper are summarized as follows:

(1) The reason why historical cases in low-density areas should not be included in the case reuse step is analyzed from the perspective of safety and reliability requirements in industrial operational optimization problems.
(2) A novel abnormal case removal method, which could effectively remove the abnormal cases before case reuse, is proposed on the basis of the Local Outlier Factor (LOF), and properly integrated into the case retrieval step.
(3) The effectiveness and superiority of the newly proposed local density-based abnormal case removal method is verified by a numerical optimization case study and an industrial operational optimization case study.

The rest of this paper is organized as follows. Some preliminaries of the CBR framework and the distance-based similarity measurements are briefly reviewed in Section 2, then the motivations, principles, and procedures of the local density-based abnormal case removal method are systematically presented in Section 3. Section 4 exhibits the operational optimization results of a numerical case study and an industrial case study. Finally, conclusions are given in Section 5.

2. Preliminaries

In this section, some basic knowledge on the CBR framework and the distance-based similarity measurements is introduced. Unlike the model-based methods, CBR solves the target problem with several related cases stored in the case base. To be specific, the case base should be constructed with as many historical cases as possible. Each case consists of a problem description and a case solution. Figure 1 gives the basic framework of CBR (also known as the CBR cycle).

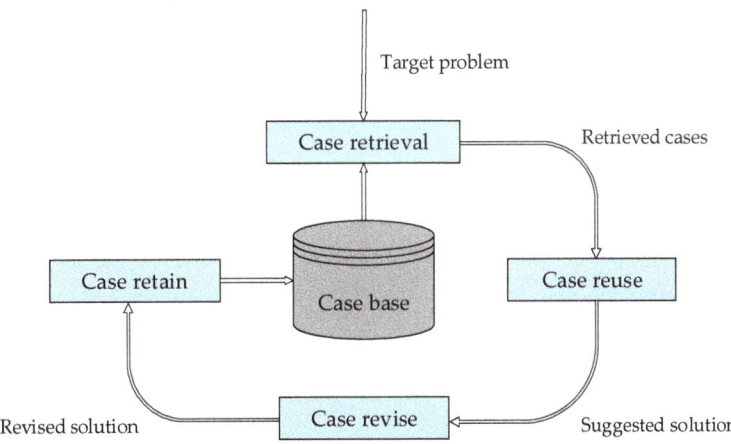

Figure 1. Basic framework of CBR.

It could be seen from Figure 1 that case retrieval is the first step of the CBR cycle. The task of case retrieval is to retrieve several valuable cases from the constructed case base. Supposing the number of retrieved cases is fixed as k, the retrieved cases are the first k cases with the most similar problem descriptions to the target problem. After the case retrieval step, the case reuse is performed to obtain a suggested solution according to the retrieved cases. If the suggested solution is not applicable to the target problem, the suggested solution needs revising to adapt to the target problem. In the last step, the experience of solving this target problem is stored to update the case base, which enable CBR to constantly learn during the CBR cycle.

In general, CBR solves the target problem by learning from historical cases with similar problem descriptions to the target problem. Therefore, case retrieval is the foundation of CBR, and the retrieval accuracy directly affects the performance of CBR [29–31]. In previous studies, most case retrievals are based on distance-based similarity. Table 1 lists five most commonly used distances for similarity measurement in CBR.

Table 1. The most commonly used distances for similarity measurement in CBR.

Name	Formula
Euclidean Distance	$D(X_1, X_2) = \sqrt{(X_1 - X_2)(X_1 - X_2)^T}$
Mahalanobis Distance	$D(X_1, X_2) = \sqrt{(X_1 - X_2)^T \Sigma^{-1} (X_1 - X_2)}$
Cosine angle Distance	$D(X_1, X_2) = \frac{X_1 X_2^T}{\|X_1\| \|X_2\|}$
Manhattan Distance	$D(X_1, X_2) = \sum_{i=1}^{N} |X_{1,i} - X_{2,i}|$
Chebyshev Distance	$D(X_1, X_2) = \max(|X_{1,i} - X_{2,i}|), i = 1, \cdots, N$

As shown in Table 1, several distances can be applied to measure the similarity between two cases. Under the CBR framework, great attention has been paid to measure the similarity between the target problem and historical problems in the case base. However, due to the complexity of industrial processes, it is still hard to choose an appropriate similarity index that only retrieves valuable cases when facing gross measuring error and multiple working conditions. Therefore, it is necessary to develop an abnormal case removal method so as to obtain the most valuable cases in industrial operational optimization.

3. Methods

3.1. Analysis of Case Retrieval in Industrial Operational Optimization

To improve product quality and enhance economic benefits, operational optimization has been widely implemented in industrial processes. CBR can find the optimal operational settings by learning from the historical optimal operational settings in the case base, so it has been widely studied in the industrial operational optimization community. Suppose that there are k cases overall retrieved from the case base, and $X_i (i = 1, 2, \cdots, k)$ and $Y_i (i = 1, 2, \cdots, k)$ represent the problem descriptions and the optimal solutions of the ith retrieved case, respectfully. Under the CBR framework, the suggested solution $\widetilde{Y}_t$ of the target problem X_t can be determined as follows:

$$\widetilde{Y}_t = \frac{\sum_{i=1}^{k} S(X_i, X_t) Y_i}{\sum_{i=1}^{k} S(X_i, X_t)} \tag{1}$$

where $S(X_i, X_t)$ represents the similarity between the target problem X_t and the problem description of the ith historical case X_i. In fact, the suggested solution $\widetilde{Y}_t$ is a weighted sum of historical optimal solutions. Concretely, k historical cases are selected by the case retrieval step according to their similarity to the target problem. Moreover, Equation (1) shows that the weight of the suggested solution is only determined by the similarity between the target problem and the problem description of the selected historical case. In other words, the case retrieval step not only provides some helpful candidates for the suggested solution, but also determines their weights in the suggested solution. Hence, the accuracy of case retrieval is vital to the performance of industrial operational optimization.

Since CBR assumes that similar problem descriptions always have similar case solutions [32], most of the previous studies tend to discover the most similar cases to the target problem. Although the classic case retrieval methods have been proved effective in many fields, the accuracy of case retrieval is still inevitably affected by measuring error and by multiple working conditions. As a result, not all retrieved cases are helpful for solving the target problem. The concrete reasons are as follows.

(a) Accuracy of case retrieval would be influenced by the measuring error

Industrial data are collected by various kinds of sensors installed in the factory. Since perturbations and noises are inevitable in industrial processes, measuring error is naturally introduced in the case base. Consequently, the descriptions of historical cases are not accurate. For the ith case, its measured description $\hat{X}_i$ can be represented as follows:

$$\hat{X}_i = X_i + W_i \tag{2}$$

where X_i and W_i are the accurate description and the measuring error of the ith case, respectively. Considering the measuring error in its corresponding measured description, the true Euclidean distance between X_i and X_t are calculated as follows:

$$D(X_i, X_t) = \sqrt{\left(\left(\hat{X}_i - W_i\right) - \left(\hat{X}_t - W_t\right)\right)\left(\left(\hat{X}_i - W_i\right) - \left(\hat{X}_t - W_t\right)\right)^T} \tag{3}$$

Then the similarity between X_i and X_t can be calculated as follows:

$$S(X_i, X_t) = \frac{1}{1 + D(X_i, X_t)} \tag{4}$$

Obviously, the measuring error in industrial data would degrade the accuracy of case retrieval and make it hard to evaluate the importance of historical cases in solving the target problem. Therefore, it is necessary to eliminate negative impacts from historical cases that have gross measuring error.

(b) Accuracy of case retrieval would be influenced by the multiple working conditions

Industrial processes always run in many working conditions, which leads to some undesirable results if the number of retrieved cases is not appropriate. That is to say, not only the similarity $S(X_i, X_t)$ but also the number k have an impact on the accuracy of case retrieval. Therefore, an appropriate parameter k is crucial for the success of industrial operational optimization under the CBR framework. However, for a particular process, there are different numbers of historical cases in different working conditions, suggesting that the case base is imbalanced. There are a larger number of cases in common working conditions and a smaller number of cases in uncommon working conditions. Therefore, it is easy to retrieve enough cases from a common working condition, yet difficult to do the same from an uncommon working condition. Since the parameter k is fixed as a constant in classic CBR, it may perform well for some working conditions but perform poorly for others. The reason why classic CBR has a different performance in different working conditions is that some irrelevant cases from other working conditions may be retrieved if the target problem belongs to uncommon working conditions. Thus, the suggested solution may be inapplicable.

In summary, both the measuring error and the multiple working conditions would decrease the accuracy of case retrieval, which is going to affect the performance of operational optimization under the CBR framework. To decrease the negative impact from these abnormal cases, a local density-based abnormal case removal method for the case retrieval step is proposed in the following subsection.

3.2. Local Density-Based Abnormal Case Removal

Most of the previous studies on case retrieval have only focused on similarity measurement, while the distribution of retrieved cases was neglected. The goal of case retrieval is, in essence, to search the case base for valuable cases in order to solve the target problem. In Section 3.1, the reasons as to why abnormal cases commonly exist in industry are thoroughly analyzed. Consequently, the retrieval results may not be reliable and the accuracy of case retrieval needs enhancing. In contrast to the model-based methods, CBR directly uses the operational information in retrieved cases, so the accuracy of retrieved cases is vital to the performance of CBR. In another words, abnormal cases are harmful for the industrial operational optimization, so they must be removed before the case reuse. In this paper, it is believed that the distribution of retrieved cases can reflect their reliability. By eliminating low-reliability cases, the quality of the retrieved cases can be significantly enhanced. Figure 2 presents a demonstration of the relationship between the distribution and the reliability of cases.

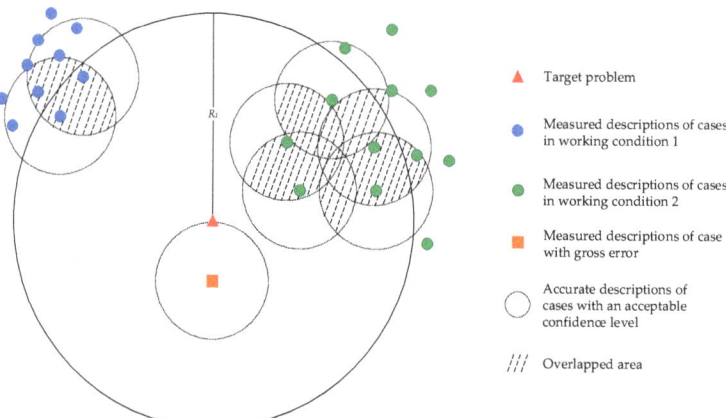

Figure 2. Distribution and reliability of the retrieved cases in industrial processes.

As shown in Figure 2, the retrieved cases are not uniformly distributed in the whole space. Moreover, the accurate descriptions of historical cases are uncertain due to the existence of measuring error. In this paper, the measuring error is assumed to follow the Gaussian distribution. With a certain confidence level, accurate descriptions of historical cases lie in dashed circles centered in their corresponding measured descriptions. Since the similarity is usually calculated according to the measured descriptions, cases with the highest similarity are not necessarily the most helpful cases for the target problem. However, there exist some overlaps in the area with high-density cases, showing cases in the high-density area have higher reliability than other cases since the accurate descriptions are more likely to lie in the overlaps. Therefore, although cases in the low-density area may have a higher similarity to the target problem, they should not proceed to the case reuse step due to their lower reliability.

Another issue that impacts the accuracy of case retrieval is the multiple working conditions of industrial processes. For a target problem that lies on the edge of a working condition, its nearest neighbors probably include cases from other working conditions. Obviously, these cases will not help to solve the target problem and should not be included in the retrieved cases. This issue can be partly solved by assigning different number of retrieved cases to every working condition, but it requires identifying the working conditions in advance and setting a different k parameter for every working condition. Consequently, it demands more priori knowledge and becomes much more complicated. Considering the working condition identification problem can be transformed into a classic classification problem, the K-Nearest Neighbors (KNN) classifier believes that the target problem belongs to the working condition that the majority of its nearest neighbors belongs to. That is to say, the number of retrieved cases from other working conditions is less than the number of retrieved cases from the working condition that the target problem belongs to. Since all retrieved cases belong to the same neighborhood, cases from other working conditions are more likely to be in the low-density area, so they can be identified by calculating the density of retrieved cases.

To conclude, measuring error and multiple working conditions are two inevitable problems affecting the accuracy of case retrieval and degrading the performance of CBR. Therefore, developing an abnormal case removal method is urgent and necessary. Since cases in a high-density area are more reliable than those in a low-density area, the latter should be removed from the retrieved cases. In this subsection, a local density-based abnormal case removal algorithm is designed based on the Local Outlier Factor (LOF), which is a common index showing how isolated a data point is comparing with its nearest data points. The LOF of historical case X_i is defined as follows:

$$LOF(X_i) = \frac{1}{m}\sum_{q=1}^{m} \frac{lrd(X_q)}{lrd(X_i)} \quad (5)$$

where m is an adjustable parameter; $lrd(X_q)$ and $lrd(X_i)$ stand for the local reachability density of case X_q and X_i, respectively; X_q is the qth similar cases in the retrieved cases. Particularly, the $lrd(X_i)$ can be represented as follows:

$$lrd(X_i) = \left(\frac{1}{m}\sum_{q=1}^{m} D(X_i, X_q)\right)^{-1} \quad (6)$$

where $D(X_i, X_q)$ is the Euclidean distance between X_q and X_i.

As shown in Equation (5), LOF reflects the average ratio of $lrd(X_q)$ to $lrd(X_i)$. Therefore, a bigger LOF indicates a smaller local density, and the corresponding case should be removed. Normally, the threshold of LOF is determined after the whole dataset has been analyzed, while in this paper, the threshold of LOF can be adaptively adjusted. To automatically eliminate the retrieved cases in a low-density area, the threshold of the local density-based abnormal case removal algorithm is designed as follows:

$$\zeta = \mu + \alpha\sqrt{\frac{\sum_{i=1}^{k}(X(i) - \mu)^2}{k-1}} \quad (7)$$

where α is an adjustable parameter of the threshold ζ, and μ is the average LOF of the retrieved cases, which can be calculated as follows:

$$\mu = \frac{1}{k}\sum_{i=1}^{k} LOF(X_i) \quad (8)$$

In this paper, k is optimized according to the mean absolute error of the training set; m and α are optimized determined by orthogonal experiments. With the optimal parameter k, m, α, pseudo-codes of the designed local density-based abnormal case removal algorithm are shown in Algorithm 1.

Algorithm 1: Local density-based abnormal case removal

Input: k retrieved cases; optimal parameter m, α
Output: The retrieved cases without abnormal cases
1 Calculate the local density of every retrieved case according to Equation (6)
2 Calculate the LOF of every retrieved case according to Equation (5)
3 Calculate the threshold of the retrieved cases according to Equations (7) and (8)
4 Remove the cases whose LOF are higher than the threshold

With the aforementioned local density-based abnormal case removal algorithm, procedures of the industrial operational optimization are as follows:

Step 1: construct the case base with history data;
Step 2: for a target problem, select k most similar cases from the case base and construct the original retrieved cases $C_i = \{X_i, Y_i\}(i = 1, \cdots, k)$;
Step 3: employ the local density-based abnormal case removal algorithm to remove wrongly retrieved cases;
Step 4: acquire the suggested solution for the target problem according to Equation (1);
Step 5: revise the suggested solution, if necessary;

Step 6: store it in the case base after the target problem is solved.

4. Case Studies

In this section, the effectiveness and the superiority of the designed local density-based abnormal case removal method were validated by two case studies. Firstly, a numerical simulation was designed, where case descriptions were featured with multiple working conditions and measurement error. Then, an industrial case study, whose data were collected from a cut-made process of cigarette production, was designed to show the effectiveness and the superiority of the abnormal case removal method in industrial operation optimization under the CBR framework. In these case studies, the proposed method was compared with classic CBR and case-based fuzzy reasoning in which the fuzzy membership function and its parameters were determined according to their ability to resist measuring error [18]. The concrete hardware and software are as follows: Intel(R) Core (TM) i5-4590, ROM 8 GB, Windows 10 professional.

4.1. Numerical Simulation

In this numerical simulation, 120 operating points were generated with MATLAB 2019A to simulate the characteristics of multiple working conditions and measurement error of industrial data. Particularly, two working conditions were generated with different centers and deviations (the deviations followed Gaussian distribution to simulate the measurement error in industry). In detail, every working condition consisted of 60 operating points, and the centers of working condition 1 and working condition 2 were set as (1, 1) and (−1, −1), respectively. In addition, standard deviations of the two working conditions were both set as 0.5. It should be noted that the operating points with larger deviation from their corresponding centers were considered as operating points with gross error, and they should be removed before the case reuse. Figure 3 shows the distribution of the generated dataset, which can perfectly reflect the characteristics of industrial data.

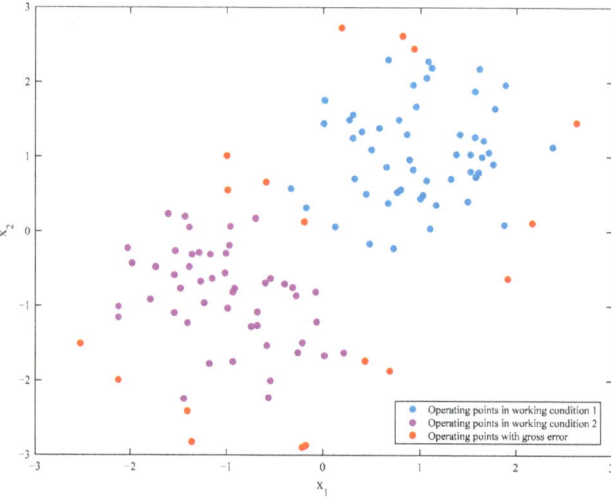

Figure 3. Distribution of the generated dataset.

As shown in Figure 3, the operating points lying on the edge of working condition 1 and working condition 2 were considered as operating points with gross error in this study. Moreover, the case solutions of working condition 1 and working condition 2 were designed as Equations (9) and (10), respectively.

$$Y_1(i) = 0.2(x_1(i) - 1)^2 + 0.3(x_2(i) - 1)^2 + (x_1(i) - 1) + 4 \qquad (9)$$

$$Y_2(i) = -0.2(x_1(i) + 1)^2 + 0.5(x_2(i) + 1) - 4 \tag{10}$$

Their parameters were designed differently to reflect diverse operating experience in different working conditions. Furthermore, Equations (9)–(12) were designed as quadratic polynomials to represent the nonlinearity in the operating experience. For operating points with gross error, their measured descriptions were heavily deviated from their accurate descriptions. Consequently, their case solutions are less helpful for operational optimization than those of normal cases. For this reason, the case solutions of working condition 1 and working condition 2 with gross error were designed as Equations (11) and (12), respectively.

$$Y_{1e}(i) = 0.2(x_1(i) - 1)^2 + 0.3(x_2(i) - 1)^2 + (x_1(i) - 1) + 8 \tag{11}$$

$$Y_{2e}(i) = -0.2(x_1(i) + 1)^2 + 0.5(x_2(i) + 1) - 8 \tag{12}$$

In this numerical simulation, 60 operating points were randomly chosen from the generated dataset as a case base, while the rest of 60 operating points were equally divided into two datasets. To be specific, the first was used as training dataset to pick out the optimal parameters including k, m, and α, and the last was chosen as a testing dataset to evaluate the performance of the designed abnormal case removal method with the selected optimal parameters. The concrete evaluation criterion was Mean Absolute Error (MAE).

$$MAE = \frac{\sum_{i=1}^{n} |Y_i - Y_{i,suggested}|}{n} \tag{13}$$

where n is the number of cases in the testing dataset. Y_i and $Y_{i,suggested}$ are the optimal solution and the suggested solution of the ith cases, respectively.

Since k is a crucial parameter for case retrieval and its value directly affects the performance of CBR, sensitivity analysis was firstly carried out to find the best parameter k. Figure 4 presents the MAE of the training dataset when k changed from 1 to 15.

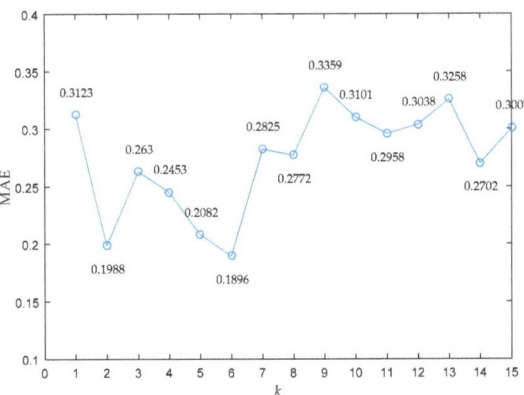

Figure 4. MAE of the training dataset with different parameter k.

As shown in Figure 4, the tendency of MAE firstly decreases with k changing from 1 to 6, and then generally increases with k changing from 6 to 15. The minimal MAE was 0.1896 when the parameter k was chosen as 6. Therefore, the number of retrieved cases was set as 6 both in classic CBR and the improved CBR with the proposed abnormal case removal method. In addition, in order to find out the best parameters m and α for the abnormal case removal algorithm, orthogonal experiments were designed with the training dataset. In particular, the parameter m was set from 1 to 5 while the parameter α was set from 0.2 to 2.2. Table 2 shows the MAE of the training dataset with different combination of parameter m and parameter α.

Table 2. MAE of the training dataset with different parameter combination. Bold shows the optimal number.

MAE	$m=1$	$m=2$	$m=3$	$m=4$	$m=5$
$\alpha = 0.2$	0.1858	0.2378	0.1937	0.1984	0.2070
$\alpha = 0.4$	0.1872	0.2380	0.1622	0.2001	0.1948
$\alpha = 0.6$	0.1870	0.2078	0.1651	0.1675	0.1526
$\alpha = 0.8$	0.1742	0.1967	0.1671	0.1475	0.1521
$\alpha = 1.0$	0.1741	0.1930	0.1956	**0.1457**	0.1470
$\alpha = 1.2$	0.1647	0.2032	0.1950	0.1458	0.1478
$\alpha = 1.4$	0.1935	0.2016	0.1882	0.1541	0.1478
$\alpha = 1.6$	0.1930	0.2018	0.1873	0.1893	0.1602
$\alpha = 1.8$	0.1859	0.1840	0.1840	0.1877	0.1877
$\alpha = 2.0$	0.1797	0.1896	0.1896	0.1896	0.1896
$\alpha = 2.2$	0.1896	0.1896	0.1896	0.1896	0.1896

As shown in Table 2, the minimal MAE of the training dataset was 0.1457 when the parameter m and α were set as 4 and 1, respectively. The reason as to why m and α could influence the MAE of the training dataset were analyzed as follows:

(1) Supposing the parameter m was fixed as a constant, if the selected parameter α was too small, it would result in a lower threshold ξ and more normal cases would be removed by mistake in the retrieved cases. This would increase the MAE.
(2) Supposing the parameter m was fixed as a constant, if the selected parameter α was too big, it would result in a larger threshold ξ and more abnormal cases would be preserved in the retrieved cases. This would increase the MAE.
(3) Supposing the parameter α was fixed as a constant, if the selected parameter m was too small, fewer nearest neighbors would be included in the calculation of LOF. This would make the LOF more vulnerable to uncertainty so as to increase the MAE.
(4) Supposing the parameter α was fixed as a constant, if the selected parameter m was too big, more nearest neighbors would be included in the calculation of LOF. This would reduce the distinguish ability of LOF so as to increase the MAE.

In the end, the best parameters of the designed abnormal case removal algorithm were set as $k=6$, $m=4$ and $\alpha=1$, respectively. With the aforementioned parameter combination, the testing dataset was finally used to show the effectiveness and the superiority of our method. Additionally, Cauchy fuzzy membership function was selected for the case-based fuzzy reasoning and its optimal parameters were 0.725 and 0.837, based on its performance against measuring error. The concrete fuzzy membership functions evaluation method and parameters optimization method can be found in reference [18]. Figure 5 presents the concrete results.

According to Figure 5, it can be found that the set values of our method are closer to their corresponding optimal set values than that of the other two methods. Specifically, there are overall five operating points (marked with red boxes) in which our method outperformed the classic CBR and case-based fuzzy reasoning. As an average, the abnormal case removal method improved the setting accuracy in the testing dataset by 20.3% compared with classic CBR, and by 8.5% compared with case-based fuzzy reasoning. The reason why our method can obtain better results is that some abnormal cases retrieved by the classic case retrieval step could be removed with Equations (5) and (7). By eliminating these abnormal cases whose LOFs are higher than the threshold, the impacts of these cases can be removed in the case reuse step, thus improving the quality of the retrieved cases. Naturally, the MAE of the testing dataset would be decreased, and the performance of operational optimization would be improved under the CBR framework.

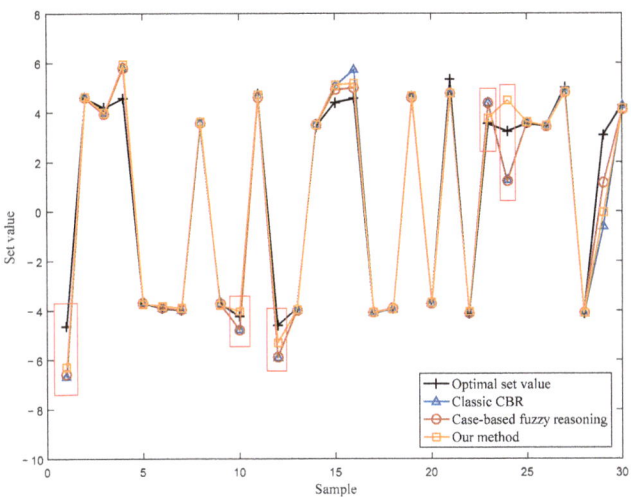

Figure 5. Set values of the testing dataset for numerical simulation.

4.2. Operational Optimization of an Industrial Cut-Made Process of Cigarette Production

In this case study, the designed abnormal case removal method was tested with industrial data collected from a cut-made process of cigarette production. In this production, the operator aims to keep the moisture content of leaf-silk close to the desirable value, and the operational optimality has an impact on the quality of cigarettes. Specifically, the studied cut-made process includes the following three procedures: (1) the leaf-silk drying procedure, (2) the blending procedure, and (3) the spicing procedure. Since many operating experiences were stored in the production data, the set value of the moisture content of the leaf-silk drying procedure could be determined with historical optimal cases. Table 3 presents the basic structure of historical cases for the operational optimization of cut-made process of cigarette production.

Table 3. Structure of historical case for the operational optimization of cut-made process.

Case Description	Case Solution
Average ambient temperature at the drying machine	
Average ambient moisture at the drying machine	
Average leaf-silk moisture content of production line B	
Average leaf-silk moisture content of production line C	The optimal set value of leaf-silk drying machine in production line A
Tobacco stems moisture content	
Expanded leaf-silk moisture content	
Blending time	
Average ambient temperature at spicing	
Average ambient moisture at spicing	

After data preprocessing, a total of 200 cases were extracted for having valuable operating experience from the production data. Then, 100 cases were randomly chosen from the 200 cases as the case base, while the rest were equally divided into two datasets. The first was used as training dataset while the last was chosen as testing dataset. Similar to the numerical simulation, MAE was chosen to evaluate its operational optimization performance, and an orthogonal experiment was conducted to find the best parameter combination for the abnormal case removal algorithm and CBR. By trial and error, the best parameters of the proposed abnormal case removal algorithm were set as $k = 8$, $m = 5$ and $\alpha = 0.6$, based on which the operational optimization performance in the training dataset was improved by 22.3% compared with classic CBR. Furthermore, the Gaussian membership function was selected, and the optimized parameters were displayed in Table 4. Figure 6 exhibits the set values provided by these methods for the industrial cut-made process in the testing dataset.

Table 4. Optimized parameters of Gaussian membership function in the industrial case study.

Case Description	Optimized Parameters
Average ambient temperature at the drying machine	0.4317
Average ambient moisture at the drying machine	0.3811
Average leaf-silk moisture content of production line B	0.5302
Average leaf-silk moisture content of production line C	0.3529
Tobacco stems moisture content	0.4173
Expanded leaf-silk moisture content	0.5513
Blending time	0.5556
Average ambient temperature at spicing	0.4098
Average ambient moisture at spicing	0.4885

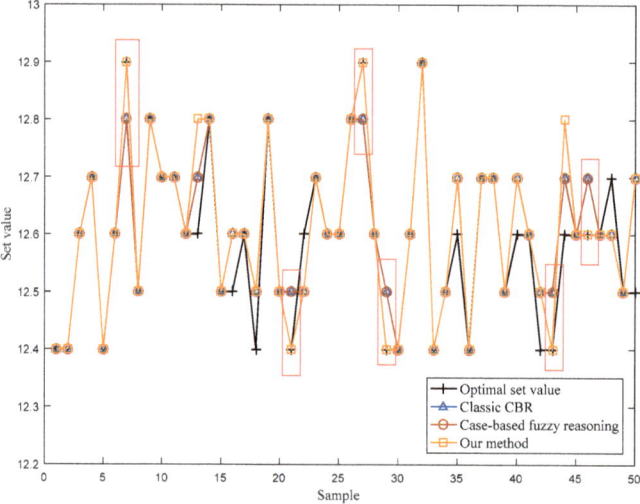

Figure 6. Set values of the testing dataset for industrial cut-made process.

As shown in Figure 6, CBR with the designed abnormal case removal method (our method) can obtain better results in the operational optimization of moisture content of leaf-silk drying machine in production line A. In particular, overall, there are six operating points (marked with red boxes) in which our method outperformed the classic CBR and case-based fuzzy reasoning. This is due to some abnormal cases being removed by the proposed case removal method in the case retrieval step. Furthermore, the influence of multiple working conditions was not considered in the case-based fuzzy reasoning, and thus the performance of CBR with the designed abnormal case removal method was better. In summary, the MAE of classic CBR in testing dataset was 0.034 and the MAE of case-based fuzzy reasoning was 0.03, while the MAE of our method in the testing dataset was 0.026. The proposed abnormal case removal method improved the MAE by 23.5% compared to classic CBR, and by 13.3% compared to case-based fuzzy reasoning. Therefore, the effectiveness and the superiority of the local density-based abnormal case removal method was proven, and it is suitable for the operational optimization of industrial processes.

5. Conclusions

This paper proposed a local density-based abnormal case removal method for the industrial operational optimization problem. Particularly, the reason as to why abnormal cases should be removed from the case set retrieved by traditional method was analyzed in view of the safety and reliability requirements of industrial operational optimization. Then, historical cases whose LOF exceeded the corresponding threshold were removed by the designed local density-based abnormal case removal algorithm. The simulation results showed that, compared with classic CBR, the local density-based abnormal case removal method could improve the performance of operational optimization by 20.3% in the numerical case and 23.5% in the industrial case study, while improving the performance of operational optimization by 8.5% in the numerical case and 13.3% in the industrial case study compared with case-based fuzzy reasoning. In this paper, the calculation of local density increased computation cost, thus, how to obtain the local density of retrieved cases with lower computation burden would be an interesting topic in the future.

Author Contributions: Conceptualization, X.P. and Y.W.; Data curation, X.P. and L.G.; Formal analysis, X.P. and Y.X.; Funding acquisition, Y.W.; Investigation, X.P., Y.X. and L.G.; Methodology, X.P. and Y.X.; Project administration, Y.W.; Software, X.P.; Supervision, Y.W.; Validation, X.P. and L.G.; Visualization, X.P. and L.G.; Writing—original draft, X.P.; Writing—review & editing, X.P., Y.W. and Y.X. All authors have read and agreed to the published version of the manuscript.

Funding: This research was supported in part by National Natural Science Foundation of China (NSFC) (U1911401), in part by the National Key Research and Development Program of China (2020YFB1713800), and the Science and Technology Innovation Program of Hunan Province (2021RC4054).

Data Availability Statement: The data set used in the numerical case was generated with the MATLAB 2019A.

Conflicts of Interest: The authors declare no conflict of interest. The funders had no role in the design of the study; in the collection, analyses, or interpretation of data; in the writing of the manuscript, or in the decision to publish the results.

References

1. Chen, H.T.; Chai, Z.; Dogru, O.; Jiang, B.; Huang, B. Data-Driven Fault Detection for Dynamic Systems with Performance Degradation: A Unified Transfer Learning Framework. *IEEE Trans. Instrum. Meas.* **2020**, *70*, 3504712. [CrossRef]
2. Sun, B.; Yang, C.H.; Zhu, H.Q.; Li, Y.G.; Gui, W.H. Modeling, Optimization, and Control of Solution Purification Process in Zinc Hydrometallurgy. *IEEE/CAA J. Autom. Sin.* **2018**, *5*, 564–576. [CrossRef]
3. Xue, Y.F.; Wang, Y.L.; Sun, B.; Peng, X.Y. An Efficient Computational Cost Reduction Strategy for the Population-Based Intelligent Optimization of Nonlinear Dynamical Systems. *IEEE Trans. Ind. Inf.* **2021**, *17*, 6624–6633. [CrossRef]
4. Xie, R.; Liu, W.H.; Chen, M.Y.; Shi, Y.J. A Robust Operation Method with Advanced Adiabatic Compressed Air Energy Storage for Integrated Energy System under Failure Conditions. *Machines* **2022**, *10*, 51. [CrossRef]
5. Chen, Q.D.; Ding, J.L.; Chai, T.Y.; Pan, Q.K. Evolutionary Optimization Under Uncertainty: The Strategies to Handle Varied Constraints for Fluid Catalytic Cracking Operation. *IEEE Trans. Cybern.* **2022**, *52*, 2249–2262. [CrossRef]
6. Yang, C.E.; Ding, J.L.; Jin, Y.C.; Wang, C.Z.; Chai, T.Y. Multitasking Multiobjective Evolutionary Operational Indices Optimization of Beneficiation Processes. *IEEE Trans. Autom. Sci. Eng.* **2019**, *16*, 1046–1057. [CrossRef]
7. Boggs, P.T.; Tolle, J.W. Sequential Quadratic Programming for Large-Scale Nonlinear Optimization. *J. Comput. Appl. Math.* **2000**, *124*, 123–137. [CrossRef]
8. Liu, Q.Y.; Zha, Y.W.; Liu, T.; Lu, C. Research on Adaptive Control of Air-Borne Bolting Rigs Based on Genetic Algorithm Optimization. *Machines* **2021**, *9*, 240. [CrossRef]
9. Zheng, X.Y.; Su, X.Y. Sliding Mode Control of Electro-Hydraulic Servo System Based on Optimization of Quantum Particle Swarm Algorithm. *Machines* **2021**, *9*, 283. [CrossRef]
10. Chen, H.T.; Chai, Z.; Dogru, O.; Jiang, B.; Huang, B. Data-Driven Designs of Fault Detection Systems via Neural Network-Aided Learning. *IEEE Trans. Neural Netw. Learn. Syst.* **2021**. [CrossRef]
11. Chai, T.Y.; Ding, J.L.; Wang, H. Multi-Objective Hybrid Intelligent Optimization of Operational Indices for Industrial Processes and Application. *IFAC Proc. Vol.* **2011**, *44*, 10517–10522. [CrossRef]
12. Ran, G.T.; Liu, J.; Li, C.J.; Lam, H.-K.; Li, D.Y.; Chen, H.T. Fuzzy-Model-Based Asynchronous Fault Detection for Markov Jump Systems with Partially Unknown Transition Probabilities: An Adaptive Event-Triggered Approach. *IEEE Trans. Fuzzy Syst.* **2022**. [CrossRef]

13. Pan, Z.F.; Chen, H.T.; Wang, Y.L.; Huang, B.; Gui, W.H. A New Perspective on AE- and VAE-Based Process Monitoring. *TechRxiv* **2022**. [CrossRef]
14. Wang, Y.J.; Li, H.G. A Novel Intelligent Modeling Framework Integrating Convolutional Neural Network with An Adaptive Time-Series Window and Its Application to Industrial Process Operational Optimization. *Chemom. Intell. Lab. Syst.* **2018**, *179*, 64–72. [CrossRef]
15. Ding, J.; Modares, H.; Chai, T.; Lewis, F.L. Data-Based Multiobjective Plant-Wide Performance Optimization of Industrial Processes Under Dynamic Environments. *IEEE Trans. Ind. Inf.* **2016**, *12*, 454–465. [CrossRef]
16. Li, Y.N.; Wang, X.L.; Liu, Z.J.; Bai, X.W.; Tan, J. A Data-Based Optimal Setting Method for the Coking Flue Gas Denitration Process. *Can. J. Chem. Eng.* **2018**, *97*, 876–887. [CrossRef]
17. Ding, J.L.; Chai, T.Y.; Wang, H.F.; Wang, J.W.; Zheng, X.P. An Intelligent Factory-Wide Optimal Operation System for Continuous Production Process. *Enterp. Inf. Syst.* **2016**, *10*, 286–302. [CrossRef]
18. Zhang, Z.P.; Chen, D.J.; Feng, Y.Z.; Yuan, Z.H.; Chen, B.Z.; Qin, W.Z.; Zou, S.W.; Qin, S.; Han, J.F. A Strategy for Enhancing the Operational Agility of Petroleum Refinery Plant using Case Based Fuzzy Reasoning Method. *Comput. Chem. Eng.* **2018**, *111*, 27–36. [CrossRef]
19. Aamodt, A. Case-Based Reasoning: Foundational Issues, Methodological Variations, and System Approaches Aicom—Artificial Intelligence Communications. *AI Commun.* **1994**, *7*, 39–59. [CrossRef]
20. Zhai, Z.; Ortega, J.; Castillejo, P.; Beltran, V. A Triangular Similarity Measure for Case Retrieval in CBR and Its Application to an Agricultural Decision Support System. *Sensors* **2019**, *19*, 4605. [CrossRef]
21. Fei, L.G.; Feng, Y.Q. A Novel Retrieval Strategy for Case-Based Reasoning Based on Attitudinal Choquet Integral. *Eng. Appl. Artif. Intell.* **2020**, *94*, 103791. [CrossRef]
22. Zhu, G.N.; Hu, J.; Qi, J.; Ma, J.; Peng, Y.H. An Integrated Feature Selection and Cluster Analysis Techniques for Case-Based Reasoning. *Eng. Appl. Artif. Intell.* **2015**, *39*, 14–22. [CrossRef]
23. López, B. Case-Based Reasoning: A Concise Introduction. In *Synthesis Lectures on Artificial Intelligence and Machine Learning*; Morgan Claypool Publishers: San Rafael, CA, USA, 2013; Volume 7, pp. 1–103.
24. Ahn, J.; Park, M.; Lee, H.S.; Ahn, S.J.; Ji, S.H.; Song, K.; Son, B.S. Covariance Effect Analysis of Similarity Measurement Methods for Early Construction Cost Estimation using Case-Based Reasoning. *Autom. Constr.* **2017**, *81*, 254–266. [CrossRef]
25. Cheng, M.Y.; Tsai, H.C.; Chiu, Y.H. Fuzzy Case-Based Reasoning for Coping with Construction Disputes. *Expert Syst. Appl.* **2009**, *36*, 4106–4113. [CrossRef]
26. Li, H.; Sun, J. Gaussian Case-Based Reasoning for Business Failure Prediction with Empirical Data in China. *Inf. Sci.* **2009**, *179*, 89–108. [CrossRef]
27. Pan, D.; Jiang, Z.H.; Chen, Z.P.; Gui, W.H.; Xie, Y.F.; Yang, C.H. Temperature Measurement and Compensation Method of Blast Furnace Molten Iron Based on Infrared Computer Vision. *IEEE Trans. Instrum. Meas.* **2019**, *68*, 3576–3588. [CrossRef]
28. Liang, S.; Zeng, J. Fault Detection for Complex System under Multi-Operation Conditions Based on Correlation Analysis and Improved Similarity. *Symmetry* **2020**, *12*, 1836. [CrossRef]
29. Chergui, O.; Begdouri, A.; Groux-Leclet, D. Integrating a Bayesian Semantic Similarity Approach into CBR for Knowledge Reuse in Community Question Answering. *Knowl. Based Syst.* **2019**, *185*, 104919. [CrossRef]
30. Smyth, B.; Mckenna, E. Competence Guided Incremental Footprint-Based Retrieval. *Knowl. Based Syst.* **2001**, *14*, 155–161. [CrossRef]
31. Zhang, Q.; Shi, C.Y.; Niu, Z.D.; Cao, L.B. HCBC: A Hierarchical Case-Based Classifier Integrated with Conceptual Clustering. *IEEE Trans. Knowl. Data Eng.* **2019**, *31*, 152–165. [CrossRef]
32. Zhong, S.S.; Xie, X.L.; Lin, L. Two-Layer Random Forests Model For Case Reuse In Case-Based Reasoning. *Expert Syst. Appl.* **2015**, *42*, 9412–9425. [CrossRef]

Article

An Improved Fault Diagnosis Approach Using LSSVM for Complex Industrial Systems

Shuyue Guan [1], Darong Huang [1,*], Shenghui Guo [2], Ling Zhao [1] and Hongtian Chen [3]

[1] College of Information Science and Engineering, Chongqing Jiaotong University, Chongqing 400074, China; 622200070034@mails.cqjtu.edu.cn (S.G.); zhaoling@cqjtu.edu.cn (L.Z.)
[2] College of Electronics and Information Engineering, Suzhou University of Science and Technology, Suzhou 215009, China; shguo@usts.edu.cn
[3] Department of Chemical and Materials Engineering, University of Alberta, Edmonton, AB T6G 1H9, Canada; hc11@uablerta.ca
* Correspondence: drhuang@cqjtu.edu.cn

Abstract: Fault diagnosis is a challenging topic for complex industrial systems due to the varying environments such systems find themselves in. In order to improve the performance of fault diagnosis, this study designs a novel approach by using particle swarm optimization (PSO) with wavelet mutation and least square support (LSSVM). The implementation entails the following three steps. Firstly, the original signals are decomposed through an orthogonal wavelet packet decomposition algorithm. Secondly, the decomposed signals are reconstructed to obtain the fault features. Finally, the extracted features are used as the inputs of the fault diagnosis model established in this research to improve classification accuracy. This joint optimization method not only solves the problem of PSO falling easily into the local extremum, but also improves the classification performance of fault diagnosis effectively. Through experimental verification, the wavelet mutation particle swarm optimazation and least sqaure support vector machine (WMPSO-LSSVM) fault diagnosis model has a maximum fault recognition efficiency that is 12% higher than LSSVM and 9% higher than extreme learning machine (ELM). The error of the corresponding regression model under the WMPSO-LSSVM algorithm is 0.365 less than that of the traditional linear regression model. Therefore, the proposed fault scheme can effectively identify faults that occur in complex industrial systems.

Keywords: fault diagnosis; PSO; wavelet mutation; LSSVM

Citation: Guan, S.; Huang, D.; Guo, S.; Zhao, L.; Chen, H. An Improved Fault Diagnosis Approach Using LSSVM for Complex Industrial Systems. *Machines* **2022**, *10*, 443. https://doi.org/10.3390/machines10060443

Academic Editor: Ahmed Abu-Siada

Received: 7 May 2022
Accepted: 31 May 2022
Published: 4 June 2022

Publisher's Note: MDPI stays neutral with regard to jurisdictional claims in published maps and institutional affiliations.

Copyright: © 2022 by the authors. Licensee MDPI, Basel, Switzerland. This article is an open access article distributed under the terms and conditions of the Creative Commons Attribution (CC BY) license (https:// creativecommons.org/licenses/by/ 4.0/).

1. Introduction

Fault diagnosis and detection for complex industrial systems has been widely investigated and rapidly developed in recent years [1–5]. In essence, fault diagnosis in industrial environments is pattern recognition based on fault features. In engineering systems, fault diagnosis is usually carried out in two aspects: model-based and data-based [6]. With the progress of science and technology, intelligent pattern recognition algorithms for fault signals have been developed vigorously, such as neural networks [7–9], K-nearest neighbor [10–12], and LSSVM [13–15]. Neural networks have the advantage of being able to approximate arbitrary complex nonlinearities and have good robustness [6,16]. For example, Xu et al. [17] proposed a fault diagnosis method based on neural networks and fuzzy theory for rotating machinery. In [4], a performance degradation and fault detection model for industrial systems was proposed based on transfer learning and federated neural networks, and the analysis illustrated its effectiveness and feasibility for industrial systems. For the purpose of fault detection, Chen et al. [9] established a data-driven fault detection scheme based on two neural networks, which can construct the optimal model adaptively. These methods demonstrate the effectiveness of neural network algorithms in fault diagnosis for dynamic industrial systems [18]. In another respect, vibration signals can be converted into two-dimensional digital images representing the patterns of the permutation entropy of those signals, as in [19], where a deep neural network was established for pattern recognition. Usually, a neural network algorithm needs a large amount of data training to establish a model with high diagnostic accuracy. However, it is difficult to obtain a large amount of fault data from complex systems in practice. K-nearest neighbor is one of the simplest algorithms based on data-driven classification technology, and it is easy to implement and requires no parameter estimations. It is widely applied in pattern recognition, fault diagnosis, and the multiple classification problem [20–22]. Ma et al. [23] proposed a multilabel learning algorithm based on the K-nearest neighbor algorithm for managing the prognostics and health of rolling bearings, and Gan et al. [24] used the K-nearest classifier to identify different rolling bearing conditions for industrial systems. Nevertheless, K-nearest neighbor is highly dependent on samples, the effect of this defect on classification accuracy cannot be neglected.

Support vector machine (SVM), as a classical pattern identification method, is widely used in various fields. For example, a temporal-based support vector machine for the detection and identification of several toxic gases in a gas mixture was proposed in [25], which also indicates the great potential of SVM. LSSVM, which is a modification of the SVM, was proposed by Suykens and Vandewalle in [26]. Inequality constraints in SVM are replaced by equality constraints in LSSVM, reducing the difficulty of calculation. Zhang et al. [27] combined a generalized frequency response function and LSSVM to achieve fault classification for a nonlinear analog circuit. The results showed that the fault diagnosis method can obtain high recognition accuracy. Product function correntropy and LSSVM were presented in [28] to improve the fault diagnosis performance for rolling bearings in varying industrial conditions. In order to further improve the effectiveness of LSSVM, Zhang et al. [29] used PSO to optimize LSSVM, and their proposed PSO-LSSVM fault diagnosis method had a high recognition rate. Similarly, a fault identification method for rolling bearings in industrial systems was proposed in [30]. In addition, Ren et al. achieved fault detection and diagnosis in complex industrial systems based on PSO-LSSVM, and their experimental results showed that this method can be applied well in the field of industry. As mentioned above, as a classical intelligent optimization algorithm, PSO is widely used due to its convenience of implementation: it does not require that extra attention be paid to parameter tuning. However, the PSO algorithm also has many disadvantages, such as a poor ability to search locally, and its tendency to fall easily into the local extremum [31–33]. To solve this problem, many scholars have made great efforts. For example, Zhang et al. [34] introduced dynamic inertia weights and gradient information to improve PSO. At the same time, a bearing fault diagnosis method via an LSSVM identification model was presented. Liu et al. [35] established a fault detection model based on a chaotic PSO algorithm and a kernel-independent component analysis, and the simulation results showed that the optimization method can avoid the phenomenon of the PSO algorithm's susceptibilty to falling into a local extremum. Furthermore, an improved PSO- and SVM-based fault diagnosis methodology was presented in [36] to predict faults in nuclear power plants.

Motivated by the above observations, the first contribution of this study is to design a novel fault diagnosis method based on WMPSO-LSSVM that can achieve a high classification accuracy. The second contribution is to solve the problems of the PSO algorithm's susceptibilty to falling into a local extremum and its low search precision. In addition, this study adopts the data-driven method to realize the fault diagnosis and prognostics of the actual complex parts in an industrial system, and a contrast experiment shows that the established joint optimization scheme has superior performance and strong robustness, which can promote the development of mechanical fault diagnosis.

The remaining parts of this study include Section 2, which introduces the signal preprocessing and feature extraction methods, which are based on an orthogonal wavelet packet algorithm (WPT); Section 3, in which the WMPSO-LSSVM-based fault diagnosis scheme is presented; Section 4, where the effectiveness of this study is verified by actual fault data and comparison experiments; and finally, Section 5, in which the conclusion is given.

2. Signal Decomposition and Feature Extraction-Based Orthogonal Wavelet Packet Transform

Wavelet transforms have been widely used for vibration signal pre-processing for industrial systems. Generally, wavelet transforms only decompose the low-frequency part of the signal, and do not treat the high-frequency portion of the signal at all. However, the detailed information that can characterize the vibration signal usually exists in the high-frequency section. Therefore, the orthogonal wavelet packet transform is introduced to solve this problem. Furthermore, the vibration signal of industrial systems can be decomposed in this way without information loss, which lays a foundation for obtaining high fault diagnosis accuracy. The theoretical basis is described as follows.

In multiresolution analysis, $L^2(R)$ is a square-integrable space and $L^2(R) = \underset{j \in Z}{\oplus} W_j$, indicating that the multiresolution analysis decomposes $L^2(R)$ into the orthogonal sum of all subspaces $W_j (j \in Z)$, according to the different scale factors j. $W_j (j \in Z)$ is the wavelet subspace of the wavelet function $\psi(t)$. Then, we hope to further subdivide $W_j (j \in Z)$ through a binary fraction. Therefore, the scale subspace V_j and the wavelet subspace W_j can be represented through a new subspace U_j^n, if there are the following conditions:

$$\begin{cases} U_j^0 = V_j j \in Z \\ U_j^1 = W_j j \in Z \end{cases} \quad (1)$$

Then, the orthogonal decomposition of the Hilbert space can be expressed as follows:

$$U_{j+1}^0 = U_j^0 \oplus U_j^1 \quad (2)$$

Suppose U_j^n is the wavelet subspace of $u_n(t)$, U_j^{2n} is the wavelet subspace of $u_{2n}(t)$, and $u_n(t)$ is:

$$\begin{cases} u_{2n}(t) = \sqrt{2} \sum_{k \in Z} h(k) u_n(2t - k) \\ u_{2n+1}(t) = \sqrt{2} \sum_{k \in Z} g(k) u_n(2t - k) \end{cases} \quad (3)$$

where $h(k)$ represents the low-pass filter coefficients and $g(k)$ represents the high-pass filter coefficients, and $g(k) = (-1)^k h(1-k)$. Then, Formula (3) can be rewritten as follows:

$$\begin{cases} u_{2n}(t) = \sqrt{2} \sum_{k \in Z} h_k u_n(2t - k) \\ u_{2n+1}(t) = \sqrt{2} \sum_{k \in Z} g_k u_n(2t - k) \end{cases} \quad (4)$$

where $u_0(t) = \phi(t)$ ($\phi(t)$ is the scale function), $u_1(t) = \psi(t)$ ($\psi(t)$ is the wavelet basis function), and the sequence $\{u_n(t)\}_{n \in Z_+}$ is the orthogonal wavelet packet basis.

Suppose $f(n)$ is the signal to be decomposed. In fact, a wavelet packet transform of $f(n)$ is a projection coefficient on the wavelet packet basis $\{u_n(t)\}_{n \in Z_+}$:

$$p_f(n, j, k) = \langle f(t), u_n(t) \rangle = \int_{-\infty}^{+\infty} f(t) \left[2^{-j/2} \bar{u}_n \left(2^{-j} \bar{t} - k \right) \right] dt \quad (5)$$

where $\{p_s(n, j, k)\}_{k \in Z}$ is the sequence of transformation coefficients of $f(n)$ on U_j^n.

Usually, the transformation coefficients $\{p_s(n, j, k)\}_{k \in Z}$ can be calculated through the Mallat algorithm:

$$\begin{cases} p_f(2n, j, k) = \sum_{l \in Z} h_{l-2k} p_f(n, j-1, l) \\ p_f(2n+1, j, k) = \sum_{l \in Z} g_{l-2k} p_f(n, j-1, l) \end{cases} \quad (6)$$

According to the above discussion, the decomposition processing of the original signal is depicted and illustrated in Figure 1.

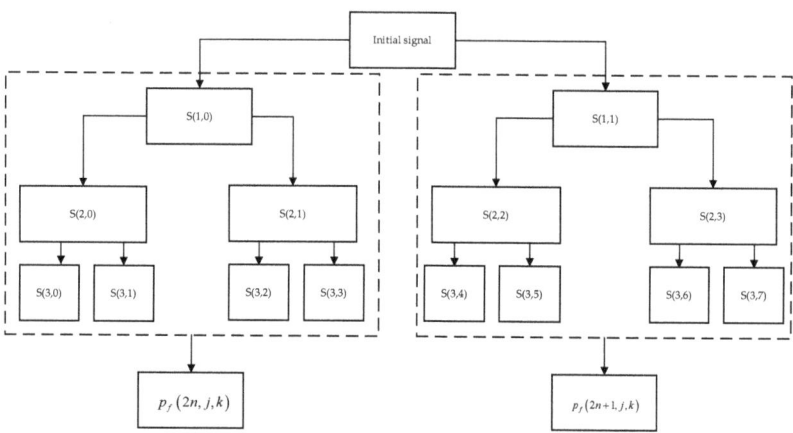

Figure 1. Wavelet decomposition for original signals.

Because of the integrity and orthogonality of the wavelet packet space, the original signal $f(n)$ is almost completely intact after wavelet decomposition, which provides conditions for analyzing signal characteristics.

According to the above definition of the orthogonal wavelet packet transform, the signal $f(n)$ has been projected adaptively into the orthogonal wavelet packet space; then, the obtained component can be regarded as the energy distributed in the corresponding space. If the energy distribution of signals in the space of each orthogonal wavelet packet can be calculated at a certain decomposition level, then the characteristics can be extracted by sorting these energies according to the frequency index of U_j^n. The energy distribution in the time-frequency localization space can be interpreted as follows:

$$E(j,n) = \sum_{k \in Z} \left[p_f(n,j,k) \right]^2 \tag{7}$$

Therefore, if the original signal $f(t)$ is decomposed by P levels, the energy feature vector extracted from the original signal can be expressed as follows:

$$E^*(P,f) = \left[E(P,0), E(P,1), \ldots, E\left(P, 2^P - 1\right) \right] \tag{8}$$

3. Improved Fault Diagnosis Approach Using WMPSO-LSSVM

3.1. Least Squares Support Vector Machine

The literature of various fields shows that the LSSVM model performs well on various datasets, so it can process the data generated under unknown working conditions in complex industrial systems well. In addition, the complete theoretical basis of LSSVM can also ensure its stability. The principle of LSSVM is as follows:

$$\min_{w,b} \frac{1}{2} \|w\|^2 + C \sum_{i=1}^{m} \zeta_i^2 \tag{9}$$

$$s.t. y_i \left(w^T x_i + b \right) = 1 - \zeta_i, i = 1, 2, \ldots, m \tag{10}$$

where $\{(x_1, y_1), (x_2, y_2), \ldots, (x_l, y_l)\}$ are the samples to be observed, w is the perpendicular vector of the line, b is the offset of the hyperplane, C is the regularization parameter, and ζ_i represents the fluctuations in the error of each sample.

To obtain an accurate solution to the above optimal problem, the Lagrange function with slack variables can be established as follows:

$$L(w,b,\zeta,\alpha,\lambda) = \frac{1}{2}\|w\|^2 + C\sum_i \zeta_i^2 + \sum_i \alpha_i\left(1 - \zeta_i y_i\left(w^T \varphi_{lssvm}(x_i) + b\right)\right) - \sum_i \lambda_i \zeta_i \quad (11)$$

where α_i is the Lagrange multiplier of the original problem, and λ_i is the Lagrange multiplier of the additional slack variables.

Take the derivative of each variable in Formulas (9) and (10) and let them be 0. The following equalities hold:

$$\begin{cases} w = \sum_i^n \alpha_i x_i y_i \\ \sum_{i=1}^n \alpha_i y_i = 0 \\ C - \alpha_i - \lambda_i = 0 \end{cases} \quad (12)$$

Thus, Formula (11) can be rewritten as follows:

$$L(\zeta,\alpha,\lambda) = \sum_i \alpha_i^2 + \sum_{i,j} \alpha_i \alpha_j y_i y_j X_i^T X_j \quad (13)$$

Therefore, the optimal problem of Formulas (9) and (10) can be expressed as follows:

$$\begin{cases} \max_\alpha W(\alpha) = \sum_{i=1}^n \alpha_i - \frac{1}{2}\sum_{i,j=1}^n y_i y_j \alpha_i \alpha_j \langle x_i, x_j \rangle \\ s.t. \sum_{i=1}^n \alpha_i y_i = 0 \end{cases} \quad (14)$$

Given the varying conditions of industrial systems, the vibration signal of equipment follows a nonlinear relationship. In order to solve the problem of linear indivisibility in primordial space, it is necessary to transform the failure samples into multi-dimensional distinguishable space by introducing kernel functions. Therefore, Formula (14) can be written as follows:

$$\begin{cases} \max_\alpha W(\alpha) = \sum_{i=1}^n \alpha_i - \frac{1}{2}\sum_{i,j}^n y^{(i)} y^{(j)} \alpha_i \alpha_j k(x_i, x_j) \\ s.t. \sum_{i=1}^n \alpha_i y_i = 0 \end{cases} \quad (15)$$

where $k(x_i, x_j)$ is the kernel function, and the selection of the kernel function has great flexibility. The common kernel functions are described as follows:

1. Linear kernel function:

$$K(x_i, x_j) = x_i \cdot x_j \quad (16)$$

2. Polynomial kernel function:

$$K(x_i, x_j) = (x_i \cdot x_j + 1)^l, l = 1, 2, \ldots \quad (17)$$

3. Gaussian kernel function:

$$K(x_i, x_j) = \exp\left[-\frac{\|x_i - x_j\|^2}{2\sigma^2}\right] \quad (18)$$

The Gaussian kernel function selected in this paper can effectively transform the data from the low-dimensional non-separable space to the high-dimensional separable space, and it can further improve the classification accuracy of the model. Another advantage of Gaussian kernels, compared to other kernels, is that the more complex the model, the stronger the performance. In addition, no matter how many dimensions are the characteristics of each sample point, each sample can be transformed into the total sample quantity dimension after processing by the Gaussian kernel function, which expands the dimension and the diversity of data.

It is natural to notice that LSSVM's classification accuracy is closely related to the penalty factor and parameter σ of the kernel functions. If the kernel function is too small, there will be an over-fitting phenomenon in the classification; otherwise, there will be an under-fitting phenomenon. Similarly, the larger the penalty factor, the more likely it is to overfit; and the smaller the penalty factor is, the more likely it is to underfit. Thus, in order to improve the accuracy of fault diagnosis for industrial systems, an optimized approach, named WMPSO-LSSVM, is proposed in the next section.

3.2. WMPSO-Based Parameters Optimization of LSSVM

As mentioned above, the regularization parameter and kernel functions play an important role in LSSVM. Thus, in this paper, we adopt the proposed WMPSO algorithm to optimize the parameters and establish a desirable model with high classification accuracy. Firstly, the basic model of PSO is as follows:

$$C_i = m \times C_i + c_1 \times rand \times (gbest - \sigma_i) + c_2 \times rand \times (qbest - \sigma_i) \tag{19}$$

$$\sigma_i = \sigma_i + C_i \tag{20}$$

where C_i is the regulation parameter of the LSSVM as well as the current velocity of PSO, and σ_i is the kernel function of the LSSVM as well as the location of particles in PSO. m indicates the weight coefficient, c_1 and c_2 are learning factors, and $rand$ is a random number between 0 and 1. Meanwhile, $gbest$ and $qbest$ store the optimal values corresponding to the penalty coefficient C and the kernel parameter σ, respectively.

Suppose there is a group of particle swarms $S = (S_1, S_2, \ldots, S_n)$ in an n-dimensional space; C and σ can be presented as follows:

$$C = (C_1, C_2, \ldots, C_i) \tag{21}$$

$$\sigma = (\sigma_1, \sigma_2, \ldots, \sigma_i) \tag{22}$$

In this paper, the wavelet function μ^* is used to conduct a random perturbation of all the dimensions of the contemporary optimal value $Q_g^m(t)$ particles, and the perturbation result is taken as the position of the particles. The calculation model is given as follows:

$$\bar{\sigma}^m(t) = \mu^* Q_g^m(t) \tag{23}$$

For the sake of the accuracy of the WMPSO algorithm, the Morlet function was selected as the wavelet base in this study, as shown in Figure 2.

The Morlet wavelet has more accurate and high-resolution spectral estimation, and has thus been widely used. Compared with the Gaussian and Cauchy variations often used in particle swarm optimizations, the Morlet wavelet searches more effectively in the solution space because there is an equal probability of producing positive and negative numbers.

In addition, the Morlet wavelet function changes the local solution more frequently in the solution space, and it is easier to obtain the optimal solution in the local optimization. The Morlet wavelet function can fine-tune the particle, so it is a remarkable choice to select the Morlet wavelet for mutation.

Thus, the wavelet function value applied is expressed as follows:

$$\mu^* = \frac{1}{\sqrt{a}} e^{-\left(\frac{\varphi^*}{a^*}\right)^2/2} \cos\left(5\left(\frac{\varphi^*}{a^*}\right)\right) \tag{24}$$

Meanwhile, the scale parameter a^* is calculated by Formula (25):

$$a^* = e^{-\ln(g) \times \left(1 - \frac{t}{T}\right)^{\gamma_{wm}}} + \ln(g) \qquad (25)$$

where γ_{wm} is the shape parameter, t is the current iteration number, T is the maximum number of iterations, and g is the limit of a^*.

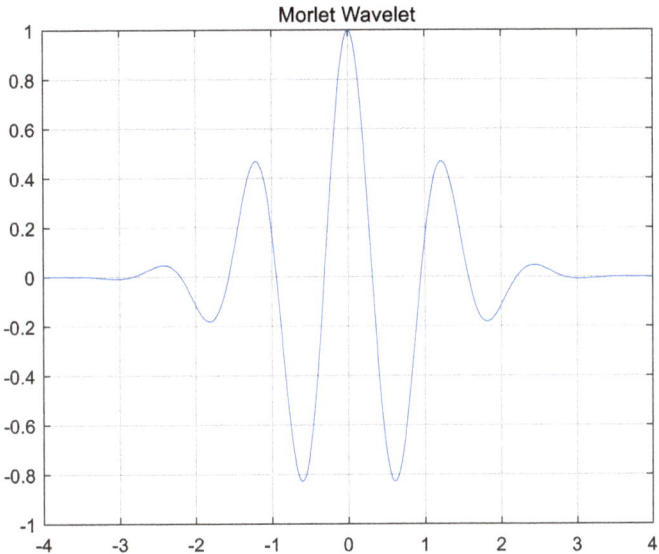

Figure 2. Morlet-wavelet function.

Therefore, after the perturbance by using the wavelet mutation function, the new positions of the particles are $\tilde{\sigma}^m = (\tilde{\sigma}_1^m, \tilde{\sigma}_2^m, \ldots, \tilde{\sigma}_n^m)$. Once the position and kernel parameter σ are determined, the regularization factor C can be confirmed according to Formula (19). The optimization process for the parameters in this study is given in Algorithm 1:

Algorithm 1 The process of the WMPSO parameters' optimization

---

Initialize σ_i \\ σ_i is the position of the $i_t h$ particle
  Calculate fitness function \\ Individual extreme values of particles can be calculated by fitness function
  while $i <= T$ **do** \\ T is the maximum number of iterations performed by the algorithm
  $i = i + 1$
    for $j = 1$ to n **do**
      Update velocity C_i based on Equation (19)
      Update position σ_i based on Equation (20)
      if $p_m > rand$ **then**
        Calculate a^* based on Equation (25)
        $\varphi^* = 2.5 * a^* * rand(1, 30)$
        Calculate μ^* based on Equation (24)
        Update position σ_i based on Equation (23)
      end if
      Calculate fitness function
      Update Q_i and Q_g
    end for
  end while

3.3. Design of WMPSO-LSSVM-Based Fault Diagnosis Scheme for Industrial Systems

Based on the above analysis, the WMPSO-LSSVM-based data-driven fault diagnosis approach is designed as follows:

1. Decompose the composite fault data of industrial systems based on the orthogonal wavelet packet algorithm and extract the fault characteristics;
2. Take the extracted characteristics as the input to the WVPSO-LSSVM identification model, training to obtain the regularization coefficient C and kernel parameter σ. The training process is summarized as follows:

   - Initialize the following parameters: the evolution algebra of the particles, the learning factors c_1 and c_2, the regularization factor C, the kernel parameter σ, and the historical optimal kernel parameter Q_σ;
   - Calculate the new information of the C and σ, and update a new generation of the particles;
   - Calculate the fitness value of the particles according to the fitness function, and update the individual and global optimal values of C and σ on this basis;
   - Evaluate whether the maximum number of iterations or searching boundaries has been reached. If so, store the C and σ, and construct the WMPSO-LSSVM-based identification model;

3. Take the extracted characteristics as the input to the WVPSO-LSSVM identification model, testing to obtain the classification result.

The corresponding flowchart is presented in Figure 3.

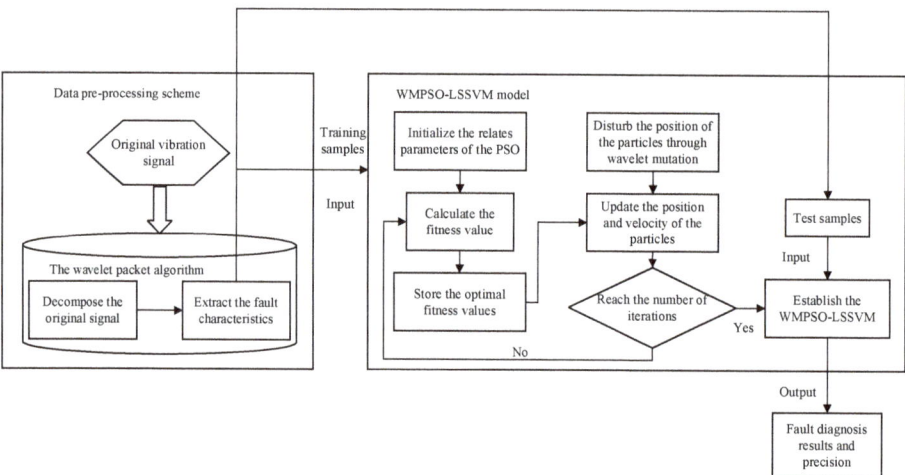

Figure 3. The flowchart of the proposed WMPSO-LSSVM algorithm.

4. Experimental Applications for Industrial Systems Based on WMPSO-LSSVM

The effectiveness and superiority of this study for industrial systems are evaluated on a database taken from the Guangdong Provincial Key Laboratory of Petrochemical Equipment Fault Diagnosis of China. Meanwhile, some comparative experiments are used to further prove the fault diagnosis performance of the proposed method.

As shown in Figure 4, the industrial system studied in this section is the main fan motor of a steam turbine, and the specific research object of this system is the gearbox containing the rolling bearings. The actual data of the gearbox and bearings are obtained from the intelligent fault diagnosis system, which consists of an acceleration sensor, a preamplifier (PMP), an explosion-proof BOX (BOX), a data collector (butylated hydroxytoluene), and a server (PC-1).

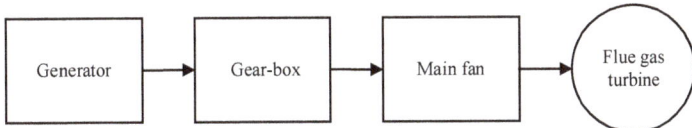

Figure 4. Schematic diagram of the main fan system.

In addition, the acceleration sensor is installed on the generator to obtain the vibration signals; the role of the BOX is to protect the preamplifier; the preamplifier is installed in the BOX for signal amplification; the data collector is installed in the steam turbine of the main fan for signal acquisition and processing; and the server is used for data storage and management.

The accelerometer used to measure the vibration acceleration mainly contains the following information. The highest amplitude is 50 g, the channel number is 6, the maximum transmission distance is 300 m, the working power supply is 18–30 VDC, and the working current is constant (2–10 mA). The actual industrial system operation environment and data collection situation are shown in Figures 5–8.

Figure 5. The on-site industrial environment.

Figure 6. The local-data acquisition system.

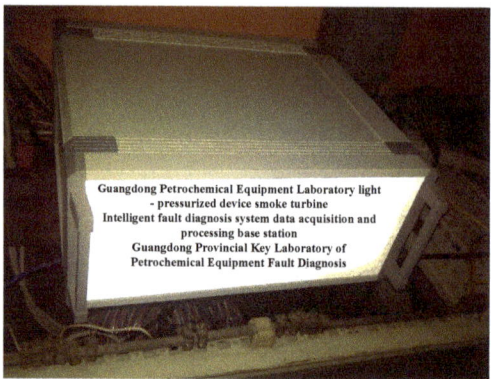

Figure 7. The data acquisition base station.

Figure 8. The data acquisition platform.

The data collected by the intelligent fault diagnosis system mainly include seven states, which are different fault combinations of gears and bearings. Their fault modes and corresponding indicators are shown in Table 1, and the waveforms of the part of the original vibration signals are shown in Figures 9–12.

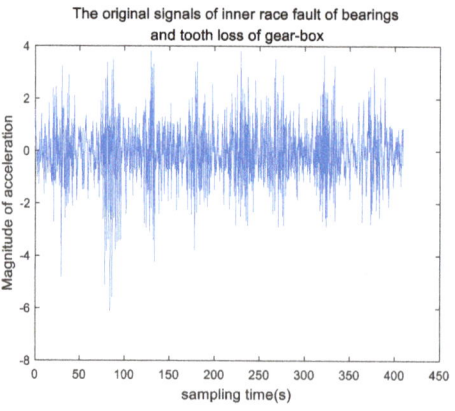

Figure 9. The original signals of the inner race fault of the bearings and the tooth loss of gear-box.

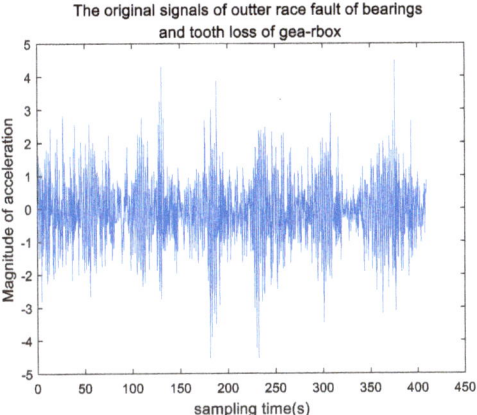

Figure 10. The original signals of the outer race fault of the bearings and the tooth loss of the gear-box.

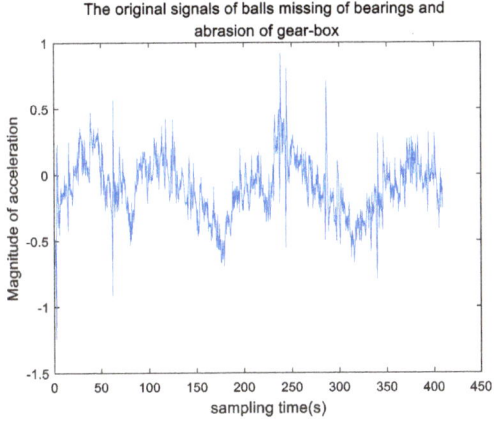

Figure 11. The original signals of balls that are missing bearings and the abrasion of the gear-box.

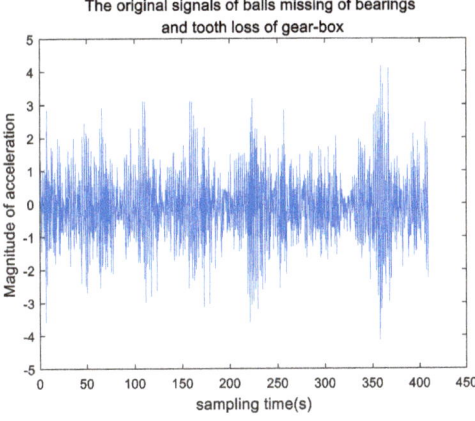

Figure 12. The original signals of balls that are missing bearings and the tooth loss of the gear-box.

Table 1. Seven fault states of the key components for entire systems.

Description of Seven States	Vibration Index	Impulsion Index	Tolerance Index	Peak Index	Kurtosis Index
State 1: missing gear teeth and outer ring wear of right bearing	1.1975 1.3132	2.5531 6.8919	2.9015 8.2115	2.1319 5.3947	3.0860 4.1036
State 2: missing gear teeth and lack of balls on left bearing	1.2293 **1.2920**	3.1451 4.9894	3.6689 5.9483	2.5414 3.9279	2.7140 **3.5757**
State 3: missing gear teeth and outer ring wear on left bearing	1.2657 1.3558	**4.3240** 7.5935	**5.1791** 9.1797	**3.3671** 5.7598	**3.4370** 5.4632
State 4: missing gear teeth and inner ring wear on right bearing	1.2438 1.3082	3.2264 5.6916	3.7968 6.8665	2.5912 4.3945	2.8526 4.3278
State 5: wear of gear and inner ring wear on left bearing	1.2252 1.3433	2.2448 **4.2110**	2.6442 **4.9972**	1.8041 **3.3652**	2.3961 4.6594
State 6: wear of gear and lack of balls on left bearing	1.2257 1.3227	2.6885 5.3905	3.3278 6.7998	2.4035 4.1221	2.7392 8.0007
State 7: wear of gear and outer ring wear on left bearing	**1.3007** 1.3742	**4.3120** 7.4453	**5.1996** 9.0964	3.3152 5.5460	**3.6755** 5.4385

The numbers in bold in the table represent the time domain index of the faulty component. Look at the numbers in the table. If the data in the table appears to be significantly asynchronous, this can be used to distinguish component failures. Taking the waveform indicator as an example, 1.2920 is obviously out of sync with all the numbers in the second row of the waveform column, and 1.3007 is also out of sync with the numbers in the first row of the waveform column, so it can be used as the basis for division.

Therefore, according to the indicators in bold in Table 1, the following analysis can be obtained.

- States 2 and 7 can be distinguished via the vibration index;
- States 3 and 5 can be distinguished via the impulsion and tolerance indices;
- States 5 and 7 can be distinguished via the impulsion and tolerance indices;
- States 3 and 5 can be distinguished via the peak index;
- States 2 and 7 can also be distinguished via the kurtosis index, as can states 2 and 3.

Then, the original signals are decomposed into three layers using the wavelet packet decomposition algorithm, and the node coefficients are calculated according to Formula (5). The corresponding results are given in Figure 13. In addition, the wavelet packet coefficients of the third layer, consisting of nodes 7 to 14 and calculated according to Formula (6), are shown in Figure 14.

The spectral distributions of the non-stationary vibration signals of the gearbox and bearings are closely related to their characteristic structures. Therefore, the energy distributions in the wavelet packet space of the original vibration signals decomposed by the wavelet packet are the fault features of the gears and bearings to be extracted. The parts of the characteristic extraction results are shown in Figure 15.

Finally, by using 75% of the extracted fault features as the input to establish the optimal WMPSO-LSSVM and by inputting the test samples into the model, the classification results can be obtained. The experimental results of LSSVM, PSO-LSSVM, and WMPSO-LSSVM are given in Figures 16–18, respectively.

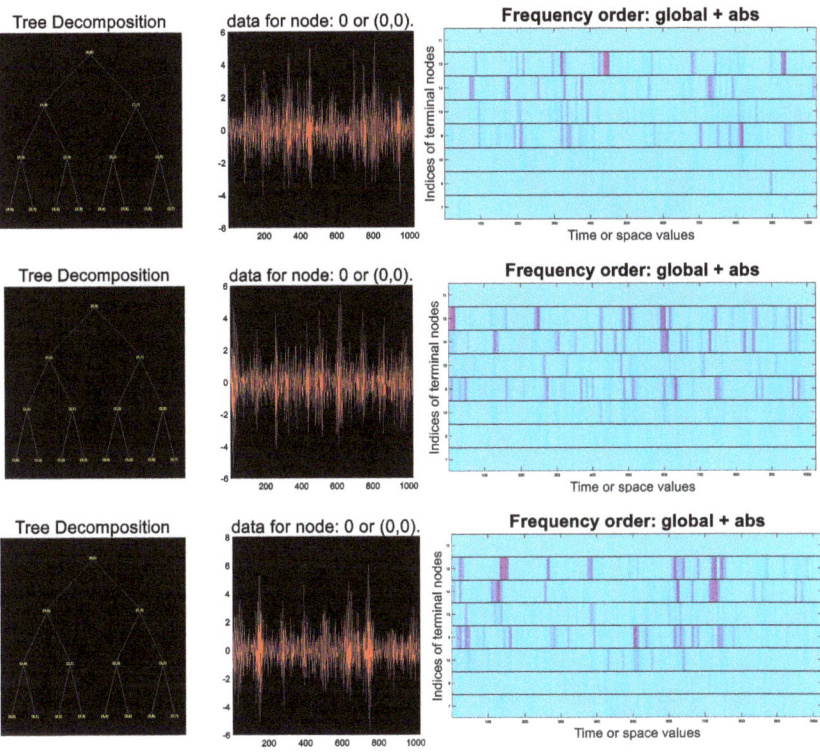

Figure 13. The decomposition results of the vibration signals.

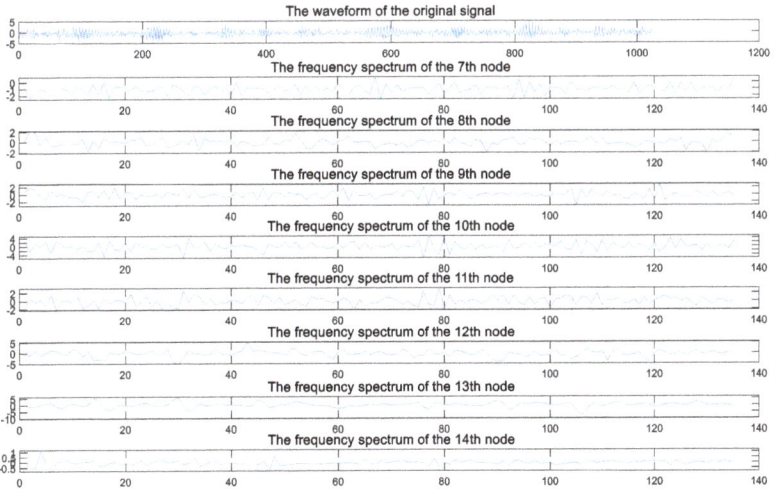

Figure 14. The node coefficients of the wavelet-packet algorithm.

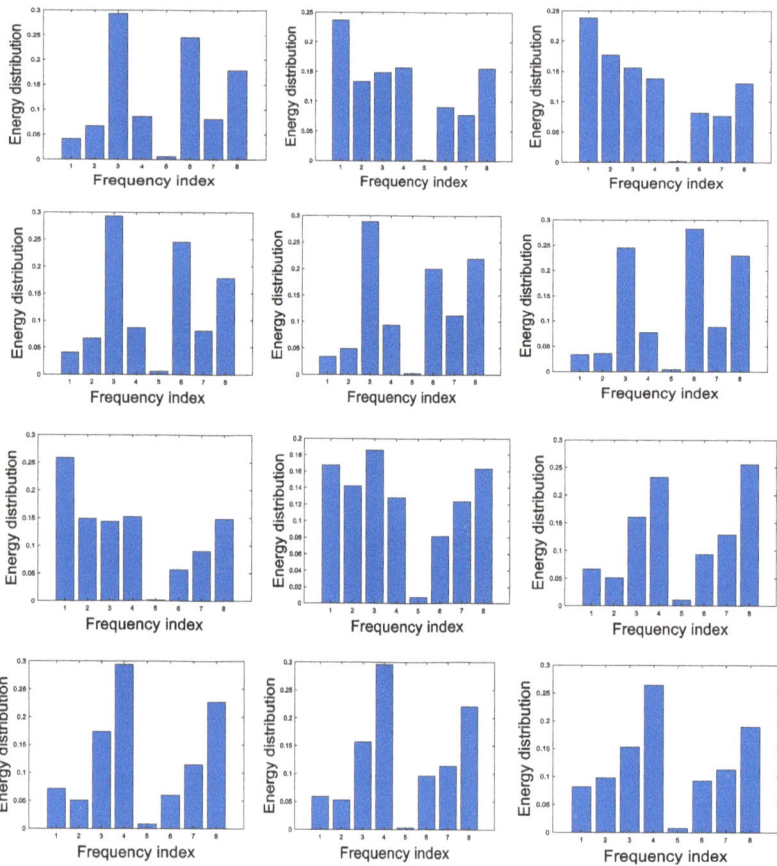

Figure 15. The fault characteristic extraction results of the gear-box and bearings.

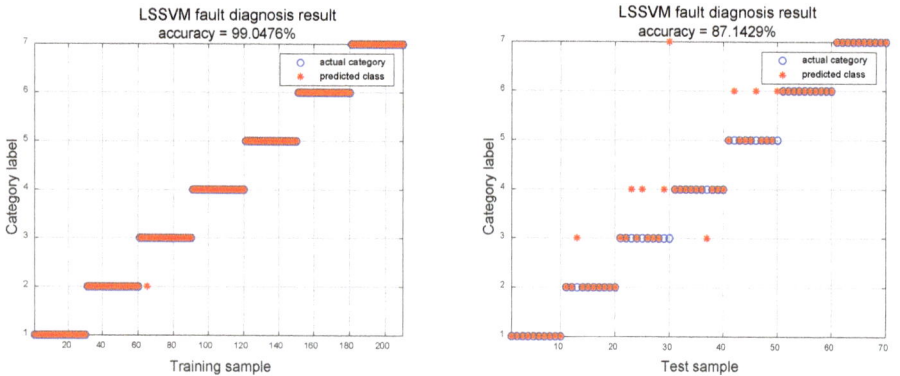

Figure 16. The classification results of LSSVM.

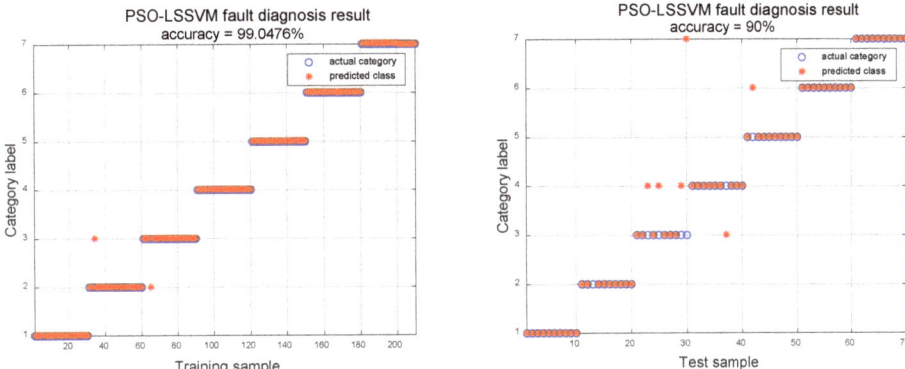

Figure 17. The classification results of PSO-LSSVM.

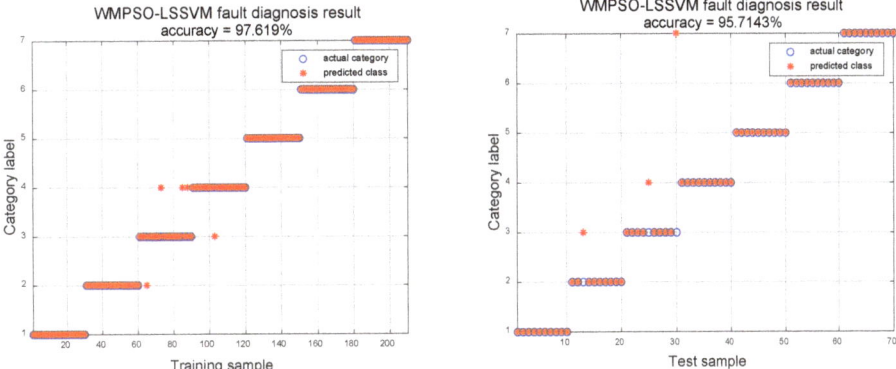

Figure 18. The classification results of WMPSO-LSSVM.

In order to further verify the superiority of the WVPSO-LSSVM classification model for key components of industrial systems, ELM and the traditional BP network are used for comparison purposes; the experimental results are shown in Table 2 and Figures 19–22.

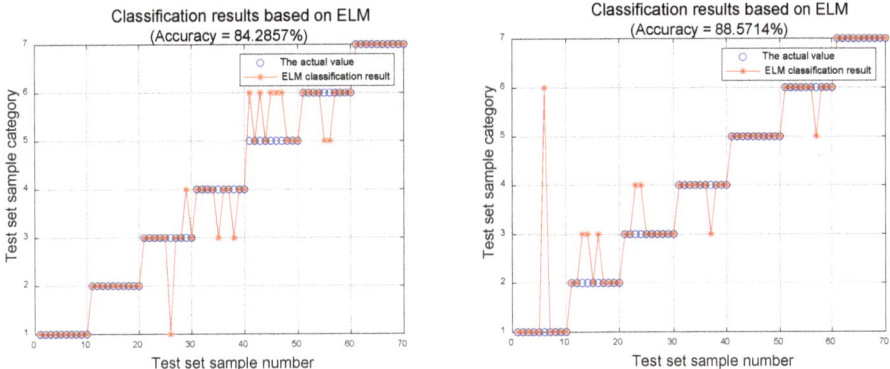

Figure 19. Fault diagnosis results based on ELM (1).

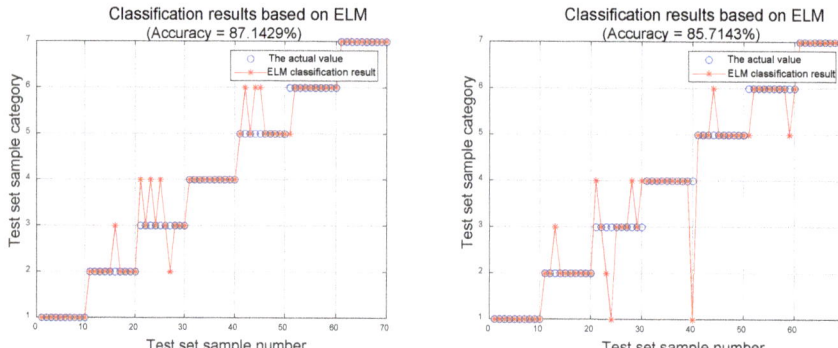

Figure 20. Fault diagnosis results based on ELM (2).

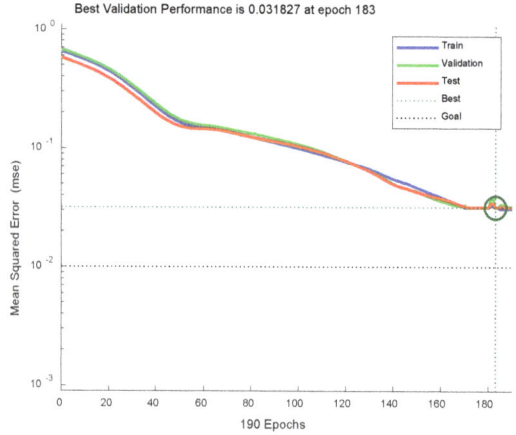

Figure 21. Training performance of the neural-networks.

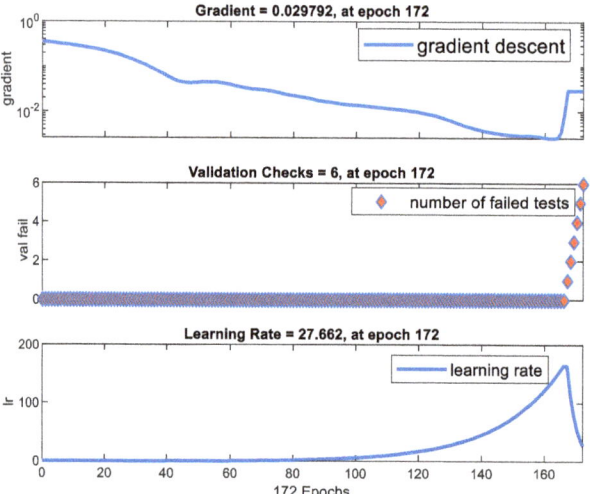

Figure 22. Training state of the neural-netwoks.

Table 2. This table contrasts the results of the three mechanisms.

Classification Method	BP	ELM	LSSVM	PSO-LSSVM	WMPSO-LSSVM
Classification accuracy (%)	64.29	86.50	84.17	90.00	95.71

To evaluate the performance of the WMPSO-LSSVM classification model, the confusion matrices of the WMPSO-LSSVM and ELM are presented, respectively, in Figures 23 and 24.

In the Figures 23 and 24, the blue square represents the number of correctly classified samples, while the pink square represents the number of incorrectly classified samples. For example, in Figure 23, there is only one incorrectly classified sample for the second type, and the remaining nine are correctly classified. The more diagonally distributed samples in the matrix, the better the performance of the model. And according to the results, the WMPSO-LSSVM has a higher precision than ELM.

In order to further verify the effectiveness of the proposed algorithm, the corresponding WVPSO-LSSVM regression model for the bearings and gearbox is established, and the composite fault characteristic trend is predicted. The comparative results are shown in Figures 25–28 and Table 3.

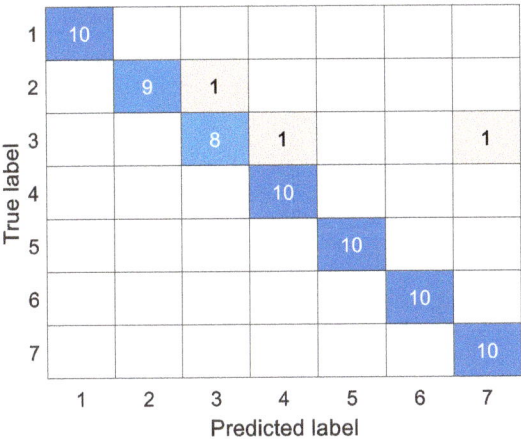

Figure 23. The confusion matrix of the WMPSO-LSSVM model.

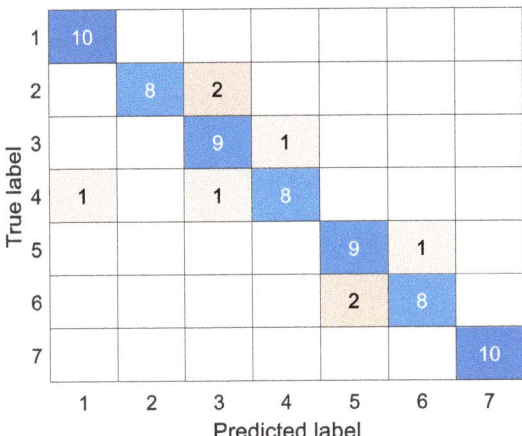

Figure 24. The confusion matrix result of the ELM model.

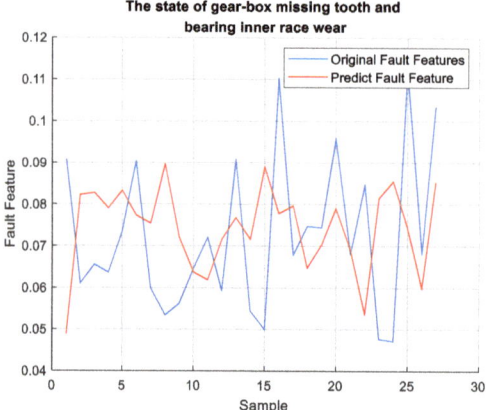

Figure 25. Bearing inner ring wear and gear tooth loss.

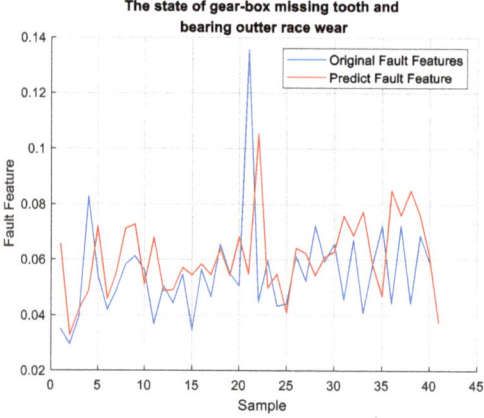

Figure 26. Bearing outer ring wear and gear tooth loss.

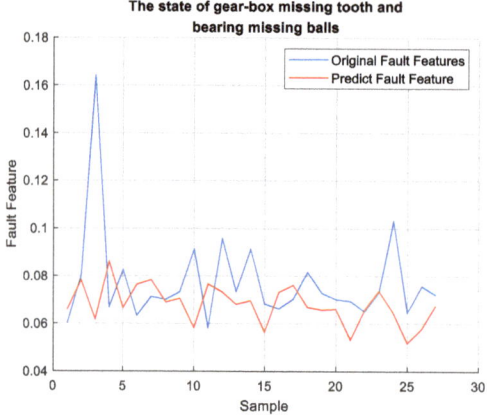

Figure 27. Bearing missing balls and gear tooth loss.

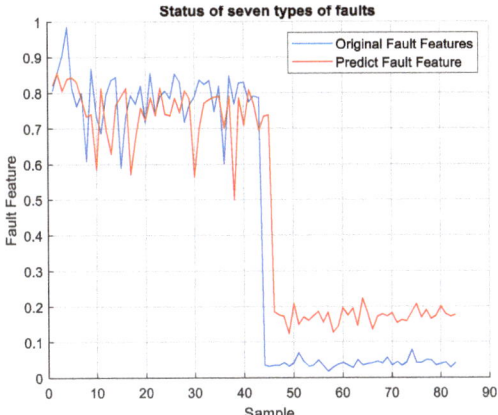

Figure 28. The state of seven types of faults.

Table 3. Comparison between the WMPSO-LSSVM regression model and the linear regression model.

Fault Types	Method WMPSO-LSSVM	Linear Regression
Bearing inner ring wear and gear tooth loss	0.0707154	0.411682
Bearing outer ring wear and gear tooth loss	0.00146932	0.2976
Bearings missing balls and gear tooth loss	0.00260635	0.545191
Seven types of fault features	0.0224879	0.304906

Since the weight and the deviation of ELM are randomly generated, the inconsistent networks generated each time will eventually lead to a large performance difference, although the learning speed of ELM is fast and its generalization performance is good. Furthermore, because the BP neural network is a gradient descent method, its optimized objective function is extremely complex, and there will be a zig-zag phenomenon in the training process, which makes the BP algorithm inefficient. The accuracy of the BP neural network also depends largely on the sample size, and the number of fault samples obtained from industrial systems is small. Thus, it is not suitable for limited fault data of complex industrial systems.

In addition, it can be seen from the comparative experimental results that the WMPSO-LSSVM model has strong performance. The introduction of the Gaussian kernel function in WMPSO-LSSVM can expand the diversity and dimension of limited data and solve the defect of traditional neural networks' unsuitability for small samples. At the same time, the model can not only classify complex fault data effectively, but can also predict the complex fault characteristic trend, which has good applicability to complex fault data in industrial systems.

5. Conclusions

In this research, aiming to address the difficulty of the low precision of fault diagnosis methods for industrial systems, a new fault diagnosis methodology, named WMPSO-LSSVM, is proposed. Based on the decomposition of fault signals for feature extraction, the gearbox and bearings derived from the composable components are taken as the specific objects, and the vibration can be decomposed without information loss based on WPT. By comparing the proposed method with the existing pattern recognition methods, the results show that the WMPSO-LSSVM method can achieve higher classification accuracy for multiple fault modes in industrial systems.

In addition, PSO optimized by the wavelet mutation is combined with the LSSVM algorithm to realize the further optimization of the regularization parameter and kernel function in the LSSVM, thereby improving the fault diagnosis accuracy. Particles that jump out of the local extreme value through the wavelet mutation algorithm will seek the optimal solution of parameters in the global space, so the optimal hyperplane of the LSSVM model can be established. As demonstrated via the comparative experiments, the accuracy of the WMPSO-LSSVM is almost 12% higher than that of the LSSVM, and is 9% higher than the ELM; moreover, the average error of the regression is 0.365 less than that of the traditional linear regression model, implying the potency of this scheme.

However, how to better select the parameters in the wavelet mutation function adaptively is not yet resolved in this work. Further research on the optimization of parameters in wavelet mutation is warranted.

In summary, the WMPSO-LSSVM proposed in this paper can significantly improve the fault diagnosis accuracy for complex industrial systems, and therefore, it offers better operability and scalability in the actual industrial environment.

Author Contributions: S.G. (Shuyue Guan): writing-original draft, methodology, investigation; D.H.: funding acquisition, supervision; S.G. (Shenghui Guo): writing—review & edtiting, methodology; L.Z.: methodology; H.C.: writing—review & editing. All authors have read and agreed to the published version of the manuscript.

Funding: This work was supported in part by the National Natural Science Foundation of P.R. China, under Grants 61663008 and 62073051, the Chong-qing Technology Innovation, Application Special Key Project, under Grant cstc2019jscx-mbdxX0015, and the 2018 Reliable control and safety maintenance of dynamic system under Grant JDDSTD2018001.

Informed Consent Statement: Not applicable.

Data Availability Statement: The data is provided by the Guangdong Provincial Key Laboratory of Petrochemical Equipment Fault Diagnosis of China for providing the dataset of the rolling bearings.

Acknowledgments: The authors would like to thank the Guangdong Provincial Key Laboratory of Petrochemical Equipment Fault Diagnosis of China for providing the dataset of the rolling bearings.

Conflicts of Interest: The authors declare no conflict of interest.

Notations

$L^2(R)$	square-intergrable space
W_j	wavelet subspace
$\psi(t)$	wavelet function in wavelet packet algorithm
V_j	scale subspace
U_j^n	Hilbert space
$u_n(t)$	orthogonal wavelet packet basis
$h(k)$	low-pass filter coefficients
$g(k)$	high-pass filter coefficients
$\phi(t)$	scale function in wavelet packet
$f(n)$	original signal
$p_f(n,j,k)$	a sequence of transformation coefficients in wavelet packet
$E(j,n)$	energy distribution
w	the perpendicular vector in LSSVM
b	an offset of the hyperplane in LSSVM
C	regularization parameter in LSSVM
ζ	the fluctuations of the error in LSSVM
α	Lagrange multiplier of the original problem
λ	Lagrange multiplier of the additional slack variables
$K(x_i, x_j)$	kernel function
C_i	the velocity of the ith particle
σ_i	the position of the ith particle

σ	kernel parameter of the Gaussian kernel function
m	weight coefficient in PSO
c_1	learning factor in PSO
c_2	learning factor in PSO
$rand$	random number uniformly distributed in [0, 1]
$gbest$	the best particle that indicates the global best
$qbest$	the best particle that indicates the local best
S	particle swarm
μ^*	wavelet function in the mutation wavelet algorithm
a^*	scale parameter in the mutation wavelet algorithm
γ_{wm}	shape parameter
t	the current iteration number
T	the maximum number of iterations
g	limit of scale parameter
$\bar{\sigma}^m$	the new position of the disturbed particle
p_m	the mutation rate
$Q_g^m(t)$	the global best of the ith particle
φ^*	wavelet function basis in Morlet
Q_g	the best particle that indicates the global best of the disturbed particle
Q_i	the best particle that indicates the individual best of the disturbed particle
Q_σ	the historical optimal kernel parameter

References

1. Yuan, H.D.; Wu, N.L.; Chen, X.Y.; Wang, Y. Fault Diagnosis of Rolling Bearing Based on Shift Invariant Sparse Feature and Optimized Support Vector Machine. *Machines* **2021**, *9*, 98. [CrossRef]
2. Espinoza-Sepulveda, N.; Sinha, J. Mathematical Validation of Experimentally Optimized Parameters Used in a Vibration-Based Machine-Learning Model for Fault Diagnosis in Rotating Machines. *Machines* **2021**, *9*, 155. [CrossRef]
3. Nguyen, V.; Hoang, D.T.; Tran, X.T.; Van M.; Kang, H.J. A Bearing Fault Diagnosis Method Using Multi-Branch Deep Neural Network. *Machines* **2022**, *9*, 345. [CrossRef]
4. Chen, H.; Chai, Z.; Jiang, B.; Huang, B. Data-driven fault detection for dynamic systems with performance degradation: A unified transfer learning framework. *IEEE Trans. Instrum. Meas.* **2020**, *70*, 3504712. [CrossRef]
5. Ran, G.; Liu, J.; Li, C.; Lam, H.-K.; Li, D.; Chen, H. Fuzzy-Model-Based Asynchronous Fault Detection for Markov Jump Systems with Partially Unknown Transition Probabilities: An Adaptive Event-Triggered Approach. *IEEE Trans. Fuzzy Syst.* **2022**, 1–10. [CrossRef]
6. Gao, Z.W.; Cecati, C.; Ding, S.X. A Survey of Fault Diagnosis and Fault-Tolerant Techniques-Part I: Fault Diagnosis with Model-Based and Signal-Based Approaches. *IEEE Trans. Ind. Electron.* **2015**, *62*, 3757–3767. [CrossRef]
7. Chadha, G.S.; Panambilly, A.; Schwung, A.; Ding, S.X. Bidirectional deep recurrent neural networks for process fault classification. *ISA Trans.* **2021**, *106*, 330–342. [CrossRef]
8. Wang, L.; Zhang, Z.J.; Long, H.; Xu, J.; Liu, R. Wind Turbine Gearbox Failure Identification with Deep Neural Networks. *IEEE Trans. Ind. Inform.* **2017**, *13*, 1360–1368. [CrossRef]
9. Chen, H.T.; Li, L.L.; Shang, C.; Huang, B. Fault Detection for Nonlinear Dynamic Systems with Consideration of Modeling Errors: A Data-Driven Approach. *IEEE Trans. Cybern.* **2022**, 1–11. [CrossRef]
10. Javaid, A.; Javaid, N.; Wadud, Z.; Saba, T.; Sheta, O.E.; Saleem, M.Q.; Alzahrani, M.E. Machine Learning Algorithms and Fault Detection for Improved Belief Function Based Decision Fusion in Wireless Sensor Networks. *Sensors* **2019**, *19*, 1334. [CrossRef]
11. Moreno, A.P.; Santiago, O.L.; de Lazaro, J.M.B.; Moreno, E.G. Comparative Evaluation of Classification Methods Used in the Fault Diagnosis of Industrial Processes. *IEEE Lat. Am. Trans.* **2013**, *11*, 682–689. [CrossRef]
12. Ebrahimi, B.M.; Faiz, J. Diagnosis and performance analysis of three-phase permanent magnet synchronous motors with static, dynamic and mixed eccentricity. *IET Electr. Power Appl.* **2010**, *4*, 53–66. [CrossRef]
13. Zuo, M.J.; Xiang, G.; Hu, S. Fault diagnosis of the constant current remote power supply system in CUINs based on the improved water cycle algorithm. *India J. Geo-Mar. Sci.* **2021**, *50*, 914–921.
14. Zhao, Y.P.; Wang, J.J.; Li, X.Y.; Peng, G.J.; Yang, Z. Extended least squares support vector machine with applications to fault diagnosis of aircraft engine. *ISA Trans.* **2020**, *97*, 189–201. [CrossRef]
15. Heidari, M.; Homaei, H.; Golestanian, H.; Heidari, A. Fault diagnosis of gearboxes using wavelet support vector machine, least square support vector machine and wavelet packet transform. *J. Vibroeng.* **2016**, *18*, 860–875.
16. Milosevic, N.; Rackovic, M. Classification based on Missing Features in Deep Convolutional Neural Networks. *Neural Netw. World* **2019**, *29*, 221–234. [CrossRef]
17. Chen, H.; Chen, Z.; Chai, Z.; Jiang, B.; Huang, B. A single-side neural network-aided canonical correlation analysis with applications to fault diagnosis. *IEEE Trans. Cybern.* **2021**, 1–13. [CrossRef]

18. Chen, H.T.; Chai, Z.; Dogru, O.; Jiang, B.; Huang, B. Data-Driven Designs of Fault Detection Systems via Neural Network-Aided Learning. *IEEE Trans. Neural Netw. Learn. Syst.* **2021**, 1–12. [CrossRef]
19. Landauskas, M.; Cao, M.; Ragulskis, M. Permutation entropy-based 2D feature extraction for bearing fault diagnosis. *Nonlinear Dyn.* **2020**, *102*, 1717–1731. [CrossRef]
20. Kherif, O.; Benmhamed, Y.; Teguar, M.; Boubakeur, A.; Ghoneim, S.S. Accuracy Improvement of Power Transformer Faults Diagnostic Using KNN Classifier with Decision Tree Principle. *IEEE Access* **2021**, *9*, 81693–81701. [CrossRef]
21. Liu, Y.W.; Cheng, Y.Q.; Zhang, Z.Z.; Wu, J. Multi-information Fusion Fault Diagnosis Based on KNN and Improved Evidence Theory. *J. Vib. Eng. Technol.* **2022**, *10*, 841–852. [CrossRef]
22. Liu, T.Y.; Luo, H.; Kaynak, O.; Yin, S. A Novel Control-Performance-Oriented Data-Driven Fault Classification Approach. *IEEE Syst. J.* **2020**, *14*, 1830–1839. [CrossRef]
23. Ma, X.; Hu, Y.; Wang, M.H.; Li, F.; Wang, Y. Degradation State Partition and Compound Fault Diagnosis of Rolling Bearing Based on Personalized Multilabel Learning. *IEEE Trans. Instrum. Meas.* **2021**, *70*, 1–11. [CrossRef]
24. Gan, X.; Lu, H.; Yang, G.Y.; Liu, J. Rolling Bearing Diagnosis Based on Composite Multiscale Weighted Permutation Entropy. *Entropy* **2018**, *20*, 821. [CrossRef] [PubMed]
25. Djeziri, M.A.; Djedidi, O.; Morati, N.; Seguin, J.L.; Bendahan, M.; Contaret, T. A temporal-based SVM approach for the detection and identification of pollutant gases in a gas mixture. *Appl. Intell.* **2021**, *52*, 6065–6078. [CrossRef]
26. Suykens, J.A.K.; Vandewalle, J. Least squares support vector machine classifiers. *Neural Process Lett.* **1999**, *9*, 293–300. [CrossRef]
27. Zhang, J.L. Fault Diagnosis of Nonlinear Analog Circuit Based on Generalized Frequency Response Function and LSSVM Classifier Fusion. *Math. Probl. Eng.* **2020**, *2020*, 8274570. [CrossRef]
28. Fu, Y.X.; Jia, L.M.; Qin, Y.; Yang, J. Product function correntropy and its application in rolling bearing fault identification. *Measurement* **2017**, *97*, 88–99. [CrossRef]
29. Zhang, K.; Su, J.P.; Sun, S.; Liu, Z.; Wang, J.; Du, M.; Liu, Z.; Zhang, Q. Compressor fault diagnosis system based on PCA-PSO-LSSVM algorithm. *Sci. Prog.* **2021**, *104*, 1–16. [CrossRef] [PubMed]
30. Xu, H.B.; Chen, G.H. An intelligent fault identification method of rolling bearings based on LSSVM optimized by improved PSO. *Mech. Syst. Signal Process.* **2013**, *35*, 167–175. [CrossRef]
31. Deng, W.; Yao, R.; Zhao, H.M.; Yang, X.; Li, G. A novel intelligent diagnosis method using optimal LS-SVM with improved PSO algorithm. *Soft Comput.* **2019**, *23*, 2445–2462. [CrossRef]
32. Yuan, H.D.; Chen, J.; Dong, G.M. Bearing Fault Diagnosis Based on Improved Locality-Constrained Linear Coding and Adaptive PSO-Optimized SVM. *Math. Probl. Eng.* **2017**, *2017*, 7257603. [CrossRef]
33. Zhang, D.D.; Xiang, W.G.; Cao, Q.W.; Chen, S. Application of incremental support vector regression based on optimal training subset and improved particle swarm optimization algorithm in real-time sensor fault diagnosis. *Appl. Intell.* **2020**, *51*, 3323–3338. [CrossRef]
34. Zhang, Q.F.; Chen, S.; Fan, Z.P. Bearing fault diagnosis based on improved particle swarm optimized VMD and SVM models. *Adv. Mech. Eng.* **2021**, *13*, 1–12. [CrossRef]
35. Liu, M.G.; Li, X.S.; Lou, C.Y.; Jiang, J. A Fault Detection Method Based on CPSO-Improved KICA. *Entropy* **2019**, *21*, 668. [CrossRef] [PubMed]
36. Wang, H.; Peng, M.J.; Hines, J.W.; Zheng, G.Y.; Liu, Y.K.; Upadhyaya, B.R. A hybrid fault diagnosis methodology with support vector machine and improved particle swarm optimization for nuclear power plants. *ISA Trans.* **2020**, *95*, 358–371. [CrossRef] [PubMed]

Article

An Adaptive Fusion Convolutional Denoising Network and Its Application to the Fault Diagnosis of Shore Bridge Lift Gearbox

Rongqiang Zhao * and Xiong Hu

Logistics Engineering College, Shanghai Maritime University, Shanghai 201306, China; huxiong@shmtu.edu.cn
* Correspondence: zhaorongqiang@stu.shmtu.edu.cn

Abstract: Traditional fault diagnosis methods are limited in the condition detection of shore bridge lifting gearboxes due to their limited ability to extract signal features and their sensitivity to noise. In order to solve this problem, an adaptive fusion convolutional denoising network (AF-CDN) was proposed in this paper. First, a novel 1D and 2D adaptive fused convolutional neural network structure is built. The fusion of both 1D and 2D convolutional models can effectively improve the feature extraction capability of the network. Then, a gradient updating method based on the Kalman filter mechanism is designed. The effectiveness of the developed method is evaluated by using the benchmark datasets and the actual data collected for the shore bridge lift gearbox. Finally, the effectiveness of the proposed algorithm is proved through the experimental validation in the paper. The main contributions of this paper are described as follows: the proposed AF-CDN can improve the diagnosis accuracy by 1.5–9.1% when compared with the normal CNN methods. The robustness of the diagnostic network can be significantly improved.

Keywords: gearbox fault diagnosis; convolution fusion; state identification

Citation: Zhao, R.; Hu, X. An Adaptive Fusion Convolutional Denoising Network and Its Application to the Fault Diagnosis of Shore Bridge Lift Gearbox. *Machines* **2022**, *10*, 424. https://doi.org/10.3390/machines10060424

Academic Editors: Hongtian Chen, Kai Zhong, Guangtao Ran and Chao Cheng

Received: 23 April 2022
Accepted: 25 May 2022
Published: 26 May 2022

Publisher's Note: MDPI stays neutral with regard to jurisdictional claims in published maps and institutional affiliations.

Copyright: © 2022 by the authors. Licensee MDPI, Basel, Switzerland. This article is an open access article distributed under the terms and conditions of the Creative Commons Attribution (CC BY) license (https://creativecommons.org/licenses/by/4.0/).

1. Introduction

With the further development of globalization, automated container terminals (ACTs) are increasingly widespread. The automatic loading and unloading of containers by ACTs ensure the orderly flow of goods. This plays an important role in the globalization of the economy. As an integral part of the ACT, the reliable working condition of the port crane ensures the efficient operation of the entire terminal. The gearbox of the port crane, as an important power component, works for long periods of time and under heavy loads. A reliable condition monitoring and fault diagnosis system for the port crane gearbox is essential for a port crane [1,2]. The failure of a port crane can lead to port blockages and unnecessary economic losses or even cause injury or death. Therefore, it is essential to ensure that it works safely and securely. In practical scenarios, it is usually experienced experts or engineers who perform the maintenance of the equipment through their previous experience. For example, an experienced expert can determine the status of a device by tapping on it and locating faults according to the feedback signal characteristics. However, some critical equipment requires effective online monitoring so that faults can be detected and handled as soon as they occur.

The development of sensors such as vibration sensors, acoustic sensors, temperature sensors, pressure sensors, etc., can provide an effective means of obtaining information for such equipment [3–5]. This provides an effective means of detecting equipment in real time. Traditional analysis methods are mainly based on manual feature extraction of the collected signals. The methods of feature extraction for signals include time domain features, frequency domain features, and time–frequency domain features. Common features of the time domain include the mean value, standard deviation, root mean square value, peak value, shape indicator, skewness, kurtosis, crest indicator, clearance indicator, impulse indicator, etc., [6]. Frequency domain features usually refer to feature signals extracted from the frequency spectrum, mainly including the mean frequency, frequency center, root mean square frequency, the standard deviation of frequency, etc. [7]. Frequency signals often better represent some of the hidden features of the signal than the time domain. The time–frequency domain features include energy entropy, which is usually extracted by wavelet transform (WT), wavelet package transform (WPT), or empirical model decomposition (EMD) [8–13].

With the development of artificial intelligence technology [14,15], machine learning methods are used to identify faults based on the features extracted, such as expert systems, ANN, and SVM [16,17]. The intelligent algorithms of the fault diagnosis model have a strong nonlinear fitting capability [18,19]. With a provided training target and an optimization algorithm, the intelligent algorithm often achieves a good diagnosis result after continuous iteration of the optimal search. However, the efficiency of signal feature extraction may have a significant impact on the diagnostic accuracy of these methods

In this paper, a 1D and 2D adaptive fusion convolutional neural network structure is proposed, while the parameters are integrated with a Kalman filter during the iterative training process. AF-CDN converts raw data into 2D data and uses the fast Fourier transform (FFT) technique to extract features from the signal. Then, the two signals are adaptively fused. At the same time, the use of Kalman filter technology can effectively eliminate the influence of noise in the raw data on the diagnostic results. The network has excellent diagnostic accuracy, while the robustness is greatly improved. Based on the historical data, we built an online condition monitoring system for port crane gearboxes. We also test our proposed algorithm on a public bearing dataset from Case Western Reserve University, and the results show that AF-CDN is well suited for different situations.

The main contributions of this paper are summarized as follows:

(1) A 1D and 2D adaptive convolutional approach is proposed, through which the feature extraction capability of the network can be greatly enhanced. We design a 1D and 2D fused convolutional signal extraction layer (perception layer). First, the FFT-processed 1D information is fed into the 1D convolution input, and then the sequence of the original signal is aligned and fed into the 2D convolution input.

(2) A Kalman filter-based method for updating network parameters is proposed. Improvements are made to the minibatch stochastic gradient descent (MSGD) method. The information within the minibatch is effectively integrated based on the Kalman filter mechanism.

The subsequent sections of this manuscript are organized as follows: Section 2 presents the preliminary work. Section 3 describes the method proposed in detail. Section 4 designs experimental validation for the proposed algorithms. Section 5 presents the conclusion and provides suggestions for future work

2. Preliminary Work

2.1. Convolutional Neural Networks

Convolutional neural networks are an important branch of neural networks [20]. However, unlike back propagation (BP) neural networks, convolutional neural networks have a strong feature extraction capability. After the convolutional operation, the network can perform feature extraction on the signal fed into the network.

A CNN has a convolutional layer, a pooling layer, and a full connection layer. Since the proposal of the convolutional neural network, a rich variety of CNN structures have been developed over the decades, including LeNET, AlexNET, VGG, GoogleLeNET, ResNET, DenseNET, etc. A typical convolutional neural network structure is shown in Figure 1.

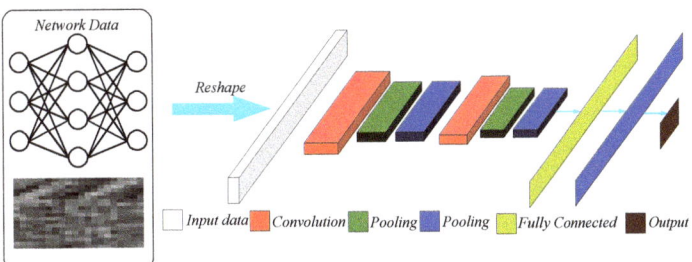

Figure 1. Typical convolutional neural network structure.

The forward computation of a CNN can be expressed as follows.

$$\mathcal{G}(X) = g^{(L)}(\ldots g^{(2)}(g^{(1)}(X, \theta^{(1)}), \theta^{(2)}) \ldots, \theta^{(L)}). \quad (1)$$

$\mathcal{G}$ is the mapping equation of the network. g is the nonlinear function of each layer inside the network. θ is the connection parameters of each layer and L is the number of layers of the neural network. $X = \{x_1, x_2 \ldots x_p \ldots x_Q\}$ is the input to the network, which can be one- or two-dimensional. Q is the number of data in the dataset.

The convolutional layer of a CNN consists of a convolution core and a bias. After the input of the network has been convolved, the bias of the layer is added, and, finally, the output of the network is obtained by passing through the nonlinear layer. The equation for the convolution layer is shown as follows [20].

$$O_i^{(l)} = g^{(l-1)} \left(\sum_{j=1}^{n^{(l-1)}} w_{ij}^{(l)} * X_j^{(l-1)} + b_i^{(l)} \right). \quad (2)$$

In the formula, $O^{(l)}$ is the output of the lth layer. $i = 1, 2 \ldots n^{(l)}$, $n^{(l)}$ is the output size of the lth layer. $j = 1, 2 \ldots n^{(l-1)}$, $n^{(l-1)}$ is the output size of the $(l-1)$th layer. w is the value of the convolution core. b is the value of bias. The pooling layer is used to further extract the information from the convolutional output. The pooling operation can be max pooling, down pooling or average pooling. After the pooling operation, the representative features in the local area are further extracted. Taking down pooling as an example, it takes the smallest value inside the pooling size range and generates a new output. As shown in Equation (3), let the pooling size be $p \times p$.

$$x_{ijl} = \min(o_{i'j'l} : i \le i' < i+p, j \le j' < j+p). \quad (3)$$

The input information to the network is passed through a number of convolution, pooling, and nonlinear computations. Then, the output value of the last pooling layer is reshaped, and this value is fed into the full connecting layer. Finally, the diagnosis result is provided after a softmax layer. We assume that the network output has K classes. $Y = \{y_1, y_2 \ldots y_q \ldots y_Q\}$ is the output set of the dataset. Q is the number of data in the dataset. If $y_q \in \{1, 2 \ldots k \ldots K\}$, the predicted output of the network is shown as Equation (4):

$$\hat{y}(\hat{y}_q = k | o^{(L-1)}; w^{(L)}) = \frac{e^{w_k^{(L)} o^{(L-1)}}}{\sum_{i=1}^{K} e^{w_i^{(L)} o^{(L-1)}}}. \quad (4)$$

where $w^{(L)}$ is the network parameter for the softmax layer. $\hat{Y} = \{\hat{y}_1, \hat{y}_2 \ldots \hat{y}_q \ldots \hat{y}_Q\}$.

The network parameters are updated using the minibatch stochastic gradient descent method after the network has completed forward propagation. J is the loss function of the network, which can be mean square error (MSE) or cross-entropy, etc. After each forward calculation is completed, the output of the network is updated by iterating backward derivation. The network parameters are updated by Equations (5) and (6). γ is the learning rate.

$$W = W - \gamma \frac{\partial J(W, B; X, \hat{Y})}{\partial W}. \tag{5}$$

$$B = B - \gamma \frac{\partial J(W, B; X, \hat{Y})}{\partial B}. \tag{6}$$

2.2. CNN-Based Fault Diagnosis

Convolutional neural networks are widely used in data-based fault diagnosis applications. Based on the different types of convolutional kernel operations, they can be divided into 1D-CNNs and 2D-CNNs. The 1D convolutional structure is proposed mainly in response to the fact that the neural network often requires manual feature extraction of the raw signal when performing recognition. One-dimensional convolutional neural networks can use the raw data directly as an input to the neural network. For example, Eren et al. proposed an adaptive 1D convolution method that can extract data features directly from the raw time-domain data [21]. An online diagnostic network based on 1D-CNN was designed for the effective diagnosis of a gearbox, where vibration sensors cannot be used, and the signal was collected by a rotary encoder [22]. A deep convolutional structure, Deep Inception Net with Atrous Convolution (ACDIN), was designed in [23] based on 1D-CNNs, which improved the feature extraction ability of the network by adding an inception layer. The 1D convolution was improved by Atrous convolution. This led to a significant increase in the diagnostic capability of the network. To address the problem of uneven distribution of samples in the dataset, Jia et al. proposed a 1D-CNN with normalized weights for one-dimensional input data [24]. Jiang et al. designed a multi-scale signal resolution method using one-dimensional convolution for signal feature extraction, which achieved a positive result [25]. Appana et al. proposed the extraction of the raw signal by CNN for the case of multiple faults and environmental influences [26]. One-dimensional CNNs have excellent environmental adaptability and can effectively resist interference.

The main issue that needs to be solved when 2D convolutional neural networks are used for fault diagnosis is how to convert the acquired 1D raw signal into 2D data that can be fed into the network. A number of approaches have been proposed to solve this problem. Guo et al. used the residual processed short-time Fourier transform (STFT)-transformed image of the original signal as the input into the CNN [27]. Long et al. proposed a signal to image conversion mechanism to transform the raw time domain signal into 2D grey images [28]. This enables feature extraction of the collected vibration signals in a similar manner to picture recognition. Han proposed a spatiotemporal convolutional neural network (ST-CNN), which extracts spatiotemporal features via the spatiotemporal pattern network (STPN) and then makes a diagnosis based on the CNN [29]. In [30], Yu et al. used a pseudo-color map to represent the data extracted by STFT and then fed the images into a CNN for training recognition. Sun et al. used the dual-tree complex wavelet transform method to extract features from the raw data, and the DTCWT wavelet sub-bands were used as multiple rows of a matrix so that a 2D signal was formed and sent to 2D convolution for processing [31]. Similarly, the Hilbert envelope demodulation spectra (HEDS) of reconstructed signals in each frequency band were also spliced into the 2D signal matrix [32]. The HEDS of the reconstructed signal for each frequency band were stitched into a 2D signal matrix to produce a 2D signal. The time domain signals were arranged row by row to form a 2D input matrix as the network input for diagnostics. Min et al. arranged the time domain signals row by row to form a 2D input matrix as the network input for diagnostics [33].

3. The Proposed Method

3.1. Adaptive Fusion Convolutional Denoising Network

The proposed AF-CDN structure diagram is shown in Figure 2. Firstly, the equipment data are collected by vibration sensors. Then, the 1D data are obtained by FFT and fed into the 1D channel of the perception layer. The raw data are arranged in order to obtain 2D data. These data are then fed into the 2D channel of the perception layer. After feature extraction in the perception layer, all feature values are flattened and then pooled. The features then go through two more inception and pooling layers. The output values are fed to the auxiliary classifier for classification after each pooling layer. The final pooling layer output is combined with the output of the auxiliary classifier to provide the final diagnosis result. The loss value is calculated from the output, and if the predetermined loss condition is satisfied, then the training is stopped; if not, then training continues.

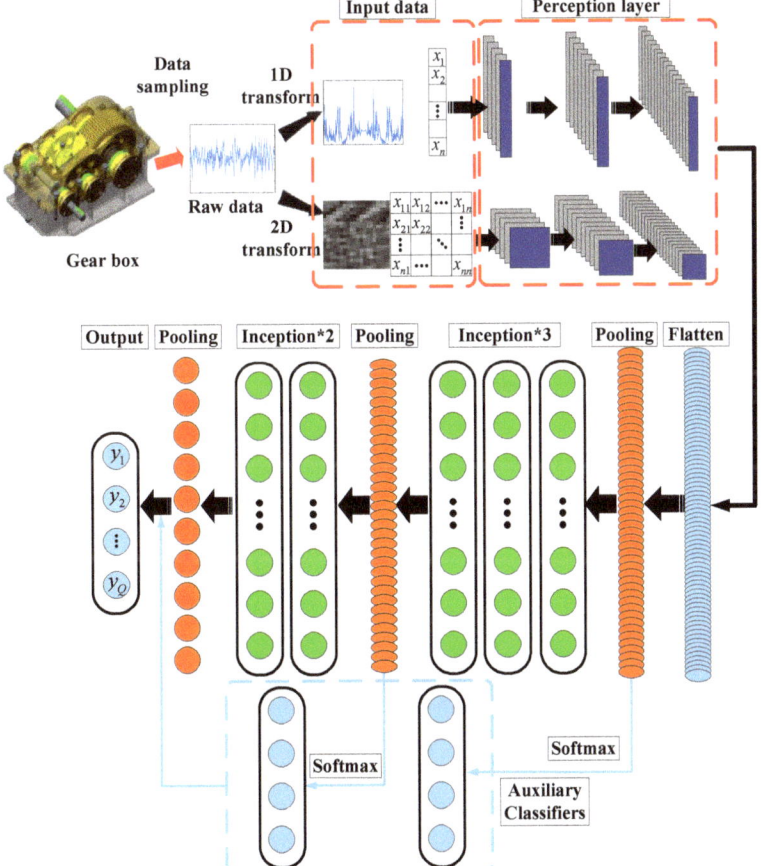

Figure 2. AF-CDN structure diagram.

Numerous research results in the field of CV have favorably illustrated the importance of 2D convolution for the accurate recognition of valuable information in 2D images. As shown in Section 2.2, the development of convolutional neural networks in the field of fault diagnosis also started with 1D-CNNs, and then researchers successively proposed various methods to convert 1D data into 2D data, thus enabling 2D-CNNs to be widely used in the field of fault diagnosis. The schematic diagram of the data processing is shown in Figure 3.

After the original data are FFT-transformed, the positive half-axis frequency data are taken and arranged sequentially to obtain 1D data. Two-dimensional data, on the other hand, are arranged with the raw data starting from the first row of the matrix, followed by the second row, until the entire matrix is filled. During the 2D conversion, the signals are arranged sequentially, so the time-series property of the signals is preserved.

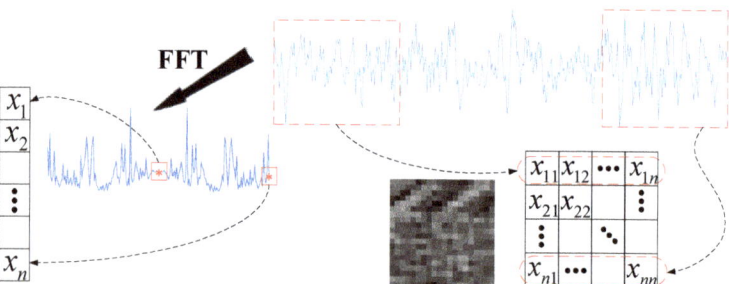

Figure 3. One-dimensional and two-dimensional data transform.

The algorithm flow chart of AF-CDN is shown in Figure 4. The main steps of the proposed algorithm are described as Algorithm 1.

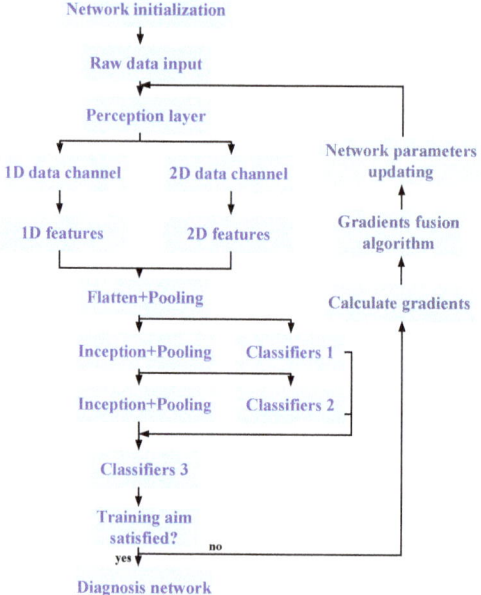

Figure 4. Algorithm flow chart of AF-CDN.

Algorithm 1: AF-CDN

Input: Network $\mathcal{G}$, Training epoch N, Input data $X = \{x_1, x_2 \ldots x_p \ldots x_Q\}$ and $Y = \{y_1, y_2 \ldots y_q \ldots y_Q\}$
Output: Trained $\mathcal{G}^{tr}$
1. Initialize network parameters
2. **For** $i = 1, 2, \ldots N$ **do**
3. Feed raw data into the network
4. Send the data to the perception layer to calculate 1D data and 2D data, respectively
5. Flatten and pool the 1D and 2D features of the perception layer output. The values after pooling are fed into classifier1 and inception layer, respectively
6. The values of the inception layer are pooled and fed into classifier2 and the next inception layer, respectively
7. The values of inception layer are pooled and fed into classifier3
8. Classifier3 combines the values of classifier1 and classifier2 to give the predicted output.
9. Calculate the loss value. Stop training if the training target is met, otherwise step forward.
10. Calculate the gradients and use the fusion algorithm to update the gradient value.
11. **End**

3.2. Gradient Fusion Algorithm

Ma et al. proposed a Kalman-filter-based fusion method for network parameters updating [6]. Based on this theory, we performed some improvements when we used the MSGD algorithm to calculate the updating gradients. Considering that after the sensor has been selected and the measurement position has been determined, noise in the acquired signal cannot be avoided. Therefore, we considered further analysis of the gradients during the gradient updating process. Signal noise is assumed to be hidden in the gradient information of each sample. Thus, the gradient information is fused inside the minibatch using a Kalman filter.

When calculating the gradient inside the batch, we fuse the gradient information using the Kalman filter on each of the batch size gradients.

We use k and $k-1$ moments as an example, and the Kalman filter-based gradient fusion process is described as follows.

$$w(k) = Fw(k-1) + \delta(k). \tag{7}$$

$$z(k) = H(k)w(k) + \gamma(k). \tag{8}$$

$w(k)$ is the network parameter at the moment k. F is the state transfer matrix. $\delta(k)$ is the state error, $\gamma(k)$ is the measurement error and $\delta(k)$, $\gamma(k)$ conform to a Gaussian distribution. Then according to the Kalman filter formula [34], the follow-up process in the iterative process is as follows.

First, an a priori estimate of the gradient is calculated.

$$\hat{w}^-(k) = F\hat{w}(k-1) + \delta(k). \tag{9}$$

$\hat{w}^-(k)$ is an a priori estimate of the moment k. $\hat{w}(k-1)$ is the optimal estimate at moment $k-1$. Next, we update the a priori estimated covariance.

$$P^-(k) = FP(k-1)F^T + Q(k). \tag{10}$$

$P^-(k)$ is the a priori estimated covariance, which will be used when calculating the Kalman gain. $P(k-1)$ is the posterior estimated covariance at moment $k-1$. $Q(k) = \delta(k)\delta^T(k)$ is the covariance of the state error. The measured value at the k moment is calculated according to Equation (8), and then the Kalman gain is updated [6].

$$K(k) = P^-(k)H(k)^T [H(k)P^-(k)H(k)^T + R(k)]^{-1}. \tag{11}$$

$R(k) = \gamma(k)\gamma^T(k)$ is the covariance of the measurement error. Then, the optimal estimate of the gradient at moment k can be found.

$$\begin{aligned}\hat{w}(k) &= \hat{w}^-(k) + K(k)[z(k) - H(k)\hat{w}^-(k)] \\ &= [1 - K(k)H(k)]\hat{w}^-(k) + K(k)z(k).\end{aligned} \quad (12)$$

$[1 - K(k)H(k)]$ is the confidence level of the estimate. $K(k)$ is the confidence level of the measured value.

Finally, the posterior estimated covariance is updated.

$$P(k) = [I - K(k)H(k)]P^-(k). \quad (13)$$

In specific applications, let the batch size be N_B, so $k = \{1, 2 \ldots N_B\}$.

$w(k)$ is the updated gradient value corresponding to the N_B samples. F is the state transfer matrix, it is set to I in the paper.

$H(k)$ is the measurement matrix with the measurement values matching the state values. $H(k)$ is set to I. We update the Kalman gain as in Equation (14).

$$\begin{aligned}K(k) &= P^-(k)[P^-(k) + R(k)]^{-1} \\ &= [P(k-1) + Q(k)][P(k-1) + Q(k) + R(k)]^{-1}.\end{aligned} \quad (14)$$

In practical terms, the value of R can be set to a smaller value if the measurement error of the sensor is small, which means that we are more likely to believe that the gradient of each sample is the true value. Conversely, the value of $R(k)$ can be increased appropriately. The value of $Q(k)$ is adjusted according to the range of the gradient values during actual optimization.

4. Experimental Verification

In order to verify the effectiveness of the algorithm proposed in this paper, in Case One, data acquisition, fault classification, and diagnosis results are introduced in detail on the port crane built by NetCMAs. The feasibility of AF-CDN is verified. In Case Two, we verify the algorithm on the open-source rolling bearing fault dataset from Case Western Reserve University. AF-CDN can achieve excellent diagnostic results. The universality of the proposed algorithm is illustrated. The experiments in this paper were implemented on an Intel(R) Core (TM) i7-8550U CPU PC (1.80 GHz, 8 GB RAM) NVIDIA Geforce MX 150 GPU (4 GB) 64 Bit Windows 10 operating system in a Python environment.

4.1. Case One

4.1.1. Dataset Preparation and Parameter Settings

The NetCMAs system is installed with vibration sensors, stress sensors, temperature sensors, etc., which can effectively detect the status of the whole port crane system in real time. The whole system has 32 sampling channels, and sampling information points are distributed in the T frame of the upper beam area, beam rod load area, lifting motor, gearbox, car motor, etc. The collection positions of the gearbox are as follows, V-directional and H-directional vibration on the left side of the high-speed shaft, the temperature on both the left and right side of the high-speed shaft, and V-directional vibration on the left side of the low-speed shaft. The sampling frequency of the detection system is 2.5 kHz. The sampling time is 0.8 s, and the sampling interval is 10 s. In order to avoid continuous data transmission and storage consumption, the valid values of each sampling period are calculated and saved. Figure 5 shows the driving part of the port crane lifting mechanism. Figure 6 shows the reduction gearbox and bearings. Figure 7 shows the installation position of the vibration sensor and the damaged bearing.

Figure 5. Lifting mechanism of port crane.

(a)

(b)

Figure 6. Gear box of lifting mechanism of port crane. (**a**) Drive gear set for installation. (**b**) Gear bearing.

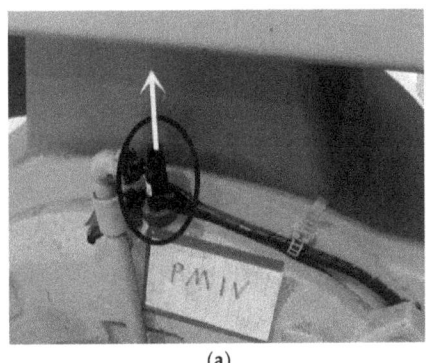
(a)

(b)

Figure 7. Experimental settings. (**a**) Installation position of vibration sensor. (**a**) Installation position of vibration sensor.

The dataset used in this case comprises four years of data on the No. 8114 lifting bridge of a port crane. The data were recorded from the time when the gearbox was first equipped until the time of failure. Due to extremely large volumes of data, only representative data are shown in Figure 8. It can be seen that there are some shock components in the wave. According to our practical application experience, all data are classified into four categories: healthy (H), sub-healthy (SH), failure (F), run-in period (R), and health (H).

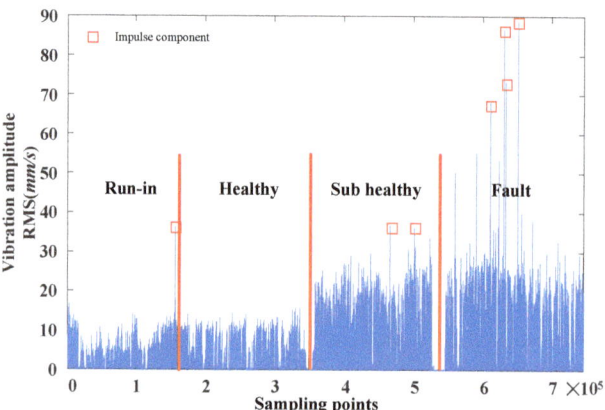

Figure 8. V-direction vibration waveform of the gearbox.

After each new gearbox is re-installed, it will run through a run-in period of time before it enters into a healthy state. After a period of operation, the equipment will be in a sub-healthy state due to the occurrence of wear of the equipment. Eventually, the damage was so significant that the equipment entered a fault state. Table 1 describes in detail the label and quantity information of the four state data in the experiment. Sample labels are onehot encoded, and each state contains 100 samples. The dimension of each sample is 1600, so the dimensions after the 2D conversion are 40 × 40. In order to ensure the speed of iteration, the batch size is set to between 20 and 40.

Table 1. Four types of fault status information.

Fault Type	Fault Diameter (inch)	Label	Sample Size
H	Healthy	0001	100
SH	Sub-Healthy	0010	100
F	Failure	0100	100
R	Run-in	1000	100

4.1.2. Experiment and Analysis

In order to demonstrate the effectiveness of AF-CDN, a simulation comparison is performed. Table 2 shows the comparison between the algorithm proposed in this paper and the 1D-CNN algorithm extracted by the FFT signal, and the 2D-CNN algorithm based on raw data.

Table 2. Comparison of accuracy.

Number of Experiments	1D-CNN	Rawdata-CNN	AF-CDN
1	91.25%	96.75%	100.00%
2	89.58%	98.75%	99.25%
3	92.50%	98.5%	99.00%
4	90.83%	98.25%	100%
5	93.75%	98.75%	100%
6	89.58%	98.50%	99.50%
7	89.17%	98.00%	100.00%
8	93.75%	98.75%	99.75%
9	90.83%	98.75%	100.00%
10	88.33%	98.50%	99.00%
Mean accuracy	90.56%	98.35%	99.65%

As can be seen from the results of the 10 experiments in the table, AF-CDN provides better diagnostic results than both the algorithm that performs signal spectral analysis alone and the algorithm based on raw signal feature extraction. Figure 9 shows the statistical information of 10 experimental results. It can be seen that AF-CDN can not only effectively improve the diagnosis accuracy but also can greatly reduce the error of each diagnosis result. The performance of AF-CDN is 10.09% higher than that of FFT alone and is 1.3% higher than that of original signal feature extraction. The average execution time of AF-CDN was 0.196 s. The 1D-CNN was 0.083 s and Rawdata-CNN was 0.154 s. The improvement in accuracy compared to 1D-CNN is significant, so the computational consumption is worthwhile. Compared to Rawdata-CNN, the computational consumption is 0.042 s higher. In a practical scenario, the computational complexity can be significantly reduced by using a high-performance GPU device, so it is worthwhile to increase the computational consumption slightly.

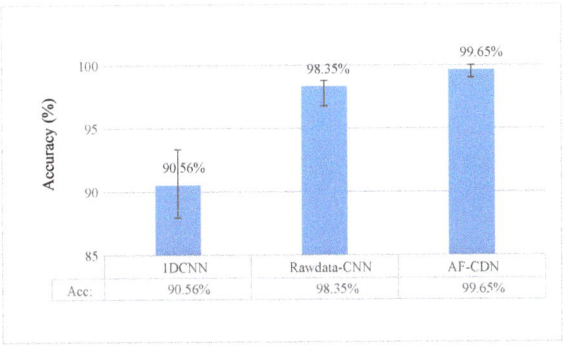

Figure 9. Results of the 10 experiments.

Figure 10 uses T-SNE visualization technology to visually display the experimental classification results. In Figure 10a,c the visualization and error matrix of the AF-CDN are presented, respectively. In Figure 10b,d the visualization and error matrix of 1D-CNN algorithm are presented, respectively.

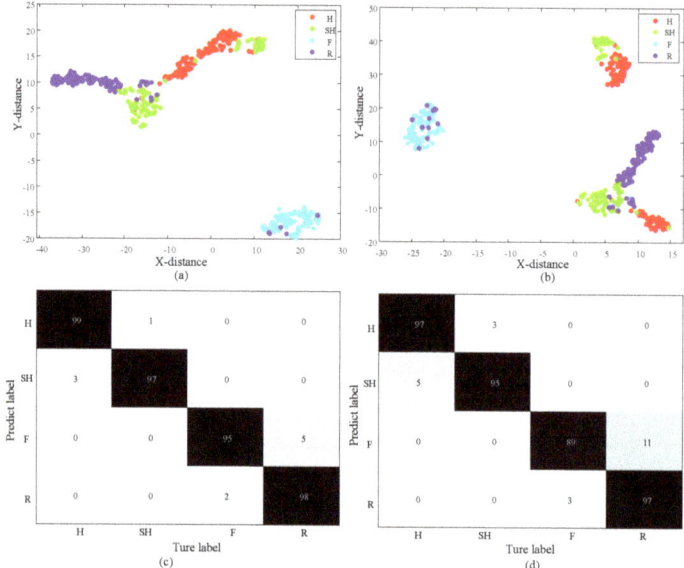

Figure 10. Visualization of classification results and error matrix: (**a**) visualization of AF-CDN; (**b**) error matrix of AF-CDN; (**c**) 1D-CNN visualization; (**d**) 1D-CNN error matrix.

At the same time, it should be pointed out that, compared with the method of feature extraction based on raw data for diagnosis, the AF-CDN in this paper demonstrates great improvement in reducing the probability of two kinds of misdiagnosis. This is also intuitive in the visualization and error matrix. The main reasons for this result can be summarized as follows: (1) compared with the single FFT signal or the original 2D signal, and the AF-CDN fuses the two signals. (2) Compared with SGD, the gradient fusion algorithm based on the Kalman filter has a stronger parameter integration ability in feature fusion of gradient information. In this way, the diagnostic ability of the network can be better improved. As can be seen from the visual figures of the two algorithms, there are more cases of misdiagnosis between H and SH and between S and F. At the same time, it should be pointed out that compared with the method of feature extraction based on original data for diagnosis, the AF-CDN provides a great improvement in reducing the probability of misdiagnosis.

4.2. Case Two

4.2.1. Dataset Preparation and Parameter Settings

The AF-CDN is also verified on the CWRU Bearing Dataset [33]. CWRU set up a variety of fault types on motor drive equipment. The vibration signals of the different positions of the drive end were collected by vibration sensors under different load conditions.

According to the different fault locations, the fault can be divided into three categories: the ball fault (BF), inner ring fault (IF), and outer ring fault (OF). Each type of fault can be further subdivided into three different fault levels according to different fault severity. Therefore, a total of nine types of faults were set up in this experiment, and the sampling method was carried out according to the paper [6,35]. For details about fault information, see Table 3. The size of each sample is 400, so the dimensions after the 2D conversion are 20×20. In order to ensure the speed of iteration, the batch size is set to between 20 and 40.

The continuous time signal collected by the vibration sensor is shown in Figure 11.

Table 3. Status information of 9 types of fault.

Fault Type	Fault Diameter(inch)	Label	Sample Size
BFI	0.007	000000001	300
BFII	0.014	000000010	300
BFIII	0.021	000000100	300
IFI	0.007	000001000	300
IFII	0.014	000010000	300
IFIII	0.021	000100000	300
OFI	0.007	001000000	300
OFII	0.014	010000000	300
OFIII	0.021	100000000	300

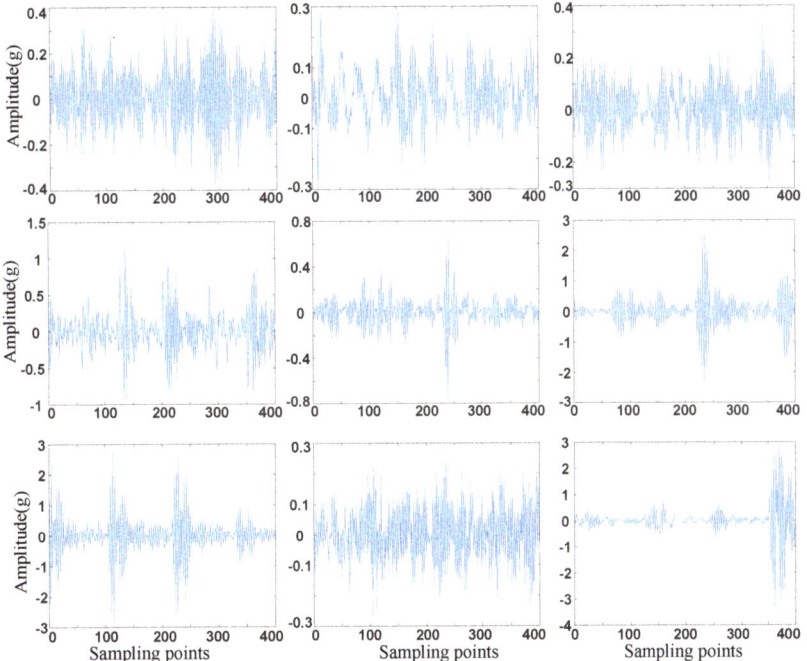

Figure 11. Continuous time signal diagram of 9 kinds of faults.

4.2.2. Experiment and Analysis

The AF-CDN is compared with some popular methods at present. As summarized in the previous chapter, the existing methods are mainly based on two signal extraction methods. One involves performing frequency domain transformation in the original time-continuous signal to obtain frequency domain features. The other mainstream approach is based on raw signals.

In this case, the existing methods and AF-CDN are compared in detail. The processing methods in the frequency domain mainly include wavelet transform, wavelet packet transform, statistical locally linear embedding, and other methods. Raw signal processing involves converting raw signals into 2D images and establishing Spectrogram methods.

Table 4 shows the comparison of experimental results between the existing mainstream methods and AF-CDN.

Table 4. Comparison of diagnostic accuracy of different methods.

Means of Classification	Features	Accuracy Rate
KNN [9]	HOCs and WT	91.2%
SVM [8]	WP	62.5–98.7%
ANN [10]	DWT (Morlet)	96.7%
ANN [10]	DWT (Daubechies 10)	93.3%
SVM [11]	Statistical Locally Linear Embedding	77.8–94.1%
2D-CNN [7]	Raw data	98.35%
2D-CNN [12]	Spectrogram	98.1–99.5%
AF-CDN	FFT + Raw data	99.44–99.78%

Compared with the traditional diagnosis methods, the proposed algorithm effectively integrates the frequency domain characteristics of vibration signals and their original signal characteristics. The accuracy of network diagnosis is further improved.

The main feature extraction methods used in the signal-based feature analysis approach include WP, WT, DWT, and statistical methods based on these. After obtaining these frequency features, KNN, SVM, ANN, etc., can be used for analysis. As shown in Table 4, such diagnosis methods based on "frequency domain signals + neural networks" have a diagnostic accuracy of 62.5–98.7%. The accuracy of the analysis method based on the raw signal is 98.1% to 99.5%. The diagnostic accuracy of the proposed method is 99.44–99.78%. The average execution time of AF-CDN is about 0.115 s. Compared with the traditional diagnosis methods, AF-CDN effectively integrates the frequency domain characteristics of vibration signals and their raw signal characteristics. The feature extraction capability of the network is excellent compared to the rest of the network structure. As analyzed in Case One, AF-CDN combines information on the frequency domain characteristics of the signal with the raw time domain information. Thus AF-CDN is able to have an excellent diagnostic result. The accuracy of network diagnosis is further improved.

5. Conclusions

This paper presents an adaptive fusion convolutional denoising network for the health monitoring of port crane speed gearboxes. At the same time, a Kalman filter is used to update the network parameters during the training process. Compared with traditional diagnosis methods, the main advantages of the method proposed in this paper are that the accuracy of diagnosis is improved greatly. The robustness of the diagnostic network can be significantly improved. Monitoring the status of port cranes' gearbox systems provides support for equipment health care. When the equipment is in the running-in state and subhealth state, workers can perform maintenance on bearing equipment in a timely manner so that the service time of the equipment can be effectively delayed. At the same time, once the equipment is in a sub-health state, it is necessary to monitor the system state at all times; once the system indicators reach the fault state, it is necessary to immediately shut down and perform the maintenance. This ensures the safety and reliability of the entire port machine. The results show that AF-CDN also has excellent diagnostic performance on public data sets. However, the shortcomings of this paper are that we have not integrated the other sensors well, including temperature sensors and vibration signals from other locations that the hardware system contains.

The system we have built in this paper is a four-stage bearing diagnosis system based on a single vibration sensor. In future work, we will further study the fusion of stress, temperature, and multi-directional vibration signals collected from multiple locations of the whole system so as to extract the accurate overall health status of the crane and fuse it with multi-sensor signals. This makes it possible to establish a whole health management system and enables effective real-time monitoring of the full lifecycle of a port crane. Secondly, a whole life cycle inspection system for the equipment should be established based on the multi-sensor information fusion technology described above. Meanwhile, distributed learning techniques will be focused on in order to fuse the data from the multi-location port machines. Meanwhile, distributed learning techniques will be focused on in order to fuse the data from the multi-location port machines.

Author Contributions: Conceptualization, R.Z.; Data curation, X.H.; Formal analysis, R.Z.; Funding acquisition, X.H.; Methodology, R.Z.; Supervision, X.H.; Writing—original draft, R.Z.; Writing—review and editing, R.Z. and X.H. All authors have read and agreed to the published version of the manuscript.

Funding: This research was funded by the National Natural Science Foundation of China grant number 31300783.

Institutional Review Board Statement: No applicable.

Informed Consent Statement: No applicable.

Data Availability Statement: The original data contributions presented in the study are included in the article; further inquiries can be directed to the corresponding authors.

Conflicts of Interest: The authors declare no conflict of interest.

References

1. Xu, L.; Chatterton, S.; Pennacchi, P. A Novel Method of Frequency Band Selection for Squared Envelope Analysis for Fault Diagnosing of Rolling Element Bearings in a Locomotive Powertrain. *Sensors* **2018**, *18*, 4344. [CrossRef] [PubMed]
2. Lei, Y.; Yang, B.; Jiang, X.; Jia, F.; Li, N.; Nandi, A.K. Applications of machine learning to machine fault diagnosis: A review and roadmap. *Mech. Syst. Signal Process.* **2020**, *138*, 106587. [CrossRef]
3. Li, H.; Liu, J.; Wu, K.; Yang, Z.; Liu, R.W.; Xiong, N. Spatio-Temporal Vessel Trajectory Clustering Based on Data Mapping and Density. *IEEE Access* **2018**, *6*, 58939–58954. [CrossRef]
4. Gao, K.; Han, F.; Dong, P.; Xiong, N.; Du, R. Connected Vehicle as a Mobile Sensor for Real Time Queue Length at Signalized Intersections. *Sensors* **2019**, *19*, 2059. [CrossRef] [PubMed]
5. Wu, M.; Tan, L.; Xiong, N. A Structure Fidelity Approach for Big Data Collection in Wireless Sensor Networks. *Sensors* **2014**, *15*, 248–273. [CrossRef]
6. Ma, X.; Wen, C.; Wen, T. An Asynchronous and Real-Time Update Paradigm of Federated Learning for Fault Diagnosis. *IEEE Trans. Ind. Inform.* **2021**, *17*, 8531–8540. [CrossRef]
7. Xia, M.; Li, T.; Xu, L.; Liu, L.; De Silva, C.W. Fault Diagnosis for Rotating Machinery Using Multiple Sensors and Convolutional Neural Networks. *IEEE/ASME Trans. Mechatron.* **2017**, *23*, 101–110. [CrossRef]
8. Hu, Q.; He, Z.; Zhang, Z.; Zi, Y. Fault diagnosis of rotating machinery based on improved wavelet package transform and SVMs ensemble. *Mech. Syst. Signal Process.* **2007**, *21*, 688–705. [CrossRef]
9. Yaqub, M.F.; Gondal, I.; Kamruzzaman, J. Inchoate Fault Detection Framework: Adaptive Selection of Wavelet Nodes and Cumulant Orders. *IEEE Trans. Instrum. Meas.* **2011**, *61*, 685–695. [CrossRef]
10. Konar, P.; Chattopadhyay, P. Bearing fault detection of induction motor using wavelet and Support Vector Machines (SVMs). *Appl. Soft Comput.* **2011**, *11*, 4203–4211. [CrossRef]
11. Wang, X.; Zheng, Y.; Zhao, Z.; Wang, J. Bearing Fault Diagnosis Based on Statistical Locally Linear Embedding. *Sensors* **2015**, *15*, 16225–16247. [CrossRef] [PubMed]
12. Verstraete, D.; Ferrada, A.; Droguett, E.L.; Meruane, V.; Modarres, M. Deep Learning Enabled Fault Diagnosis Using Time-Frequency Image Analysis of Rolling Element Bearings. *Shock Vib.* **2017**, *2017*, 5067651. [CrossRef]
13. Torrence, C.; Compo, G.P. A practical guide to wavelet analysis. *Bull. Am. Meteorol. Soc.* **1998**, *79*, 61–78. [CrossRef]
14. Chen, H.; Chen, Z.; Chai, Z.; Jiang, B.; Huang, B. A Single-Side Neural Network-Aided Canonical Correlation Analysis With Applications to Fault Diagnosis. *IEEE Trans. Cybern.* **2021**, 1–13. [CrossRef]
15. Chen, H.; Li, L.; Shang, C.; Huang, B. Fault Detection for Nonlinear Dynamic Systems With Consideration of Modeling Errors: A Data-Driven Approach. *IEEE Trans. Cybern.* **2022**, 1–11. [CrossRef]
16. Haykin, S. Neural networks expand SP's horizons. *IEEE Signal Process. Mag.* **1996**, *13*, 24–49. [CrossRef]
17. Guyon, I.; Weston, J.; Barnhill, S.; Vapnik, V. Gene Selection for Cancer Classification using Support Vector Machines. *Mach. Learn.* **2002**, *46*, 389–422. [CrossRef]
18. Huang, S.; Liu, A.; Wang, T.; Xiong, N.N. BD-VTE: A Novel Baseline Data Based Verifiable Trust Evaluation Scheme for Smart Network Systems. *IEEE Trans. Netw. Sci. Eng.* **2020**, *8*, 2087–2105. [CrossRef]
19. Yang, P.; Xiong, N.N.; Ren, J. Data Security and Privacy Protection for Cloud Storage: A Survey. *IEEE Access* **2020**, *8*, 131723–131740. [CrossRef]

20. Lecun, Y.; Bottou, L.; Bengio, Y.; Haffner, P. Gradient-based learning applied to document recognition. *Proc. IEEE* **1998**, *86*, 2278–2324. [CrossRef]
21. Eren, L.; Ince, T.; Kiranyaz, S. A Generic Intelligent Bearing Fault Diagnosis System Using Compact Adaptive 1D CNN Classifier. *J. Signal Process. Syst.* **2018**, *91*, 179–189. [CrossRef]
22. Jiao, J.; Zhao, M.; Lin, J.; Zhao, J. A multivariate encoder information based convolutional neural network for intelligent fault diagnosis of planetary gearboxes. *Knowl. Based Syst.* **2018**, *160*, 237–250. [CrossRef]
23. Chen, Y.; Peng, G.; Xie, C.; Zhang, W.; Li, C.; Liu, S. ACDIN: Bridging the gap between artificial and real bearing damages for bearing fault diagnosis. *Neurocomputing* **2018**, *294*, 61–71. [CrossRef]
24. Jia, F.; Lei, Y.; Lu, N.; Xing, S. Deep normalized convolutional neural network for imbalanced fault classification of machinery and its understanding via visualization. *Mech. Syst. Signal Process.* **2018**, *110*, 349–367. [CrossRef]
25. Jiang, G.; He, H.; Yan, J.; Xie, P. Multiscale Convolutional Neural Networks for Fault Diagnosis of Wind Turbine Gearbox. *IEEE Trans. Ind. Electron.* **2018**, *66*, 3196–3207. [CrossRef]
26. Appana, D.K.; Prosvirin, A.; Kim, J.-M. Reliable fault diagnosis of bearings with varying rotational speeds using envelope spectrum and convolution neural networks. *Soft Comput.* **2018**, *22*, 6719–6729. [CrossRef]
27. Guo, D.; Zhong, M.; Ji, H.; Liu, Y.; Yang, R. A hybrid feature model and deep learning based fault diagnosis for unmanned aerial vehicle sensors. *Neurocomputing* **2018**, *319*, 155–163. [CrossRef]
28. Wen, L.; Li, X.; Gao, L.; Zhang, Y. A New Convolutional Neural Network-Based Data-Driven Fault Diagnosis Method. *IEEE Trans. Ind. Electron.* **2017**, *65*, 5990–5998. [CrossRef]
29. Han, T.; Liu, C.; Wu, L.; Sarkar, S.; Jiang, D. An adaptive spatiotemporal feature learning approach for fault diagnosis in complex systems. *Mech. Syst. Signal Process.* **2018**, *117*, 170–187. [CrossRef]
30. Xin, Y.; Li, S.; Cheng, C.; Wang, J. An intelligent fault diagnosis method of rotating machinery based on deep neural networks and time-frequency analysis. *J. Vibroengineering* **2018**, *20*, 2321–2335. [CrossRef]
31. Sun, W.; Yao, B.; Zeng, N.; Chen, B.; He, Y.; Cao, X.; He, W. An Intelligent Gear Fault Diagnosis Methodology Using a Complex Wavelet Enhanced Convolutional Neural Network. *Materials* **2017**, *10*, 790. [CrossRef] [PubMed]
32. Cao, X.-C.; Chen, B.-Q.; Yao, B.; He, W.-P. Combining translation-invariant wavelet frames and convolutional neural network for intelligent tool wear state identification. *Comput. Ind.* **2019**, *106*, 71–84. [CrossRef]
33. Case Western Reserve University Bearing Data Center Website. Available online: http://csegroups.case.edu/bearingdatacenter/home (accessed on 15 March 2017).
34. Arulampalam, M.S.; Maskell, S.; Gordon, N.; Clapp, T. A tutorial on particle filters for online nonlinear/non-Gaussian Bayesian tracking. *IEEE Trans. Signal Process.* **2002**, *50*, 174–188. [CrossRef]
35. Chen, H.; Chai, Z.; Dogru, O.; Jiang, B.; Huang, B. Data-Driven Designs of Fault Detection Systems via Neural Network-Aided Learning. *IEEE Trans. Neural Netw. Learn. Syst.* **2021**, 1–12. [CrossRef] [PubMed]

Article

Fault Detection for Interval Type-2 T-S Fuzzy Networked Systems via Event-Triggered Control

Zhongda Lu [1,2], Chunda Zhang [1], Fengxia Xu [1,2], Zifei Wang [1] and Lijing Wang [1,2,3,*]

[1] School of Mechanical and Electrical Engineering, Qiqihar University, Qiqihar 161006, China; luzhongda@163.com (Z.L.); 2019911186@qqhru.edu.cn (C.Z.); xufengxia_hit@163.com (F.X.); 2019911187@qqhru.edu.cn (Z.W.)

[2] Collaborative Innovation Center of Intelligent Manufacturing Equipment Industrialization of Heilongjiang Province, Qiqihar University, Qiqihar 161006, China

[3] School of Electrical and Electronic Engineering, Harbin University of Science and Technology, Harbin 150080, China

* Correspondence: vipjing1002@163.com

Abstract: This paper investigates the event-triggered fault diagnosis (FD) problem for interval type-2 (IT2) Takagi–Sugeno (T-S) fuzzy networked systems. Firstly, an FD fuzzy filter is proposed by using IT2 T-S fuzzy theory to generate a residual signal. This means that the FD filter premise variable needs to not be identical to the nonlinear networked systems (NNSs). The evaluation functions are referenced to determine the occurrence of system faults. Secondly, under the event-triggered mechanism, a fault residual system (FRS) is established with parameter uncertainty, external disturbance and time delay, which can reduce signal transmission and communication pressure. Thirdly, the progressive stability of the fault residual system is guaranteed by using the Lyapunov theory. For the energy bounded condition of external noise interference, the performance criterion is established using linear matrix inequalities. The matrix parameters of the target FD filter are obtained by the convex optimization method. A less conservative fault diagnosis method can be obtained. Finally, the simulation example is provided to illustrate the effectiveness and the practicalities of the proposed theoretical method.

Keywords: fault diagnosis; event-triggered control; interval type-2 Takagi–Sugeno fuzzy model; nonlinear networked systems; filter

1. Introduction

The networked systems have been widely used because of these many advantages, their simple physical structure, reduced integration costs, resource sharing, suitable for installation, expansion and maintenance [1,2]. In order to satisfy the development of aerospace and smart manufacturing, the networked systems have increasingly strong nonlinearity, uncertainty and complexity [3,4]. New challenges are brought to the control field to deal with problems such as delay, data packet loss and network bandwidth limitation caused by network introduction [5–9]. With the development of nonlinear networked systems, it needs new performance indexes including standard interface modularization, high reliability, high stability, and so on [10,11].

Fuzzy control is an effective tool for solving nonlinear problems linearization [7,12]. Fault diagnosis (FD) technology plays a vital role in improving the reliability and safety of complex engineering systems [13,14]. The task of fault diagnosis of the networked system is to transmit the input and output data of the system to the fault diagnosis unit through the network, so as to ensure that the stable operation of the system without fault occurs [6]. The FD methods of networked systems are proposed based on the fuzzy model [7,15,16]. However, there are bad situations under time-triggered FD such as unnecessary data transmission, increased network burden, data loss, and greater network delay [8]. The event-triggered mechanism has irreplaceable advantages in the network resource-constrained system. The research on fault diagnosis technology of networked systems with event-triggered mechanisms has received extensive attention from international scholars, which has become a hot research issue in the academic community of automatic control and produced many valuable research results [8,9,15–32].

Different event triggering methods are studied such as the adaptive event-triggering mechanism [5,10,22,23,27,29,32], the dynamic event triggering mechanism containing internal dynamic variables [19,28,31], the event triggering mechanism designed by improving constant thresholds [8,15,16,20,21,25,26]. The fault filtering problem of NCSs with interval time-varying time lags is studied by using the fuzzy fault detection filter with a generic structure [17]. The authors in [22] propose a novel adaptive event-triggered fault detection approach for Markov jump systems, wherein the transition probabilities are not required to be fully known. The problem of troubleshooting networked systems subject to multiple factors is discussed [21,23,25,28,30]. The problem of fault detection for stochastic nonlinear generalized networked systems is studied, which is subject to network delay, packet loss, and asynchronous premise variables [23]. Fault diagnosis problems of NNSs with communication channels are subject to limited bandwidth and random data loss are investigated. Time-varying delay, dynamic event triggering mechanism, random nonlinearity and simultaneous packet loss are considered in building a unified fault detection dynamic model moment, which is used to solve the fault detection problem [28]. The dissipative stabilization problem is solved by considering the delay and external disturbance [30].

The existing research has been extensive. However, the complexity of real systems can no longer be described by simple models. For instance, the membership functions approaches have been proposed based on the restriction that the membership functions of the descriptive model of the systems [15,16,21]. When this issue is considered, the general T-S fuzzy modeling scheme cannot achieve the desired results [15]. The IT2 fuzzy model was developed because of its good proxy for nonlinear systems with parameter uncertainty [29–37]. The problem of the FD filtering method is proposed with event-based, which is the application in IT2 fuzzy theory under the framework of networked time-delay control systems [29]. Event-triggered dissipation-based control is investigated by using the IT2 T-S fuzzy theory to describe uncertain nonlinear networked systems [30]. The nonlinear networked system with parameter uncertainty is studied under the event-triggered mechanism with adaptive discrete H_∞ fuzzy filtering described by IT2 T-S fuzzy model [32]. In [33–38], the FD fighting design, impulse control and discrete control based on the IT2 fuzzy model are studied. Interval two-type theory is being recognized and studied by more and more scholars [39,40]. Expanding the application scope of event-driven technology in the IT2 fuzzy control system is the first motivation for writing this paper.

Then, the FD methods for fuzzy systems have been proposed without considering the problems of nonlinear perturbation and transmission-limited [13,14]. Reducing the conservativeness of existing results and redundancy in design is a difficult issue of academic concern. In summary, solutions to event-driven FD problems are important for NNSs subject to uncertainties, perturbation, and network-induced delays. The main contributions of the paper as follows:

(1) A new FD fuzzy filter is designed by using IT2 T-S fuzzy model for generating a residual signal, which means that the designed FD filter premise variable could be different from NNSs.
(2) A fault residual system is established by integrating the IT2 fuzzy theory, external disturbance, event-triggered scheme, time delays and parameter uncertainty.
(3) The stability conditions and the existence conditions of the FD filter are derived by the form of linear matrix inequalities, as a result of the Lyapunov–Krasovskii generalized function method providing the basis. Matrix decoupling implements the transformation of the filter existence conditions with stability analysis.

The rest of this paper is structured as follows. An IT2 fuzzy fault residual system is given based on the IT2 fuzzy networked control system model, event-triggered scheme, and fault diagnosis mechanism in Section 2. Section 3 is the focus of the article and is intended to discuss and clarify the stability analysis and the design of the filter for the fault residual system. Section 4 conducts simulations and discusses the validity of the proposed method. The full paper is summarized, and further research directions are given in Section 5.

Table 1 shows the abbreviations and notations used in this paper.

Table 1. Explanation of abbreviations and notations.

Symbols	Explanatory Notes
FD	fault diagnosis
IT2	interval type-2
T-S	Takagi–Sugeno
NNSs	nonlinear networked systems
FRS	Fault Residual System
LMIs	Linear matrix inequalities
ZOH	Zero-order hold
R^n	n-dimensional Euclidean space
P^{-1}	The inverse of matrix P
P^T	Transpose of matrix P
$P < 0 (\leq 0)$	Negative (semi-negative)-definite matrix
$P > 0 (\geq 0)$	Positive (semi-positive)-definite matrix
$diag\{P, Q, R\}$	Diagonal matrix of P, Q and R
$*$	Symmetric term in the matrix
$\|\cdot\|$	Euclidean norm
$L_2[0, \infty)$	The space of square summable infinite vector sequences

2. Problem Formulation

2.1. IT2 T-S Nonlinear Networked Systems

An NNSs is modeled by IT2 T–S fuzzy rules by using state-space representation, its parameter uncertainty and external perturbations are described.

Plant rule i: IF $\iota_1(x(t))$ is $\widetilde{G}_{i1}$, $\iota_2(x(t))$ is $\widetilde{G}_{i2}$, ... , and $\iota_p(x(t))$ is $\widetilde{G}_{ip}$, THEN

$$\begin{cases} \dot{x}(t) = A_i x(t) + B_i \omega(t) + B_{fi} f(t) \\ y(t) = C_i x(t) + D_i \omega(t) \end{cases} \quad (1)$$

In the IT2 T-S NNSs, A_i, B_i, B_{fi}, C_i, and D_i are system matrices. Separately, $x(t) \in R^{n_x}$, $y(t) \in R^{n_y}$, $f(t) \in R^{n_f}$ represents the state vector, measured output, and the fault signal waiting to be detected, in particular, $\omega(t) \in R^{n_\omega}$ is the external disturbance which belongs to $L_2[0, \infty)$. Define $\iota(x(t)) = [\iota_1(x(t)), \iota_2(x(t)), \ldots, \iota_p(x(t))]^T$ stands for premise variable, the number of fuzzy sets is p, the IT2 fuzzy set is described as $\widetilde{G}_{i\alpha}$, where $i = 1, 2, \ldots, r$, and $\alpha = 1, 2, \ldots, p$, the firing strength of ith rule is defined as follows [39]:

$$W_i(x(t)) = [\underline{\omega}_i(x(t)), \bar{\omega}_i(x(t))] \quad (2)$$

where $\underline{\omega}_i(x(t)) = \prod_{\alpha=1}^{p} \underline{\mu}_{\widetilde{G}_{i\alpha}}(\iota_\alpha(x(t))) \geq 0$, $\bar{\omega}_i(x(t)) = \prod_{\alpha=1}^{p} \bar{\mu}_{\widetilde{G}_{i\alpha}}(\iota_\alpha(x(t))) \geq 0$, $\bar{\mu}_{\widetilde{G}_{i\alpha}}(\iota_\alpha(x(t))) \geq \underline{\mu}_{\widetilde{G}_{i\alpha}}(\iota_\alpha(x(t))) \geq 0$, $\bar{\omega}_i(x(t)) \geq \underline{\omega}_i(x(t)) \geq 0$. We can get the IT2 fuzzy model after weighting, as follows:

$$\begin{cases} \dot{x}(t) = \sum_{i=1}^{r} \widetilde{\rho}_i(x(t))[A_i x(t) + B_i \omega(t) + B_{fi} f(t)] \\ y(t) = \sum_{i=1}^{r} \widetilde{\rho}_i(x(t))[C_i x(t) + D_i \omega(t)] \end{cases} \quad (3)$$

where $\tilde{\rho}_i(x(t)) = \overline{\rho}_i(x(t))\overline{\omega}_i(x(t)) + \underline{\rho}_i(x(t))\underline{\omega}_i(x(t)) \geq 0$, meanwhile $\sum_{i=1}^{r} \tilde{\rho}_i(x(t)) = 1$, $\underline{\rho}_i(x(t))$ and $\overline{\rho}_i(x(t))$ are greater than zero, which represent the weighting functions and satisfying:

$$\underline{\rho}_i(x(t)) + \overline{\rho}_i(x(t)) = 1 \tag{4}$$

Obviously, in the process of NNSs modeling, we define a fuzzy set for the membership function to describe its uncertainty, which provides a basis for the subsequent design of a low conservation fault diagnosis filter.

2.2. Event-Triggered FD Filter

Next, an event-triggering mechanism is introduced within the system, which is between the considered system and FD Filter as shown in Figure 1.

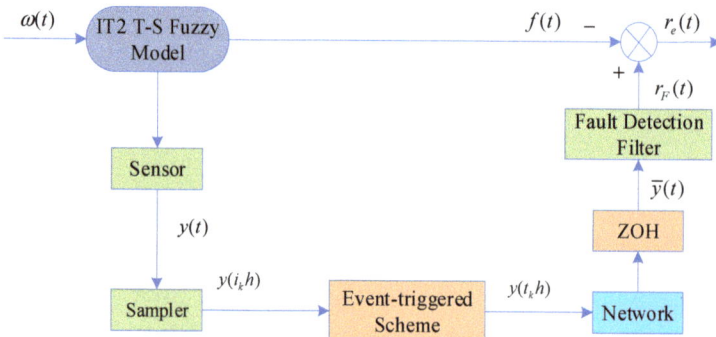

Figure 1. Framework of IT2 T-S NNSs with event-triggered scheme.

The current sampled signal must reach the trigger threshold of the event monitoring terminal before it can be transmitted to the next node. Similar to [34], we can define the event-triggering mechanism as:

$$e_k^T(t)\Lambda e_k(t) > \varepsilon y^T(i_k h)\Lambda y(i_k h) \tag{5}$$

where $\varepsilon \in [0,1)$, $e_k(t)$ is the threshold error, which is the key factor that determines whether the event trigger mechanism occurs, and is obtained by subtracting current sampled data $y(t_k h)$ from the latest transmitted data $y(i_k h)$. Λ denotes the positive triggering parameters.

ZOH provides information about the last transmitted data continuously, the input signal received by the filter can be described as

$$\overline{y}(t_k h) = y(t_k h), \ t \in [t_k h + \tau_{t_k}, t_{k+1} h + \tau_{t_{k+1}}) \tag{6}$$

The system can be transformed into a new time lag system, which can be directly analyzed with time lag system theory. Without loss of generality, the holding region of ZOH is expressed as:

$$\Omega = [t_k h + \tau_{t_k}, t_{k+1} h + \tau_{t_{k+1}}) = \bigcup_{0}^{m} \Omega_l \tag{7}$$

$$\begin{cases} \Omega_0 = [t_k h + \tau_{t_k}, t_k h + h + \overline{\tau}) \\ \Omega_i = [t_k h + ih + \overline{\tau}, t_k h + (i+1)h + \overline{\tau}), \ i = 1, 2 \ldots, m-1 \\ \Omega_m = [t_k h + mh + \overline{\tau}, t_{k+1} h + \tau_{t_{k+1}}) \end{cases} \tag{8}$$

Define $\tau(t) = t - i_k h$, where $i_k h = t_k h + lh$, $l = 0, 1, \ldots, m$, and then we can obtain:

$$0 < \tau_m \leq \tau(t) \leq h + \overline{\tau} = \tau_M \tag{9}$$

Based on the above, $\bar{y}(t)$ can be rewritten as:

$$\bar{y}(t_k h) = [y(t - \tau(t)) - e_k(t)] \qquad (10)$$

Remark 1. *The introduction of the event triggering mechanism (5) reduces redundant transmission data and saves network resources.*

Summarizing the previous discussion, the IT2 fuzzy FD filter is modeled by IT2 T–S fuzzy rules:

Filter Rule j: IF $\varphi_1(x(t))$ is $\tilde{O}_{j1}$, $\varphi_2(x(t))$ is $\tilde{O}_{j2}$, ..., and $\varphi_q(x(t))$ is $\tilde{O}_{jq}$, THEN

$$\begin{cases} \dot{x}_F(t) = \hat{A}_j x_F(t) + \hat{B}_j \bar{y}(t) \\ r_F(t) = \hat{C}_j x_F(t) + \hat{D}_j \bar{y}(t) \end{cases} \qquad (11)$$

in which, $\hat{A}_j, \hat{B}_j, \hat{C}_j$, and $\hat{D}_j$ are FD filter gain matrices. $x_F(t) \in R^{n_x}$, $\bar{y}(t) \in R^{n_y}$, and $r_F(t) \in R^{n_r}$ represent the state vector, the output, and residual output vector of the event-triggered FD filter. The fuzzy set is $\tilde{O}_{j\beta}$, $j = 1, 2, \ldots, s$, $\beta = 1, 2, \ldots, q$, q is the number of fuzzy sets. $\varphi(x(t)) = [\varphi_1(x(t)), \varphi_2(x(t)), \ldots, \varphi_q(x(t))]^T$ are the premise variables. The firing strength of jth rule is expressed by interval sets:

$$K_j(x(t)) = [\underline{\kappa}_j(x(t)), \bar{\kappa}_j(x(t))] \qquad (12)$$

with $\underline{\kappa}_j(x(t)) = \prod_{\beta=1}^{q} \underline{\mu}_{\tilde{O}_{j\beta}}(\varphi_\beta(x(t))) \geq 0$, $\bar{\kappa}_j(x(t)) = \prod_{\beta=1}^{q} \bar{\mu}_{\tilde{O}_{j\beta}}(\varphi_\beta(x(t))) \geq 0$, $\bar{\mu}_{\tilde{N}_{j\lambda}}(\varphi_\lambda(x(t))) \geq \underline{\mu}_{\tilde{N}_{j\lambda}}(\varphi_\lambda(x(t))) \geq 0$, $\bar{\kappa}_j(x(t)) \geq \underline{\kappa}_j(x(t)) \geq 0$, $\underline{\kappa}_j(x(t))$ and $\bar{\kappa}_j(x(t))$ represent, the bounds of membership, where $\underline{\mu}_{\tilde{N}_{j\lambda}}(\varphi_\lambda(x(t)))$ and $\bar{\mu}_{\tilde{N}_{j\lambda}}(\varphi_\lambda(x(t)))$ represent the bounds of the membership function, respectively. The event-triggered FD filter is designed as:

$$\begin{cases} \dot{x}_F(t) = \sum_{j=1}^{r} \tilde{\phi}_j(x(t))[\hat{A}_j x_F(t) + \hat{B}_j \bar{y}(t)] \\ r_F(t) = \sum_{j=1}^{r} \tilde{\phi}_j(x(t))[\hat{C}_j x_F(t) + \hat{D}_j \bar{y}(t)] \end{cases} \qquad (13)$$

where $\tilde{\phi}_j(x(t)) = \underline{\phi}_j(x(t))\bar{\kappa}_j(x(t)) + \bar{\phi}_j(x(t))\underline{\kappa}_j(x(t)) \geq 0$, $\sum_{j=1}^{r} \tilde{\phi}_j(x(t)) = 1$, while $\bar{\phi}_j(x(t)) \geq 0$ and $\underline{\phi}_j(x(t)) \geq 0$ are nonlinear functions used to represent the uncertainty of the FD filter, satisfying

$$\underline{\phi}_j(x(t)) + \bar{\phi}_j(x(t)) = 1 \qquad (14)$$

For the convenience of the following writing, using $\tilde{\rho}_i, \tilde{\phi}_j$ instead of $\tilde{\rho}_i(x(t)), \tilde{\phi}_j(x(t))$.

Remark 2. *The FD filter (13) proposed has two advantages. Firstly, the model has higher accuracy by using IT2 T-S fuzzy theory to describe uncertainty effectively. Secondly, the FD filter is more general as the object's affiliation function and the fuzzy rules are not shared with the FD filter.*

2.3. Fault Residual System (FRS)

In this section, the fault residual system is developed based on models (3) and (13). The fault diagnosis problem is simplified to the problem of asymptotic tracking of residuals and faults. Combination the ETS (5), and defining with $\xi(t) = \begin{bmatrix} x^T(t) & x_F^T(t) \end{bmatrix}^T$,

$\overline{\omega}(t) = \begin{bmatrix} \omega^T(t) & f^T(t) & \omega^T(t-d(t)) \end{bmatrix}^T$, $r_e(t) = r_F(t) - f(t)$, the FRS can be represented as:

$$\begin{cases} \dot{\xi}(t) = \sum_{i=1}^{r}\sum_{j=1}^{r} \widetilde{\rho}_i \widetilde{\phi}_j [\overline{A}_{ij}\xi(t) + \overline{B}_{ij}H\xi(t-\tau(t)) + \overline{B}_{\omega ij}\overline{\omega}(t) - \overline{B}_{eij}e_k(t)] \\ r_e(t) = \sum_{i=1}^{r}\sum_{j=1}^{r} \widetilde{\rho}_i \widetilde{\phi}_j [\overline{C}_{ij}\xi(t) + \overline{D}_{ij}H\xi(t-\tau(t)) + \overline{D}_{\omega ij}\overline{\omega}(t) - \overline{D}_{eij}e_k(t)] \end{cases} \quad (15)$$

$\overline{A}_{ij} = \begin{bmatrix} A_i & 0 \\ 0 & \hat{A}_j \end{bmatrix}, \overline{B}_{ij} = \begin{bmatrix} 0 \\ \hat{B}_j C_i \end{bmatrix}, \overline{B}_{\omega ij} = \begin{bmatrix} B_i & B_{fi} & 0 \\ 0 & 0 & \hat{B}_j D_i \end{bmatrix}, \overline{B}_{eij} = \begin{bmatrix} 0 \\ \hat{B}_j \end{bmatrix}, \overline{C}_{ij} = \begin{bmatrix} 0 & \hat{C}_j \end{bmatrix},$

$\overline{D}_{ij} = \begin{bmatrix} \hat{D}_j C_i \\ 0 \end{bmatrix}, \overline{D}_{\omega ij} = \begin{bmatrix} 0 & -I & \hat{D}_j D_i \end{bmatrix}, \overline{D}_{eij} = \hat{D}_j, H = \begin{bmatrix} I & 0 \end{bmatrix}.$

The target of this section is to design the FD filter (13) and triggering mechanism (5) such that the FRS (15) satisfies asymptotically stable with the H_∞ performance indicators. In the meantime, the following conditions are satisfied:

(1) When $\widetilde{\omega}(t) = 0$, the FRS (15) is considered to be asymptotically stable.
(2) Under the condition of zero initial, $r_e(t)$ contents $\|r_e(t)\|_2 < \gamma \|\widetilde{\omega}(t)\|_2$, where $\gamma > 0$ bring about H_∞ performance level.

2.4. FD Mechanism

Define the following FD mechanism.

$$\begin{aligned} J(t) &= \left\{ \int_0^t r_F^T(s) r_F(s) ds \right\}^{\frac{1}{2}} \\ J_{th} &= \sup_{w \in L_2, f=0} \left\{ \int_0^{T_d} r_F^T(s) r_F(s) ds \right\}^{\frac{1}{2}} \end{aligned} \quad (16)$$

where $J(t)$ is the residual evaluation function, and J_{th} is the threshold, T_d represents the limited length of evaluation time. The fault detection mechanism is as follows:

$$\begin{cases} J(t) > J_{th} \Rightarrow \text{with faults} \Rightarrow \text{alarm} \\ J(t) \leq J_{th} \Rightarrow \text{no faults}. \end{cases} \quad (17)$$

Lemma 1. *(Schur complement) [41] For the given matrix* $S = \begin{bmatrix} S_{11} & S_{12} \\ S_{21} & S_{22} \end{bmatrix} < 0$, *where* $S \in R^{r*r}, S_{21} = S_{12}^T$, *the following three sets of conditions and inequalities hold and are equivalent:*

(1) $S < 0$;
(2) $S_{11} < 0, S_{22} - S_{12}^T S_{11}^{-1} S_{12} < 0$;
(3) $S_{22} < 0, S_{11} - S_{12}^T S_{22}^{-1} S_{12} < 0$.

Lemma 2. *[42] For real matrices Z, X, Y with appropriate dimensions, in which the is symmetric, then*

$$Z + XK(t)Y + Y^T K(t) X^T < 0 \quad (18)$$

for all $K^T(t)K(t) \leq I$, there exists $\varepsilon > 0$, such that:

$$Z + \varepsilon XX^T + \varepsilon^{-1} Y^T Y < 0$$

Lemma 3. *[43] Given a symmetric and positive matrix $\widetilde{R}$, inequality (18) holds:*

$$-\int_{t-\overline{\tau}}^{t} \dot{\theta}^T(s) \widetilde{W} \dot{\theta}(s) ds \leq \frac{1}{\overline{\tau}} \begin{bmatrix} \theta(t) \\ \theta(t-\tau(t)) \\ \theta(t-\overline{\tau}) \end{bmatrix}^T \begin{bmatrix} -\widetilde{R} & \widetilde{R} & 0 \\ * & -2\widetilde{R} & \widetilde{R} \\ * & * & -\widetilde{R} \end{bmatrix} \begin{bmatrix} \theta(t) \\ \theta(t-\tau(t)) \\ \theta(t-\overline{\tau}) \end{bmatrix} \quad (19)$$

Remark 3. *It is worth noting that the fault residual system is built via IT2 T-S fuzzy model, considering the event-triggered communication mechanisms, disturbances and network time delays. In the existing work, there is less research on the IT2 T-S fuzzy network control system FD filtering with event triggering, which is one of the innovative points in this section.*

3. Main Conclusion

3.1. Stability Analysis

In this subsection, the following improvements will be made in the stability analysis process to reduce the system conservativeness. First, a new Lyapunov–Krasovskii function with fourfold integration is constructed; second, Wirtinger's inequality is applied to process the integral term, which is in the time derivative of the Lyapunov–Krasovskii function; third, a relaxation matrix is introduced to deal with the premise variable mismatch problem.

Theorem 1. *For given scalars $0 < \varepsilon < 1$, $0 < \tau_m \leq \tau_M$, $\gamma > 0$, and the membership functions satisfying $\tilde{w}_j - \psi_j \tilde{m}_j \geq 0 (0 < \psi_j \leq 1)$, if IT2 FRS (15) is asymptotically stable, and achieving the expected H_∞ performance level γ, then there exists parameter matrix $P > 0$, $Q_i \ (i=1,2)$, $S_i > 0 \ (i=1,2)$, $R_i > 0 \ (i=1,2,3)$, $T_i > 0 \ (i=1,2)$, $\Lambda_i > 0 \ (i=1,2)$, $\hat{A}_j, \hat{B}_j, \hat{C}_j, \hat{D}_j$ and $W_i > 0$, $(i=1,2,\ldots,r)$, meanwhile, the following inequalities exist in the appropriate dimensions:*

$$\Xi_{ij} - W_i < 0 \tag{20}$$

$$\psi_i \Xi_{ii} - \psi_i W_i + W_i < 0 \tag{21}$$

$$\psi_j \Xi_{ij} + \psi_i \Xi_{ji} - \psi_i W_j - \psi_j W_i + W_i + W_j < 0, \ i < j \tag{22}$$

for $\Xi_{ij} = \begin{bmatrix} \Xi_{ij}^{11} & \Xi_{ij}^{12} \\ * & \Xi_{ij}^{22} \end{bmatrix}$,

in which $\Xi_{ij}^{11} = \begin{bmatrix} \Phi_{ij}^{11} & \Phi_{ij}^{12} \\ * & \Phi_{ij}^{22} \end{bmatrix}$, where $\Phi_{ij}^{11} = \begin{bmatrix} \Phi_{11} & H^T R_1 & 0 & 0 & H^T R_3 \\ * & -2R_1 & R_1 & 0 & 0 \\ * & * & \Phi_{33} & R_2 & 0 \\ * & * & * & -2R_2 & 0 \\ * & * & * & * & -2R_3 \end{bmatrix}$,

$\Phi_{ij}^{12} = \begin{bmatrix} 0 & P\overline{B}_{ij} & -P\overline{B}_{eij} & -P\overline{B}_{\omega ij} \\ 0 & 0 & 0 & 0 \\ 0 & 0 & 0 & 0 \\ R_2 & 0 & 0 & 0 \\ R_3 & 0 & 0 & 0 \end{bmatrix}$, $\Phi_{ij}^{22} = \begin{bmatrix} \Phi_{66} & 0 & 0 & 0 \\ * & \varepsilon C_i^T \Lambda_2 C_i & 0 & \Phi_{79} \\ * & * & -\Lambda_1 & 0 \\ * & * & * & \Phi_{99} \end{bmatrix}$,

$\Phi_{11} = P\overline{A}_{ij} + \overline{A}_{ij}^T P + H^T(Q_1 + Q_2)H - H^T(R_1 + R_3)H$, $\Phi_{33} = -Q_1 - R_1 - R_2$, $\Phi_{66} = -Q_2 - R_3$, $\Phi_{79} = \varepsilon C_i^T \Lambda_2 \begin{bmatrix} 0 & 0 & D_i \end{bmatrix}$, $\Delta \bar{\tau} = \tau_M - \tau_m$, $\Phi_{99} = -\gamma^2 I + \varepsilon \begin{bmatrix} 0 & 0 & D_i \end{bmatrix}^T \Lambda_2 \begin{bmatrix} 0 & 0 & D_i \end{bmatrix}$,

$\Xi_{ij}^{12} = \begin{bmatrix} \frac{\tau_m}{\sqrt{2}} S_1^T \varphi_1 & \frac{\Delta \bar{\tau}}{\sqrt{2}} S_2^T \varphi_1 & \tau_m R_1^T \varphi_1 & \Delta \bar{\tau} R_2^T \varphi_1 & \tau_M R_3^T \varphi_1 & \frac{\tau_m^2}{\sqrt{6}} T_1^T \varphi_1 & \frac{\Delta \bar{\tau}^2}{\sqrt{6}} T_2^T \varphi_1 & \overline{C}_{ij} \\ 0 & 0 & 0 & 0 & 0 & 0 & 0 & 0 \\ 0 & 0 & 0 & 0 & 0 & 0 & 0 & 0 \\ 0 & 0 & 0 & 0 & 0 & 0 & 0 & 0 \\ 0 & 0 & 0 & 0 & 0 & 0 & 0 & 0 \\ \frac{\tau_m}{\sqrt{2}} S_1^T \varphi_2 & \frac{\Delta \bar{\tau}}{\sqrt{2}} S_2^T \varphi_2 & \tau_m R_1^T \varphi_2 & \Delta \bar{\tau} R_2^T \varphi_2 & \tau_M R_3^T \varphi_2 & \frac{\tau_m^2}{\sqrt{6}} T_1^T \varphi_2 & \frac{\Delta \bar{\tau}^2}{\sqrt{6}} T_2^T \varphi_2 & \overline{D}_{ij} \\ -\frac{\tau_m}{\sqrt{2}} S_1^T \varphi_3 & -\frac{\Delta \bar{\tau}}{\sqrt{2}} S_2^T \varphi_3 & -\tau_m R_1^T \varphi_3 & -\Delta \bar{\tau} R_2^T \varphi_3 & -\tau_M R_3^T \varphi_3 & -\frac{\tau_m^2}{\sqrt{6}} T_1^T \varphi_3 & -\frac{\Delta \bar{\tau}^2}{\sqrt{6}} T_2^T \varphi_3 & -\overline{D}_{eij} \\ \frac{\tau_m}{\sqrt{2}} S_1^T \varphi_4 & \frac{\Delta \bar{\tau}}{\sqrt{2}} S_2^T \varphi_4 & \tau_m R_1^T \varphi_4 & \Delta \bar{\tau} R_2^T \varphi_4 & \tau_M R_3^T \varphi_4 & \frac{\tau_m^2}{\sqrt{6}} T_1^T \varphi_4 & \frac{\Delta \bar{\tau}^2}{\sqrt{6}} T_2^T \varphi_4 & \overline{D}_{\omega ij} \end{bmatrix}$

$\Xi_{ij}^{22} = \text{diag}\{ -S_1 \ -S_2 \ -R_1 \ -R_2 \ -R_3 \ -T_1 \ -T_2 \ -I \}$,

$\varphi_1 = H\overline{A}_{ij}, \varphi_2 = H\overline{B}_{ij}, \varphi_3 = H\overline{B}_{eij}, \varphi_4 = H\overline{B}_{\omega ij}$.

Proof. For the FRS (15), construct the following Lyapunov–Krasovskii function:

$$V(t) = V_1(t) + V_2(t) + V_3(t) + V_4(t) + V_5(t) \qquad (23)$$

where

$V_1(t) = \xi^T(t) P \xi(t),$

$V_2(t) = \int_{t-\tau_m}^{t} \xi^T(s) H^T Q_1 H \xi(s) ds + \int_{t-\tau_M}^{t} \xi^T(s) H^T Q_2 H \xi(s) ds,$

$V_3(t) = \tau_m \int_{t-\tau_m}^{t} \int_{s}^{t} \dot{\xi}^T(v) H^T R_1 H \dot{\xi}(v) dv ds + (\tau_M - \tau_m) \int_{t-\tau_M}^{t-\tau_m} \int_{s}^{t} \dot{\xi}^T(v) H^T R_2 H \dot{\xi}(v) dv ds$
$\quad + \tau_M \int_{t-\tau_M}^{t} \int_{s}^{t} \dot{\xi}^T(v) H^T R_3 H \dot{\xi}(v) dv ds$

$V_4(t) = \int_{-\tau_m}^{0} \int_{\theta}^{0} \int_{t+\lambda}^{t} \dot{\xi}^T(s) H^T S_1 H \dot{\xi}(s) ds d\lambda d\theta + \int_{-\tau_M}^{-\tau_m} \int_{\theta}^{0} \int_{t+\lambda}^{t} \dot{\xi}^T(s) H^T S_2 H \dot{\xi}(s) ds d\lambda d\theta,$

$V_5(t) = \tau_m \int_{-\tau_m}^{0} \int_{\theta}^{0} \int_{\lambda}^{0} \int_{t+k}^{t} \dot{\xi}^T(s) H^T T_1 H \dot{\xi}(s) ds dk d\lambda d\theta$
$\quad + (\tau_M - \tau_m) \int_{-\tau_M}^{-\tau_m} \int_{\theta}^{0} \int_{\lambda}^{0} \int_{t+k}^{t} \dot{\xi}^T(s) H^T T_2 H \dot{\xi}(s) ds dk d\lambda d\theta$

and $P = P^T > 0, Q_i > 0, S_i > 0, T_i > 0, i = 1, 2, R_j > 0, j = 1, 2, 3$.

Along the trajectory of the FRS (15), the time derivative of $V(t)$ is:

$$\dot{V}(t) = \dot{V}_1(t) + \dot{V}_2(t) + \dot{V}_3(t) + \dot{V}_4(t) + \dot{V}_5(t) \qquad (24)$$

where

$\dot{V}_1(t) = 2\xi^T(t) P \dot{\xi}(t),$

$\dot{V}_2(t) = \xi^T(t) H^T(Q_1 + Q_2) H \xi(t) - \xi^T(t-\tau_m) H^T Q_1 H \xi(t-\tau_m) - \xi^T(t-\tau_M) H^T Q_2 H \xi(t-\tau_M),$

$\dot{V}_3(t) = \dot{\xi}^T(t) H^T \left[\tau_m^2 R_1 + (\tau_M - \tau_m)^2 R_2 + \tau_M^2 R_3 \right] H \dot{\xi}(t) - \tau_m \int_{t-\tau_m}^{t} \dot{\xi}^T(s) H^T R_1 H \dot{\xi}(s) ds$
$\quad - (\tau_M - \tau_m) \int_{t-\tau_M}^{t-\tau_m} \dot{\xi}^T(s) H^T R_2 H \dot{\xi}(s) ds - \tau_M \int_{t-\tau_M}^{t} \dot{\xi}^T(s) H^T R_3 H \dot{\xi}(s) ds$

$\dot{V}_4(t) = \frac{\tau_m^2}{2} \dot{\xi}^T(t) H^T S_1 H \dot{\xi}(t) + \frac{(\tau_M - \tau_m)^2}{2} \dot{\xi}^T(t) H^T S_2 H \dot{\xi}(t) - \int_{-\tau_m}^{0} \int_{t+\theta}^{t} \dot{\xi}^T(s) H^T S_1 H \dot{\xi}(s) ds d\theta$
$\quad - \int_{-\tau_M}^{-\tau_m} \int_{t+\theta}^{t} \dot{\xi}^T(s) H^T S_2 H \dot{\xi}(s) ds d\theta$

$\dot{V}_5(t) = \frac{\tau_m^4}{6} \dot{\xi}^T(t) H^T T_1 H \dot{\xi}(t) + \frac{(\tau_M - \tau_m)^4}{6} \dot{\xi}^T(t) H^T T_2 H \dot{\xi}(t)$
$\quad - \tau_m \int_{-\tau_m}^{0} \int_{\theta}^{0} \int_{t+\lambda}^{t} \dot{\xi}^T(s) H^T T_1 H \dot{\xi}(s) ds d\lambda d\theta - (\tau_M - \tau_m) \int_{-\tau_M}^{-\tau_m} \int_{\theta}^{0} \int_{t+\lambda}^{t} \dot{\xi}^T(s) H^T T_2 H \dot{\xi}(s) ds d\lambda d\theta$

The integral term in $\dot{V}_3(t)$, which we treat by applying Lemma 3, yields

$$-\tau_m \int_{t-\tau_m}^{t} \dot{\xi}^T(s) H^T R_1 H \dot{\xi}(s) ds \leq \begin{bmatrix} H\xi(t) \\ H\xi(t-\tau_1(t)) \\ H\xi(t-\tau_m) \end{bmatrix}^T \begin{bmatrix} -R_1 & R_1 & 0 \\ * & -2R_1 & R_1 \\ * & * & -R_1 \end{bmatrix} \begin{bmatrix} H\xi(t) \\ H\xi(t-\tau_1(t)) \\ H\xi(t-\tau_m) \end{bmatrix} \qquad (25)$$

$$-(\tau_M - \tau_m) \int_{t-\tau_M}^{t-\tau_m} \dot{\xi}^T(s) H^T R_2 H \dot{\xi}(s) ds \leq \begin{bmatrix} H\xi(t-\tau_m) \\ H\xi(t-\tau_2(t)) \\ H\xi(t-\tau_M) \end{bmatrix}^T \begin{bmatrix} -R_2 & R_2 & 0 \\ * & -2R_2 & R_2 \\ * & * & -R_2 \end{bmatrix} \begin{bmatrix} H\xi(t-\tau_m) \\ H\xi(t-\tau_2(t)) \\ H\xi(t-\tau_M) \end{bmatrix} \qquad (26)$$

$$-\tau_M \int_{t-\tau_M}^{t} \dot{\xi}^T(s) H^T R_3 H \dot{\xi}(s) ds \leq \begin{bmatrix} H\xi(t) \\ H\xi(t-\tau_3(t)) \\ H\xi(t-\tau_M) \end{bmatrix}^T \begin{bmatrix} -R_3 & R_3 & 0 \\ * & -2R_3 & R_3 \\ * & * & -R_3 \end{bmatrix} \begin{bmatrix} H\xi(t) \\ H\xi(t-\tau_3(t)) \\ H\xi(t-\tau_M) \end{bmatrix} \qquad (27)$$

Furthermore, in a bid to obtain stability conditions with low conservativeness, the following slack matrix is introduced:

$$\sum_{i=1}^{r} \sum_{j=1}^{r} \tilde{m}_i (\tilde{m}_j - \tilde{w}_j) W_i = 0, W_i = W_i^T, (i = 1, 2, \ldots, r) \qquad (28)$$

From (23) to (28), we can obtain

$$\sum_{i=1}^{r}\sum_{j=1}^{r}\widetilde{m}_i\widetilde{w}_j\Xi_{ij}$$
$$=\sum_{i=1}^{r}\sum_{j=1}^{r}\widetilde{m}_i(\widetilde{m}_j-\widetilde{w}_j+\psi_j\widetilde{m}_j-\psi_j\widetilde{m}_j)W_i+\sum_{i=1}^{r}\sum_{j=1}^{s}\widetilde{m}_i\widetilde{w}_j\Xi_{ij}$$
$$=\sum_{i=1}^{r}\widetilde{m}_i^2(\psi_i\Xi_{ii}-\psi_iW_i+W_i)$$
$$+\sum_{i=1}^{r-1}\sum_{j=i+1}^{r}\widetilde{m}_i\widetilde{m}_j(\psi_j\Xi_{ij}-\psi_jW_i+W_i+\psi_i\Xi_{ji}-\psi_iW_j+W_j)+\sum_{i=1}^{r}\sum_{j=1}^{r}\widetilde{m}_i(\widetilde{w}_j-\psi_j\widetilde{m}_j)(\Xi_{ij}-W_i)$$
(29)

under $\widetilde{w}_j - \psi_j\widetilde{m}_j \geq 0$ for all j. Combined with the event-triggering mechanism (5), we can derive

$$\dot{V}(t)+r_e^T(t)r_e(t)-\gamma^2\widetilde{\omega}^T(t)\widetilde{\omega}(t)\leq\sum_{i=1}^{r}\sum_{j=1}^{r}\widetilde{m}_i\widetilde{w}_j\zeta^T(t)\Xi_{ij}\zeta(t)$$
(30)

where

$\zeta^T(t) = [\ \eta_1(t)\ \ \eta_2(t)\ ], \zeta_1(t) = [\ \xi^T(t)\ \ \xi^T(t-\tau_1(t))\ \ \xi^T(t-\tau_m)H^T\ \ \xi^T(t-\tau_2(t))\ ],$
$\zeta_2(t) = [\ \xi^T(t-\tau_3(t))\ \ \xi^T(t-\tau_M)H^T\ \ \xi^T(t-\tau(t))H^T\ \ e_k^T(t)\ \ \widetilde{\omega}^T(t)].$

By using Schur complement, $\Xi_{ij} \leq 0$, hence, we have

$$\dot{V}(t)+r_e^T(t)r_e(t)-\gamma^2\widetilde{\omega}^T(t)\widetilde{\omega}(t)\leq 0$$
(31)

Integrating from 0 to ∞ simultaneously on the left and right sides of (30), we can obtain:

$$\int_0^{\infty}r_e^T(t)r_e(t)dt<\gamma^2\int_0^{\infty}\widetilde{\omega}^T(t)\widetilde{\omega}(t)dt$$
(32)

Equation (32) representative $\|r_e(t)\|_2 < \gamma\|\widetilde{\omega}(t)\|_2$ holds for any nonzero $\widetilde{\omega}(t) \in L_2[0,\infty)$. Thus, the FRS (15) is under the restriction of Theorem 1 is asymptotically stable and satisfies the given H_∞ performance index γ. □

Remark 4. *The Lyapunov–Krasovskii function (23) constructed contains multiple integrals, such as triple, quadruple integrals. The more system and time delay information are considered, and the amplification of the integral term processing is avoided effectively. Convergence of global asymptotic stability is guaranteed. Moreover, more recently, the introduction of the relaxation matrix (28) makes the obtained stability criterion with less conservative.*

3.2. Fault Diagnosis Filter Design

In this section, solving the parameters of the FD filter is transformed into the problem of matrix convex optimization, which can be solved by MATLAB. Using the matrix transformation and deformation, the proposed filter design method is implemented.

Theorem 2. *For given scalars $0 < \varepsilon < 1$, $0 < \tau_m \leq \tau_M$, $\gamma > 0$, and the membership functions satisfying $\widetilde{w}_j - \psi_j\widetilde{m}_j \geq 0$, $(0 < \psi_j \leq 1)$, if the IT2 FRS (15) is asymptotically stable and meets the expected H_∞ performance level γ, then there exists parameter matrix $P > 0$, $Q_i > 0$ $(i = 1, 2)$, $S_i > 0$ $(i = 1, 2)$, $R_i > 0$ $(i = 1, 2, 3)$, $T_i > 0$ $(i = 1, 2)$, $\Lambda_i > 0$ $(i = 1, 2)$, $\widetilde{A}_j$, $\widetilde{B}_j$, $\widetilde{C}_j$, $\widetilde{D}_j$ and $\widetilde{W}_i^T = \widetilde{W}_i$ have suitable dimensions satisfying the following inequality:*

$$\widetilde{\Xi}_{ij} - \widetilde{W}_i < 0$$
(33)

$$\psi_i\widetilde{\Xi}_{ii} - \psi_i\widetilde{W}_i + \widetilde{W}_i < 0$$
(34)

$$\psi_j\widetilde{\Xi}_{ij} + \psi_i\widetilde{\Xi}_{ji} - \psi_i\widetilde{W}_j - \psi_j\widetilde{W}_i + \widetilde{W}_i + \widetilde{W}_j < 0,\ i < j$$
(35)

for $\widetilde{\Xi}_{ij} = \begin{bmatrix} \widetilde{\Xi}_{ij}^{11} & \widetilde{\Xi}_{ij}^{12} \\ * & \widetilde{\Xi}_{ij}^{22} \end{bmatrix}$,

in which $\widetilde{\Xi}_{ij}^{11} = \begin{bmatrix} \widetilde{\Phi}_{ij}^{11} & \widetilde{\Phi}_{ij}^{12} \\ * & \widetilde{\Phi}_{ij}^{22} \end{bmatrix}$, where $\widetilde{\Phi}_{ij}^{11} = \begin{bmatrix} \widetilde{\Phi}_{11} & \widetilde{\Phi}_{12} & R_1 & 0 & 0 & R_3 \\ * & \widetilde{\Phi}_{22} & 0 & 0 & 0 & 0 \\ * & * & -2R_1 & R_1 & 0 & 0 \\ * & * & * & \widetilde{\Phi}_{44} & R_2 & 0 \\ * & * & * & * & -2R_2 & 0 \\ * & * & * & * & * & -2R_3 \end{bmatrix}$,

$\widetilde{\Phi}_{ij}^{12} = \begin{bmatrix} 0 & \widetilde{B}_j C_i & -\widetilde{B}_j & P_1 B_i & P_1 B_{fi} & \widetilde{B}_j D_i \\ 0 & \widetilde{B}_j C_i & -\widetilde{B}_j & Y B_i & Y B_{fi} & \widetilde{B}_j D_i \\ 0 & 0 & 0 & 0 & 0 & 0 \\ 0 & 0 & 0 & 0 & 0 & 0 \\ R_2 & 0 & 0 & 0 & 0 & 0 \\ R_3 & 0 & 0 & 0 & 0 & 0 \end{bmatrix}$, $\widetilde{\Phi}_{ij}^{22} = \begin{bmatrix} \widetilde{\Phi}_{77} & 0 & 0 & 0 & 0 & 0 \\ * & \widetilde{\Phi}_{88} & 0 & 0 & 0 & \widetilde{\Phi}_{812} \\ * & * & -\Lambda_1 & 0 & 0 & 0 \\ * & * & * & -\gamma^2 I & 0 & 0 \\ * & * & * & * & -\gamma^2 I & 0 \\ * & * & * & * & * & \widetilde{\Phi}_{1212} \end{bmatrix}$,

$\widetilde{\Phi}_{11} = P_1 A_i + P_1 \Delta A + A_i^T P_1 + A_i^T \Delta P + Q_1 + Q_2 - R_1 - R_3$, $\widetilde{\Phi}_{12} = A_i^T Y + \widetilde{A}_j + \Delta A^T Y$,

$\widetilde{\Phi}_{22} = \widetilde{A}_j + \widetilde{A}_j^T$, $\widetilde{\Phi}_{44} = -Q_1 - R_1 - R_2$, $\widetilde{\Phi}_{77} = -Q_2 - R_3$, $\widetilde{\Phi}_{88} = \varepsilon C_i^T \Lambda_2 C_i$,

$\widetilde{\Phi}_{812} = \varepsilon C_i^T \Lambda_2 D_i$, $\widetilde{\Phi}_{1212} = -\gamma^2 I + \varepsilon D_i^T \Lambda_2 D_i$.

$\Xi_{ij}^{12} = \begin{bmatrix} \frac{\tau_m}{\sqrt{2}} S_1 A_i & \frac{\Delta \bar{\tau}}{\sqrt{2}} S_2 A_i & \tau_m R_1 A_i & \Delta \bar{\tau} R_2 A_i & \tau_M R_3 A_i & \frac{\tau_m^2}{\sqrt{6}} T_1 A_i & \frac{\Delta \bar{\tau}^2}{\sqrt{6}} T_2 A_i & 0 \\ 0 & 0 & 0 & 0 & 0 & 0 & 0 & \widetilde{C}_j \\ 0 & 0 & 0 & 0 & 0 & 0 & 0 & 0 \\ \vdots \Big\}5 & \vdots \Big\}5 & \vdots \Big\}5 & \vdots \Big\}5 & \vdots \Big\}5 & \vdots \Big\}5 & \vdots \Big\}5 & \vdots \Big\}5 \\ 0 & 0 & 0 & 0 & 0 & 0 & 0 & 0 \\ 0 & 0 & 0 & 0 & 0 & 0 & 0 & \widetilde{D}_j C_i \\ 0 & 0 & 0 & 0 & 0 & 0 & 0 & -\widetilde{D}_j \\ \frac{\tau_m}{\sqrt{2}} S_1 B_i & \frac{\Delta \bar{\tau}}{\sqrt{2}} S_2 B_i & \tau_m R_1 B_i & \Delta \bar{\tau} R_2 B_i & \tau_M R_3 B_i & \frac{\tau_m^2}{\sqrt{6}} T_1 B_i & \frac{\Delta \bar{\tau}^2}{\sqrt{6}} T_2 B_i & 0 \\ 0 & 0 & 0 & 0 & 0 & 0 & 0 & -I \\ 0 & 0 & 0 & 0 & 0 & 0 & 0 & \widetilde{D}_j D_i \end{bmatrix}$,

$\widetilde{\Xi}_{ij}^{22} = \text{diag}\{ -S_1 \ -S_2 \ -R_1 \ -R_2 \ -R_3 \ -T_1 \ -T_2 \ -I \}$.

Based on the above condition for the establishment of linear matrix inequality, the filter parameter matrix is obtained as follows

$$\hat{A}_j = Y^{-1} \widetilde{A}_j, \ \hat{B}_j = Y^{-1} \widetilde{B}_j, \ \hat{C}_j = \widetilde{C}_j, \ \hat{D}_j = \widetilde{D}_j. \tag{36}$$

Proof. On the basis of Theorem 1, we set $P = \begin{bmatrix} P_1 & P_2 \\ * & P_3 \end{bmatrix}$, $J_1 = \text{diag}\{I, P_2 P_3^{-1}\}$, $J_2 = \text{diag}\{\underbrace{J_1, I \ldots I}_{18}\}$.

Then, we have to multiply the left and right sides of Equations (20)–(22) by J_2 and J_2^T. It yields that

$$\widetilde{\Xi}_{ij} - W_i + \Sigma_1^T \overline{\Delta}_f \Sigma_2 + \Sigma_2^T \overline{\Delta}_f \Sigma_1 < 0 \tag{37}$$

The application of Lemma 2 achieves the conversion of (37) to (38).

$$\widetilde{\Xi}_{ij} - W_i + \varepsilon_1^{-1} \Sigma_1^T \delta^2 \Sigma_1 + \varepsilon_1 \Sigma_2^T \Sigma_2 < 0 \tag{38}$$

To facilitate the simplification and operation of the matrix, the following expression is made:

$$\widetilde{W}_i = J_2 W_i J_2^T, \ Y = P_2 P_3^{-1} P_2^T,$$

$$\widetilde{A}_j = P_2 \hat{A}_j P_3^{-1} P_2^T, \ \widetilde{B}_j = P_2 \hat{B}_j, \ \widetilde{C}_j = \hat{C}_j P_3^{-1} P_2^T, \ \widetilde{D}_j = \hat{D}_j P_3^{-1} P_2^T.$$

Bringing them into Equations (20)–(22), we can obtain Equations (33)–(35).

By using Schur Complement Lemma, the matrix P is equivalent to $P_1 - P_2 P_3^{-1} P_2^T = P_1 - Y > 0$. Furthermore, equivalently under transformation $P_2^T P_3 x_f(t)$, the parameters of the fault detection filter can be yielded as follows:

$$\hat{A}_j = P_2^{-T} P_3 (P_2^{-1} \widetilde{A}_j P_2^{-T} P_3) P_3^{-1} P_2^T = Y^{-1} \widetilde{A}_j, \ \hat{B}_j = P_2^{-T} P_3 (P_2^{-1} \widetilde{B}_j) = Y^{-1} \widetilde{B}_j,$$

$$\hat{C}_j = (\widetilde{C}_j P_2^{-T} P_3) P_3^{-1} P_2^T = \widetilde{C}_j, \ \hat{D}_j = (\widetilde{D}_j P_2^{-T} P_3) P_3^{-1} P_2^T = \widetilde{D}_j.$$

According to Theorem 2, we determine the FD filter parameters by solving the convex optimization problems:

min γ subject to the inequalities (33)–(35).

The proof is completed. □

4. Simulation

In this section, we provide several examples to illustrate the usefulness of the designed IT2 fuzzy FD approach and to compare it with the existing results in [44,45] to show the advantages of our method.

Two rules have been considered in the following IT2 fuzzy system (system parameters are borrowed from [46])

$$\begin{cases} \dot{x}(t) = \sum_{i=1}^{2} \widetilde{\rho}_i(x(t))[A_i x(t) + B_i \omega(t) + B_{fi} f(t)] \\ y(t) = \sum_{i=1}^{2} \widetilde{\rho}_i(x(t))[C_i x(t) + D_i \omega(t)] \end{cases} \quad (39)$$

with $A_1 = \begin{bmatrix} -1 & 0.2 \\ -0.9 & 0.15 \end{bmatrix}$, $A_2 = \begin{bmatrix} -0.4 & 0.2 \\ -0.8 & -1.10 \end{bmatrix}$, $B_1 = \begin{bmatrix} 0.1 \\ 0.2 \end{bmatrix}$, $B_1 = \begin{bmatrix} 0.4 \\ 0.9 \end{bmatrix}$, $B_{f1} = \begin{bmatrix} -0.1 \\ 0.01 \end{bmatrix}$, $B_1 = \begin{bmatrix} -0.1 \\ 0.01 \end{bmatrix}$, $C_1 = \begin{bmatrix} 0.1 & 0.1 \end{bmatrix}$, $C_2 = \begin{bmatrix} 0.1 & 0.2 \end{bmatrix}$, $D_1 = D_2 = 0.01$. The membership functions of the plant and fault detection filter are depicted in Table 2. The nonlinear functions are chosen as, i.e., $\underline{\rho}_i(x_1(t)) = \sin(x_1^2(t))$, $\overline{\rho}_i(x_1(t)) = 1 - \sin(x_1^2(t))$, $i = 1, 2$, and $\underline{\phi}_j(x(t)) = \overline{\phi}_j(x(t)) = 0.5$ for $j = 1, 2$.

Table 2. Membership functions for plant and filter.

The Upper Membership Function	The Lower Membership Function
$\overline{\omega}_1(x_1(t)) = \frac{0.27 - 0.01 x_1^2(t)}{0.27}$	$\underline{\omega}_1(x_1(t)) = \frac{0.27 - 0.03 x_1^2(t)}{0.27}$
$\overline{\omega}_2(x_1(t)) = \frac{x_1^2(t)}{9}$	$\underline{\omega}_2(x_1(t)) = \frac{x_1^2(t)}{27}$
$\overline{\kappa}_1(x_1(t)) = \exp\left(-\frac{x_1^2(t)}{8}\right)$	$\underline{\kappa}_1(x_1(t)) = \exp\left(-\frac{x_1^2(t)}{4}\right)$
$\overline{\kappa}_2(x_1(t)) = 1 - \underline{\kappa}_1(x_1(t))$	$\underline{\kappa}_2(x_1(t)) = 1 - \overline{\kappa}_1(x_1(t))$

In order to derive the gain matrices of the FD filter in (7), we assume the parameter sets $(\tau_m, \tau_M, \varepsilon, \ell_l, \ell_2) = (0.01, 0.1, 0.5, 0.7, 0.5)$. Then by solving the conditions in Theorem 2, we can obtain

$$\hat{A}_1 = \begin{bmatrix} -1.6738 & 0.1545 \\ -0.5992 & -0.3587 \end{bmatrix}, \hat{A}_2 = \begin{bmatrix} -0.6885 & -0.1969 \\ 0.8140 & -2.3963 \end{bmatrix},$$

$$\hat{B}_1 = \begin{bmatrix} -2.8318 \times 10^{-12} \\ 9.3801 \times 10^{-13} \end{bmatrix}, \hat{B}_2 = \begin{bmatrix} -1.7555 \times 10^{-12} \\ -9.6037 \times 10^{-13} \end{bmatrix},$$

$$\hat{C}_1 = \begin{bmatrix} 0.1087 & -0.0306 \end{bmatrix}, \hat{C}_2 = \begin{bmatrix} 0.0980 & -0.0180 \end{bmatrix},$$
$$\hat{D}_1 = 1.2609 \times 10^{-12}, \hat{D}_2 = 1.6357 \times 10^{-12}, \Lambda = 5.3637 \times 10^{-12}.$$

Besides, the H_∞ performance is calculated as $\gamma = 2.4227$. According to the FD mechanism, we set the fault signal as

$$f(t) = \begin{cases} 2, 20 < t < 30 \\ 0, others \end{cases} \quad (40)$$

and the external disturbance $\omega(t)$ is stochastic noise that belongs to standard normal distribution. Let the initial states be $x_0 = \hat{x}_0 = \begin{bmatrix} 0 & 0 \end{bmatrix}^T$. Then, we can derive Figures 2–4. Specifically, Figure 2 depicts the actual transmission instants and intervals under the event-triggered scheme. In the simulation time (50 s) and sampling period (0.1 s), only 20.0% of sampled data are transmitted over the wireless network. Clearly, it saves many communication resources. Figures 3 and 4, respectively, show the trajectories of the error $r_e(t)$ without/with fault.

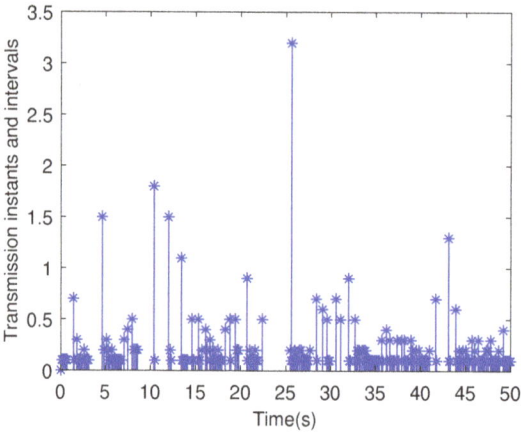

Figure 2. Transmission instants and intervals.

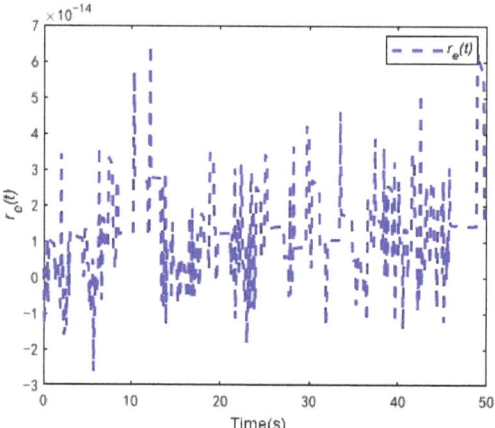

Figure 3. The trajectories of $r_e(t)$ without fault.

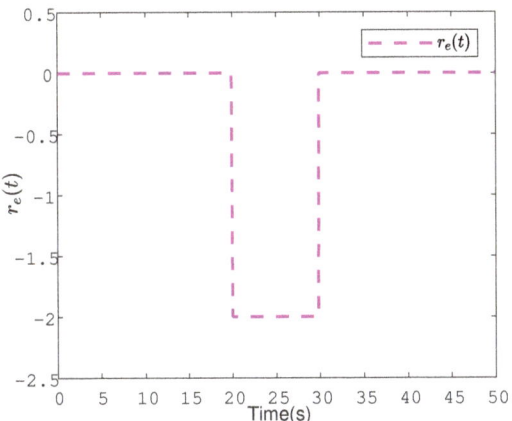

Figure 4. The trajectories of $r_e(t)$ with fault.

Moreover, the threshold J_{th} can be calculated without fault, i.e., $J_{th} = 4.0711 \times 10^{-13}$. Then, it is not hard to obtain that $J(t) = \left\{ \int_0^{24.9} r_F^T(s) r_F(s) ds \right\}^{\frac{1}{2}} = 4.0826 \times 10^{-13} > J_{th}$. This means that the fault can be detected after 4.9 s. Further, Figure 5 illustrates the fault detection results demonstrating that the proposed FD approach is effective.

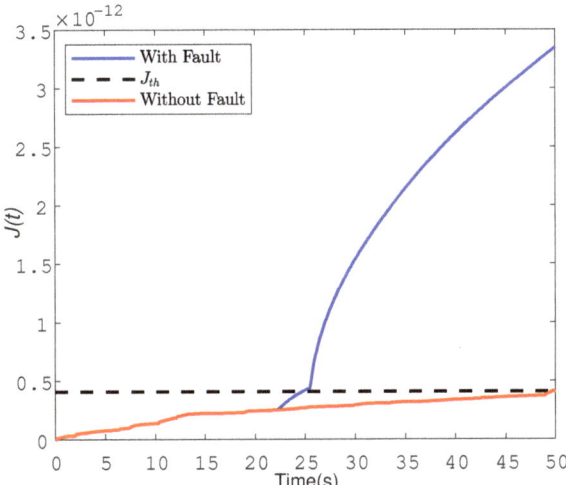

Figure 5. The trajectories of evaluation function with/without fault.

Following the above steps, considering the different types of faults, we performed three sets of simulations. Then, we produced Table 3 and derived Figures 6 and 7.

Table 3. Verification for different types of faults.

System Parameters	Fault Signal	Trigger Mechanism	Comparison of Trigger Rate (Triggering Times)		Comparison of Detection Time	
Exp a [44]	$f(t) = \begin{cases} 2\sin(t), & 30 < t < 60 \\ 0, & others \end{cases}$	cycle trigger	100% (1000)	23.9% (239)	0.5 s	0.3 s
Exp b [45]	$f(t) = \begin{cases} 1, & 1.5 < t < 2.3 \\ 0, & others \end{cases}$	adaptive Trigger	31% (31)	26% (26)	0.19 s	0.13 s
Exp c [47]	$f(t) = \begin{cases} 20\sin(t-2)(1-e^{\frac{-t+2}{4}}), & 10 < t < 30 \\ 0, & others \end{cases}$	cycle trigger	100% 3000	26.3% 789	*	0.6 s

* This is not explicitly stated in [47].

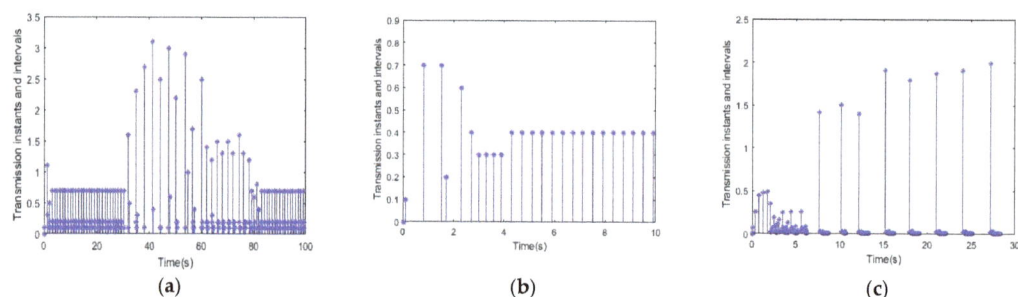

Figure 6. Transmission instants and intervals for experiment (a–c).

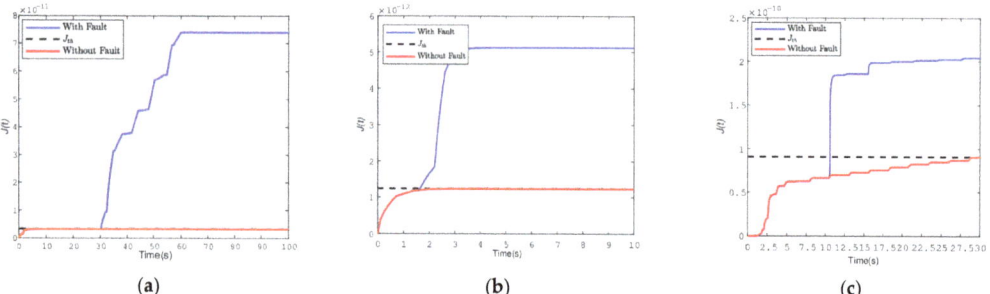

Figure 7. The trajectories of evaluation function with/without fault for experiment (a–c).

Experiment a uses the same system parameters and fault types as those the in the literature [44]. During the simulation time (100 s) with the sampling period (0.1 s), the cycle triggering time is 1000, and the events triggering time is 239. Simultaneously, the results show that the proposed method obtains a faster detection time. In experiment b, the step signal is used to represent the sudden fault. The final time is 10 s, and the sampling period is 0.1 s. With the same experimental conditions, the proposed method has fewer triggers and a faster detection speed. It can be seen that the structure of the event triggering mechanism we used is simpler. More recently, in order to discuss the effectiveness of the method for time-varying faults. Experiment c was performed by considering an inverted pendulum on a cart. It readjusts that the experimental time is 30 s and sampling period is

0.01 s, and only 26.3% of sampled data is transmitted over the wireless network. In Figure 7, one can see that the fault can be detected after 0.6 s.

5. Conclusions

The event-triggered FD problem of IT2 T-S fuzzy nonlinear networked systems has been studied in this paper. A fault residual system is established by integrating the IT2 fuzzy theory, external disturbance, event-triggered scheme, time delays and parameter uncertainties. In particular, the designed FD filter premise variable could be different from NNSs. The stability conditions and performance criterion have been proposed with the aid of the Lyapunov theory. At last, the validity has been verified by simulation experiments. The results illustrate that the proposed FD method can achieve rapid detection of faults, and the event-triggered scheme reduces the transmission rate and saves wireless communication resources. The responsiveness to different types of faults highlights its low conservativeness. The event-triggered FD problem of NNSs with random cyberattacks and packet losses will be further investigated.

Author Contributions: Conceptualization, Z.L., C.Z. and L.W.; methodology, Z.L., F.X. and L.W.; software, C.Z.; validation, Z.L., C.Z., F.X., Z.W. and L.W.; writing—review and editing, Z.L., C.Z., F.X., Z.W. and L.W.; visualization, Z.L.; supervision, L.W.; project administration, L.W. and Z.L.; funding acquisition, L.W., F.X. and Z.L. All authors have read and agreed to the published version of the manuscript.

Funding: This work was financially supported by the Heilongjiang Provincial Natural Science Foundation of China [No. LH2021F057], the Fundamental Research Funds in Heilongjiang Provincial Universities [No.135409602,135409102], the Science and Technology Project of State Grid Heilongjiang Electric Power Co., Ltd. [No.5224162000JK], and the Open project of Agricultural multidimensional sensor information Perception of Engineering and Technology Center in Heilongjiang Province [No. DWCGQKF202105].

Institutional Review Board Statement: Not applicable.

Informed Consent Statement: Not applicable.

Data Availability Statement: Not applicable.

Conflicts of Interest: The authors declare no conflict of interest.

References

1. Zhang, X.; Han, Q.; Ge, X.; Ding, D.; Ding, L.; Yue, D.; Peng, C. Networked Control Systems, A Survey of Trends and Techniques. *IEEE/CAA J. Autom. Sinica* **2020**, *7*, 1–17. [CrossRef]
2. Zhou, J.; Zhang, D. H∞ Fault Detection for Delta Operator Systems with Random Two-Channels Packet Losses and Limited Communication. *IEEE Access* **2019**, *7*, 94448–94459. [CrossRef]
3. Liu, Y.; Arunkumar, A.; Sakthivel, R.; Nithya, V.; Alsaadi, F. Finite-time Event-Triggered Non-fragile control and Fault Detection for Switched Networked Systems with Random Packet Losses. *J. Frankl. Inst.* **2019**, *357*, 11394–11420. [CrossRef]
4. Han, C.; Song, D.; Ran, G.; Yu, J. Event-Triggered Mixed Non-Fragile and Measurement Quantization Filtering Design for Interval Type-2 Fuzzy Systems. *IEEE Access* **2020**, *9*, 1533–1545. [CrossRef]
5. Wang, X.; Fei, Z.; Wang, Z.; Yu, J. Zonotopic fault detection observer design for discrete-time systems with adaptively adjusted event-triggered mechanism. *IET Control Theory Appl.* **2020**, *14*, 96–104. [CrossRef]
6. Ju, Y.; Tian, X.; Liu, H.; Ma, L. Fault detection of networked dynamical systems, a survey of trends and techniques. *Int. J. Syst. Sci.* **2021**, *52*, 3390–3409. [CrossRef]
7. Chen, H.; Liu, Z.; Alippi, C.; Huang, B.; Liu, D. Explainable Intelligent Fault Diagnosis for Nonlinear Dynamic Systems: From Unsupervised to Supervised Learning. *TechRxiv* **2022**. [CrossRef]
8. Sun, S.; Li, T.; Pang, Y.; Hua, X. Multiple delay-dependent event-triggered finite-time H∞ filtering for uncertain networked random systems against state and input constraints. *Appl. Math. Comput.* **2022**, *415*, 126711. [CrossRef]
9. Zhang, Z.; Wang, H.; Huang, M. Neural network-based event-triggered fault detection of discrete-time nonlinear uncertain systems. *J. Frankl. Inst.* **2020**, *357*, 4887–4900. [CrossRef]
10. Liu, M.; Yu, J.; Sun, Y.; Li, J. Adaptive event-triggered fault detection for Markovian jump systems with network time-delays. *Trans. Inst. Meas. Control* **2021**, *43*, 2934–2947. [CrossRef]
11. Lu, Z.; Ran, G.; Xu, F.; Lu, J. Novel mixed-triggered filter design for interval type-2 fuzzy nonlinear Markovian jump systems with randomly occurring packet dropouts. *Nonlinear Dyn.* **2019**, *97*, 1525–1540. [CrossRef]

12. Weidman, T. Comments on "Fuzzy-Model-Based Quantized Guaranteed Cost Control of Nonlinear Networked Systems". *IEEE Trans. Fuzzy Syst.* **2018**, *26*, 1086–1088. [CrossRef]
13. Guo, X.; Fan, X.; Wang, J.; Park, J. Event-triggered Switching-type Fault Detection and Isolation for Fuzzy Control Systems under DoS Attacks. *IEEE Trans. Fuzzy Syst.* **2020**, *29*, 3401–3414. [CrossRef]
14. Liu, X.; Su, X.; Shi, P.; Nguang, S.; Shen, C. Fault detection filtering for nonlinear switched systems via event-triggered communication approach. *Automatica* **2019**, *101*, 365–376. [CrossRef]
15. Qi, J.; Li, Y. Hybrid-triggered fault detection filter design for networked Takagi–Sugeno fuzzy systems subject to persistent heavy noise disturbance. *Int. J. Adapt. Control Signal Process.* **2021**, *35*, 1062–1082. [CrossRef]
16. Yi, X.; Li, G.; Liu, Y.; Fang, F. Event-triggered H∞ filtering for nonlinear networked control systems via T-S fuzzy model approach. *Neurocomoputing* **2021**, *448*, 344–352. [CrossRef]
17. Tan, Y.; Wang, K.; Su, X.; Xue, F.; Shi, P. Event-Triggered Fuzzy Filtering for Networked Systems with Application to Sensor Fault Detection. *IEEE Trans. Fuzzy Syst.* **2021**, *29*, 1409–1422. [CrossRef]
18. Ran, G.; Liu, J.; Li, C.; Chen, L.; Li, D. Event-Based Finite-Time Consensus Control of Second-Order Delayed Multi-Agent Systems. *IEEE Trans. Circuits Syst. II Express Briefs* **2021**, *68*, 276–280. [CrossRef]
19. Wan, X.; Han, T.; An, J.; Wu, M. Fault Diagnosis for Networked Switched Systems, An Improved Dynamic Event-Based Scheme. *IEEE Trans. Cybern.* **2021**, 1–12. [CrossRef]
20. Aslam, M.; Ullah, R.; Dai, X.; Sheng, A. Event-triggered scheme for fault detection and isolation of non-linear system with time-varying delay. *IET Control Theory Appl.* **2020**, *14*, 2429–2438. [CrossRef]
21. Chen, Z.; Bao, Y.; Ma, Q.; Zhang, Z. Event-Based Control for Networked T-S Fuzzy Systems via Auxiliary Random Series Approach. *IEEE Trans. Cybern.* **2020**, *50*, 2166–2175. [CrossRef]
22. Ran, G.; Liu, J.; Li, C.; Lam, H.; Li, D.; Chen, H. Fuzzy Model Based Asynchronous Fault Detection for Markov Jump Systems with Partially Unknown Transition Probabilities, An Adaptive Event-Triggered Approach. *IEEE Trans. Fuzzy Syst.* **2022**, 1–10. [CrossRef]
23. Li, R.; Yang, Y. Event-triggered fault detection for nonlinear descriptor networked control systems. *J. Frankl. Inst.* **2021**, *358*, 8715–8735. [CrossRef]
24. Huang, C.; Shen, B.; Zou, L.; Shen, Y. Event-Triggering State and Fault Estimation for a Class of Nonlinear Systems Subject to Sensor Saturations. *Sensors* **2021**, *21*, 1242. [CrossRef]
25. Chen, Z.; Zhang, B.; Zhang, Y.; Li, Y.; Zhang, Z. Event-triggered fault detection for T-S fuzzy systems subject to data losses. *International. J. Syst. Sci.* **2020**, *51*, 1162–1173. [CrossRef]
26. Liu, Q.; Long, Y.; Ju, H.; Li, T. Neural network-based event-triggered fault detection for nonlinear Markov jump system with frequency specifications. *Nonlinear Dyn.* **2021**, *103*, 1–17. [CrossRef]
27. Mishra, S.K.; Jha, A.V.; Verma, V.K.; Appasani, B.; Abdelaziz, A.Y.; Bizon, N. An Optimized Triggering Algorithm for Event-Triggered Control of Networked Control Systems. *Mathematics* **2021**, *9*, 1262. [CrossRef]
28. Ning, Z.; Wang, T.; Song, X.; Yu, J. Fault detection of nonlinear stochastic systems via a dynamic event-triggered strategy. *Signal Processing* **2019**, *167*, 107283. [CrossRef]
29. Xie, X.; Li, S.; Xu, B. Fault detection filter design for interval type-2 fuzzy systems under a novel adaptive event-triggering mechanism. *Int. J. Syst. Sci.* **2019**, *50*, 2510–2528. [CrossRef]
30. Yang, H.; Wang, X.; Park, J. Sampled-Data-Based Dissipative Stabilization of IT-2 TSFSs Via Fuzzy Adaptive Event-Triggered Protocol. *IEEE Trans. Cybern.* **2021**, *8*, 1–10. [CrossRef]
31. Guo, X.; Fan, X.; Ahn, C. Adaptive Event-Triggered Fault Detection for Interval Type-2 T-S Fuzzy Systems with Sensor Saturation. *IEEE Trans. Fuzzy Syst.* **2021**, *29*, 2310–2321. [CrossRef]
32. Xie, X.; Li, S.; Xu, B. Adaptive event-triggered H∞ fuzzy filtering for interval type-2 T-S fuzzy-model-based networked control systems with asynchronously and imperfectly matched membership functions. *J. Frankl. Inst.-Eng. Appl. Math.* **2019**, *356*, 11760–11791. [CrossRef]
33. Pan, Y.; Yang, G. Event-Driven Fault Detection for Discrete-Time Interval Type-2 Fuzzy Systems. *IEEE Trans. Syst. Man Cybern. Syst.* **2019**, *51*, 4959–4968. [CrossRef]
34. Ran, G.; Li, C.; Lam, H.; Li, D.; Han, C. Event-Based Dissipative Control of Interval Type-2 Fuzzy Markov Jump Systems Under Sensor Saturation and Actuator Nonlinearity. *IEEE Trans. Fuzzy Syst.* **2022**, *30*, 714–727. [CrossRef]
35. Rong, N.; Wang, Z. Event-Based Impulsive Control of IT2 T-S Fuzzy Interconnected System Under Deception Attacks. *Int. J. Fuzzy Syst.* **2021**, *29*, 1615–1628. [CrossRef]
36. Ran, G.; Liu, J.; Li, D.; Zhang, Y.; Huang, Y. An Event-Triggered H∞ Filter for Interval Type-2 T–S Fuzzy Nonlinear Networked Systems with Parameter Uncertainties and Delays. *Int. J. Fuzzy Syst.* **2021**, *23*, 2144–2156. [CrossRef]
37. Li, H.; Wu, L.; Lam, H.K.; Gao, Y. *Analysis and Synthesis for Interval Type-2 Fuzzy-Model-Based Systems*; Springer: Singapore, 2016.
38. Zhou, J.; Cao, J.; Chen, J.; Hu, A.; Zhang, J.; Hu, M. Dynamic Event-Triggered Predictive Control for Interval Type-2 Fuzzy Systems with Imperfect Premise Matching. *Entropy* **2021**, *23*, 1452. [CrossRef]
39. Ren, J.; Sun, J.; Fu, J. Finite-time event-triggered sliding mode control for one-sided Lipschitz nonlinear systems with uncertainties. *Nonlinear Dyn.* **2021**, *103*, 865–882. [CrossRef]
40. Lam, H.; Seneviratne, L. Stability analysis of interval type-2 fuzzy-model-based control systems. *IEEE Trans. Syst. Man Cybern. B Cybern.* **2008**, *38*, 617–628. [CrossRef]

41. Boyd, S.; Ghaoui, L.; Feron, E.; Balakrishnan, V. *Linear Matrix Inequalities in System and Control Theory*; SIAM: Philadelphia, PA, USA, 1994.
42. Petersen, I.R. A stabilization algorithm for a class of uncertain linear systems. *Syst. Control Lett.* **1987**, *8*, 351–357. [CrossRef]
43. Park, P.G.; Ko, J.W.; Jeong, C. Reciprocally convex approach to stability of systems with time-varying delays. *Automatica* **2011**, *47*, 1:235–238. [CrossRef]
44. He, Z. *Research on Fault Detection Methods of Networked Control Systems with Mixed Delays*; Northeast Petroleum University: Daqing, China, 2020; pp. 42–47.
45. Gu, Z.; Yue, D.; Park, J.H.; Xie, X. Memory-Event-Triggered Fault Detection of Networked IT2 T-S Fuzzy Systems. *IEEE Trans. Cybern.* **2022**, 1–10. [CrossRef] [PubMed]
46. Pan, Y.; Li, H.; Zhou, Q. Fault detection for interval type-2 fuzzy systems with sensor nonlinearities. *Neurocomputing* **2014**, *145*, 488–494. [CrossRef]
47. Huang, S.; Yang, G. Fault Tolerant Controller Design for T–S Fuzzy Systems with Time-Varying Delay and Actuator Faults: A K-Step Fault-Estimation Approach. *IEEE Trans. Fuzzy Syst.* **2014**, *22*, 1526–1540. [CrossRef]

Article

Health Assessment of Complex System Based on Evidential Reasoning Rule with Transformation Matrix

Zhigang Li [1], Zhijie Zhou [1], Jie Wang [1], Wei He [1,2,*] and Xiangyi Zhou [1]

1. High-Tech Institute of Xi'an, Xi'an 710025, China; analyseli@foxmail.com (Z.L.); zhouzj04@tsinghua.org.cn (Z.Z.); wjie19950921@163.com (J.W.); zhouxiangyi1995@163.com (X.Z.)
2. School of Computer Science and Information Engineering, Harbin Normal University, Harbin 150025, China
* Correspondence: hewei@hrbnu.edu.cn

Abstract: In current research of complex system health assessment with evidential reasoning (ER) rule, the relationship between the indicators reference grades and pre-defined assessment result grades is regarded as a one to one correspondence. However, in engineering practice, this strict mapping relationship is difficult to meet, and it may degrade the accuracy of the assessment. Therefore, a new ER rule-based health assessment model for a complex system with a transformation matrix is adopted. First, on the basis of the rule-based transformation technique, expert knowledge is embedded on the transformation matrix to solve the inconsistent problems between the input and the output, which keeps completeness and consistency of information transformation. Second, a complete health assessment model is established via the calculation and optimization of the model parameters. Finally, the effectiveness of the proposed model can be validated in contrast with other methods.

Keywords: evidential reasoning rule; system modelling; information transformation; parameter optimization

Citation: Li, Z.; Zhou, Z.; Wang, J.; He, W.; Zhou, X. Health Assessment of Complex System Based on Evidential Reasoning Rule with Transformation Matrix. *Machines* **2022**, *10*, 250. https://doi.org/10.3390/machines10040250

Academic Editor: Davide Astolfi

Received: 27 January 2022
Accepted: 24 March 2022
Published: 31 March 2022

Publisher's Note: MDPI stays neutral with regard to jurisdictional claims in published maps and institutional affiliations.

Copyright: © 2022 by the authors. Licensee MDPI, Basel, Switzerland. This article is an open access article distributed under the terms and conditions of the Creative Commons Attribution (CC BY) license (https://creativecommons.org/licenses/by/4.0/).

1. Introduction

A complex system, for instance, control system [1], servo system [2], energy storage system [3], is widely used in aviation, aerospace, electronics and other fields. Due to the complex structure and poor working environment, the system performance can be degraded, which affects the operation reliability of the system. Therefore, it is crucial to assess the health status of the complex system to provide decisions for management and maintenance [4].

In the current research of health assessment, there are mainly three methods called the data-based method, the qualitative knowledge-based method, and the semi-quantitative information-based method. The data-based methods assess the system performance by fitting the nonlinear relationship between the input and output of the system based on observation data, such as deep learning, neural network [5–7]. Since it is a pure black-box modelling, the assessment results cannot be explained, and there is a problem of overfitting. The qualitative knowledge-based methods provide interpretable assessment progress based on the operation mechanism of the system and expert knowledge, for example, fuzzy reasoning, belief rule base [8,9]. Due to the subjectivity of expert knowledge, the model assessment accuracy is poor. The semi-quantitative information-based methods provide both qualitative knowledge and quantitative data concurrently, providing interpretable and accurate assessment results [10]. Therefore, the health assessment based on semi-quantitative information is basically concentrated in this paper.

The evidential reasoning (ER) rule [11], as a representative semi-quantitative information-based method, originated from the Dempster-Shafer (DS) evidence theory [12], and is regard as a generalized Bayesian inference process [11]. DS evidence theory is regarded as a special case of ER, when the indicator reliability is equal to 1 [13]. In the ER rule, the quantitative data and qualitative knowledge can be effectively integrated by adopting the orthogonal operations.

Reference value is introduced to divide the input information status, then the initial evidence can be generated. To deal with the data uncertainty, the evidence weight and reliability are introduced. Particularly, the weight reflects the relative importance of multiple pieces of evidence in the aggregation of evidence. The reliability reflects the ability of the information sources to provide correct information. The reliability is influenced by the performance of the information source and external noise [14]. By clearly differentiating the two concepts, the ER rule is widely used in many fields, such as multi-attribute decision-making [15], fault diagnosis [16], health assessment [17], etc.

When using the ER rule to assess the health status of a complex system, a set of mutually exclusive and collectively exhaustive assessment result grades need to be settled in advance. First, health assessment indicators are selected, and the indicators are equivalent to evidence. Then, input indicator reference grades are introduced to conveniently collect the initial evidence pointed to the assessment grades. Finally, the initial evidence, evidence weight and reliability are integrated based on the ER rule, and the health assessment results of the complex system can be obtained. Therefore, as an important part of the assessment, indicator reference grades determine the belief distribution of initial evidence, which directly affected the assessment results.

The above assessment process determines that the input indicator reference grades and output assessment result grades strictly correspond to each other. However, in practice, the assessment result grades are determined in advance, which leads to the disaccord with input indicator reference grades. For example, the input indicator reference grades can be easily divided into "normal" and "fault" based on industry-standard, but health assessment result grades are predetermined as "health", "subhealth", "fault".

In the process of health assessment, the relationship between input indicator references grades and assessment result grades does not exactly correspond to each other. Therefore, for the sake of dealing with this problem, there are two methods to solve it. First, regarding the relationship as a one-to-one correspondence [18,19], then the input indicator references grades can be matched with assessment result grades. Second, based on expert knowledge, adding the reference grades to realize an input and output in accordance [4,8]. However, the first method neglects the consistent relationship in engineering practice, and the accuracy of the assessment results is influenced. The second method violates the prior mapping relationship, resulting in randomness and no standard of the assessment result.

To deal with the mentioned issues, a new health assessment model for a complex system based on the ER rule is proposed. First, the transformation matrix is determined according to the expert's knowledge. The input information can be converted into the initial belief distribution with regard to assessment result grades by using the rule-based information transformation technique. Thus, a general information transformation framework is constructed. Second, the evidence weight and reliability are determined by expert knowledge and the synthesis of static and dynamic characteristics separately. Then, the health assessment model of a complex system based on ER rule is constituted. Finally, to further enhance the model precision, an objective function is established to optimize the model parameters. In this paper, there are two innovations as follows:

(1) Based on transformation matrix, the mapping relationship between the antecedents and the consequent of the assessment model is established, which solves the inconsistent problem between the indicators reference grades and pre-defined assessment result grades in the engineering practice. Due to the subjectivity and limitations of the expert's knowledge, the initial values of transformation matrix may deviate from real status, hence the need to build a optimization model to further optimize the values of transformation matrix.

(2) On the basis of parameters calculation, the optimization algorithm is employed to enhance the assessment result accuracy. Then, a complete health assessment model for complex system is constituted.

This paper is organized as follows. The framework and related problems of the health assessment model are described in Section 2. In Section 3, the health assessment model based on ER for a complex system is proposed. The optimization of model parameters is presented in Section 4. In Section 5, the bus of control system and the engines are taken as examples to validate the effectiveness of the proposed model. Conclusions and future work are defined in Section 6.

2. Problem Formulation

The status of a complex system is mainly reflected by some indicators, called health status indicators. The observation information of these indicators can be obtained by placing the corresponding sensors or simulating them in the computers. Here, the health assessment of the complex system model is constructed as shown in Figure 1.

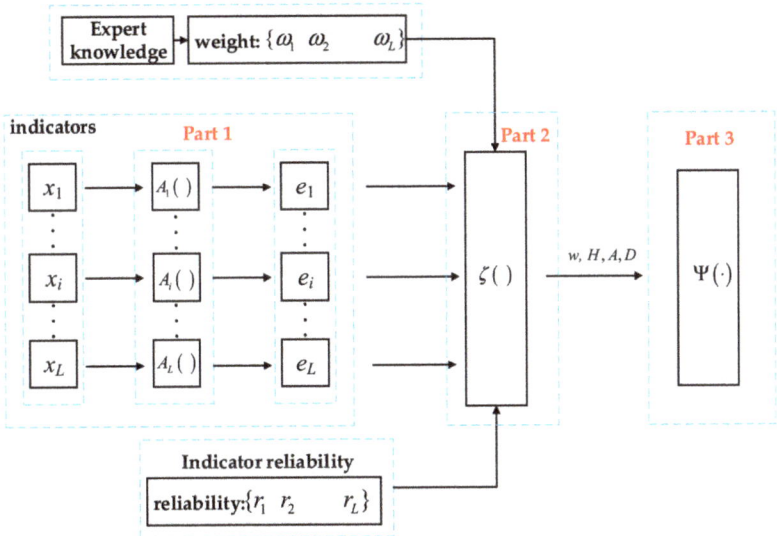

Figure 1. The structure of the health assessment model.

It can be seen from Figure 1 that the model mainly includes three parts: the first part establishes the mapping function to transform the input information into the initial evidence. The second part constitutes a complete assessment framework based on the calculation of parameters. Finally, the assessment model parameters need to be optimized in the third part.

The specific parameters of Figure 1 are as follows:

(1) x_i denotes the ith health status indicator of the complex system, where $i = 1, 2, \ldots, L$;

(2) L denotes the number of assessment indicators;

(3) A_i denotes the mapping function between the ith input indicator reference grades and assessment grades;

(4) w_i denotes the weight of the ith indicator;

(5) r_i denotes the reliability of the ith indicator;

(6) e_i denotes the initial evidence of the ith indicator.

According to the model established in Figure 1, the following two problems need to be solved in the health assessment of complex system:

(1) When assessing the health status of a complex system, the input indicators reference grades do not correspond to the assessment results grades. Therefore, Formula (1) is mainly to establish the following mapping relationship.

$$(D_1, D_2, \ldots, D_N) = A_i(H_{1,i}, H_2, \ldots, H_{N_i}) \qquad (1)$$

where $\{D_n | n = 1, 2, \ldots, N\}$ denotes N assessment result grades, $\{H_{n,i} | n = 1, 2, \ldots, N_i\}$ denotes ith input indicators N_i reference grades.

(2) The assessment model part parameters, such as indicator reference value and weight, are given by experts, which may decrease the accuracy of the assessment. Therefore, it is necessary to build an optimization model to improve the accuracy of assessment results as follows.

$$\Psi = \Psi(w, H, A, D) \qquad (2)$$

where w, H, A, D denote the indicator weight, indicator reference, transformation matrix, and assessment result grades respectively.

3. Health Assessment Method Based on the ER Rule with a Transformation Matrix

In this section, an assessment model with a transformation matrix based on the ER is adopted. The transformation of input information is conducted in Section 3.1. The calculation of model parameters is introduced in Section 3.2. The aggregation of indicators is given in Section 3.3.

3.1. Transformation Method of Input Indicators

First, it is necessary to establish an indicator system of health assessment, when assessing a complex system. There are N assessment result grades, L indicators and the numbers of ith input indicator reference grades are denoted by N_i, as shown in the Figure 2.

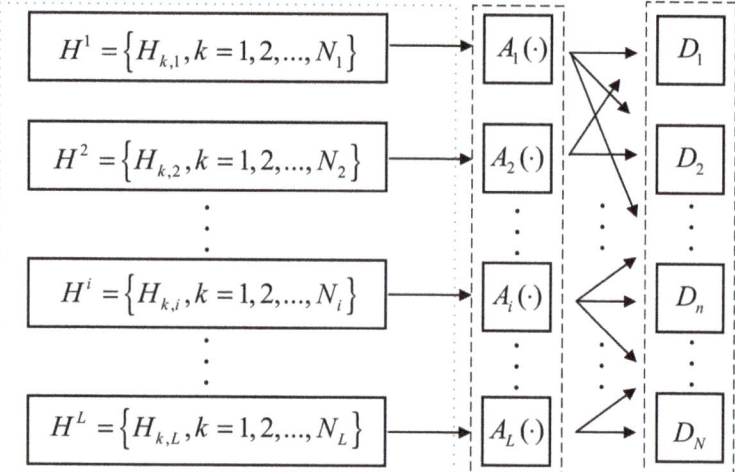

Figure 2. The transformation between the input and output.

Suppose $H^i = \{H_{1,i}, H_{2,i}, \ldots, H_{N_i,i}\}$ and $D = \{D_1, D_2, \ldots, D_N\}$ are sets of mutually exclusive and exhaustive propositions. Thus H^i and D are regarded as frames of discernment, called the discernment frame 1 and the discernment frame 2, respectively. In order to realize the transformation from discernment frame 1 to discernment frame 2, there are process of transformation as follows:

First, the mapping relationship between the kth reference grade of the ith indicator $H_{k,i}$ and assessment result grades $\{D_1, \ldots, D_N\}$ can be described as a "if-then" rule:

$$R_{k,i} : \text{if } x_i = H_{k,i}, \text{then} \{(D_1, a_{1,k}), \ldots, (D_n, a_{n,k}), \ldots, (D_N, a_{N,k})\}, \left(\sum_{n=1}^{N} a_{n,i} = 1, 0 \leq a_{n,i} \leq 1,\right) \qquad (3)$$

where $a_{n,k}$ denotes the belief degree to which D_n is regard as the consequent if, input x_i is $H_{k,i}$. $R_{k,i}$ denotes kth rule of the ith indicator. Then, the mapping relationship between the discernment frame 1 and the discernment frame 2 can be established by N_i rules. It can be described as a matrix:

$$A_i = \begin{array}{c} \\ D_1 \\ D_2 \\ \vdots \\ D_n \\ \vdots \\ D_N \end{array} \begin{array}{c} H_{1,i} \; H_{2,i} \; \cdots \; H_{k,i} \; \cdots \; H_{N_i,i} \\ \left[\begin{array}{ccccc} a_{1,1} & a_{2,2} & \cdots & a_{1,k} & \cdots & a_{1,N_i} \\ a_{2,1} & a_{2,2} & \cdots & a_{2,k} & \cdots & a_{2,N_i} \\ \vdots & \vdots & \ddots & \vdots & & \vdots \\ a_{n,1} & a_{n,2} & \cdots & a_{n,k} & \cdots & a_{n,N_i} \\ \vdots & \vdots & & \vdots & \ddots & \vdots \\ a_{N,1} & a_{N,2} & \cdots & a_{N,k} & \cdots & a_{N,N_i} \end{array} \right] \end{array} \quad (4)$$

where A_i denotes $N \times N_i$ transformation matrix, whose N_i columns are the N_i rules.

Remark 1. *The transformation matrix is established based on Formula (3), where the belief degree is allocated to any individual assessment grades and there is no ignorance left. It can be proved that transformation matrix retains the integrity and consistency of information transformation. The details of proof can be seen in the paper [20]. In other words, a belief distribution with no ignorance will not be transformed to a belief distribution with ignorance, and vice versa.*

Second, according to rule-based information transformation technique, the input information can be transformed as a belief distribution under discernment frame 1 as follows.

$$S^i(x_i^*) = \{(H_{k,i}, \gamma_{k,i}), \; k=1,2,\ldots,N_i; \; (H_\Theta, \gamma_{\Theta,i})\} \quad (5)$$

with $0 \leq \gamma_{n,i} \leq 1$ $(n = 1,\ldots,N_i, \; i=1,\ldots,L)$, where x_i^* denotes input information of the ith indicator. $H_{k,i}$ denotes the kth reference grade of the discernment frame 1, $\gamma_{k,i}$ denotes the belief degree allocated to any individual reference grade of discernment frame 1, which can be calculated as follows.

$$\begin{cases} \gamma_{k,i} = \frac{H_{k+1,i} - x_i^*}{H_{k+1,i} - H_{k,i}}, \; H_{k,i} \leq x^* \leq H_{k+1,i} \\ \gamma_{k+1,i} = 1 - \gamma_{k,i}, \; H_{k,i} \leq x^* \leq H_{k+1,i} \\ \gamma_{m,i} = 0, \; m=1,\ldots,N_i, m \neq k, k+1 \end{cases} \quad (6)$$

where $H_{k,i}$ and $H_{k+1,i}$ denote the reference values of two adjacent input indicators reference grades. If there are other information transformation techniques or qualitative indicators, the degree of global ignorance denoted by $\gamma_{\Theta,i}$ may exist.

Finally, based on transformation matrix A_i, the input information of ith indicator can be transformed as a belief distribution under discernment frame 2, as follows:

$$\widetilde{S}^i(x_i^*) = \{(D_{n,i}, \beta_{n,i}), \; n=1,2,\ldots,N; \; (D_\Theta, \beta_{\Theta,i})\} \quad (7)$$

with $0 \leq \beta_{n,i} \leq 1$ $(n=1,\ldots,N, \; i=1,\ldots,L)$, $\beta_{\Theta,i} = 1 - \sum_{n=1}^{N} \beta_{n,i}$, where $\beta_{n,i}$ and $\beta_{\Theta,i}$ denote belief degree allocated to nth individual assessment result grades and global ignorance, respectively, which can be calculated as follows:

$$b_i = A_i \times r_i \quad (8)$$

$$\beta_{\Theta,i} = 1 - \sum_{n=1}^{N} \beta_{n,i} = \gamma_{\Theta,i} \quad (9)$$

where, $b_i = [\beta_{1,i}, \beta_{2,i}, \ldots, \beta_{N,i}]$ is the belief degree under the discernment framework 2, $r_i = [\gamma_{1,i}, \gamma_{2,i}, \ldots, \gamma_{N,i}]$ is the belief degree under the discernment framework 1, A_i denotes the transformation matrix corresponding to the ith indicator.

Remark 2. *Compared with Yang's work [20], there are two contributions of this work. (1) In Yang's work, the elements of transformation matrix are only determined by the decision-makers' knowledge and experience, which may decrease the assessment accuracy. In the proposed model, the expert knowledge is used to give the initial values of the transformation matrix, and the accurate values are obtained by optimizing based on the observation data. (2) Actually, the transformation matrix makes the adjustment between different discernment frameworks realized. More importantly, this paper inherits the basic work of Yang and extends it to the field of refined health assessment.*

3.2. Calculation of Model Parameters

The indicator weight is the subjective concept that reflects the relative importance among the indicators [11,21]. Thus, the indicator weight is determined by the experts' preference to the assessment results grades. Differently, the indicator reliability is the objective concept, affected by inherent disturbance or noise when measured, resulting in the unreliability of observation data [22]. Therefore, the method that the synthesis of static and dynamic reliability is adopted, can effectively combine the expert knowledge and observation data [23].

Suppose r_i^s and r_i^d denote the statics reliability and dynamic reliability respectively. Then the indicator reliability r_i is determined as

$$r_i = \delta r_i^s + (1-\delta) r_i^d, 0 \leq \delta \leq 1 \tag{10}$$

where, δ denotes the weighting factor given by experts. r_i^s can be determined by expert experience and industry standards. r_i^d can be calculated via the method of distance, as follows.

First, the average of the ith indicator observation data is:

$$\overline{x}_i = \frac{1}{k_i} \sum_{t=1}^{k_i} x_i(k), k = 1, 2, \cdots, k_i \tag{11}$$

The distance between the ith indicator observation data and average can be expressed as:

$$d_i(x_i(k), \overline{x}_i) = |x_i(k) - \overline{x}_i| \tag{12}$$

Then, the average distance can be calculated as:

$$\overline{D}_i = \frac{1}{k_i} \sum_{k=1}^{k_i} |x_i(k) - \overline{x}_i| \tag{13}$$

The dynamic reliability is represented as:

$$r_i^d = \frac{\overline{D}_i}{\max d_i(x_i(k), \overline{x}_i)} \tag{14}$$

Remark 3. *On the one hand, the weights reflect the relative importance of indicators in the evidence aggregation process. Further, the value of the weight is strongly dependent on the decision maker. Thus, the weights can be adjusted according to actual needs. On the other hand, since the expert knowledge is limited, the initial values of the weight given by the expert may not be accurate. Thus, the weight needs to be optimized based on observation data. However, the reliability is an objective attribute of evidence, so it does not need to be optimized.*

3.3. Aggregation of Initial Evidence

Once the mapping relationship from input indicator grades to assessment grades is established based on transformation matrixes, the indicator observation data can be converted into initial evidence in the form of belief degree. The indicator weight is defined by the expert, and the indicator reliability is calculated by the above method in Section 3.2. Then, multiple indicators can be aggregated by using ER rule to obtain the health assessment results as follows:

$$\beta_{n,e(L)} = \frac{\mu\left[\prod_{i=1}^{L}\left(\tilde{\omega}_i\beta_{n,i} + 1 - \tilde{\omega}_i\sum_{n=1}^{N}\beta_{n,i}\right) - \prod_{i=1}^{L}\left(1 - \tilde{\omega}_i\sum_{n=1}^{N}\beta_{n,i}\right)\right]}{1 - \mu\prod_{i=1}^{L}(1-\tilde{\omega}_i)} \quad (15)$$

$$\beta_{\Theta,e(L)} = \frac{\mu\left[\prod_{i=1}^{L}\left(1 - \tilde{\omega}_i\sum_{n=1}^{N}\beta_{n,i}\right) - \prod_{i=1}^{L}(1-\tilde{\omega}_i)\right]}{1 - \mu\prod_{i=1}^{L}(1-\tilde{\omega}_i)} \quad (16)$$

$$\mu = \left[\sum_{n=1}^{N}\prod_{i=1}^{L}\left(\tilde{\omega}_i\beta_{n,i} + 1 - \tilde{\omega}_i\sum_{n=1}^{N}\beta_{n,i}\right) - (N-1)\times\prod_{i=1}^{L}\left(1-\tilde{\omega}_i\sum_{n=1}^{N}\beta_{n,i}\right)\right]^{-1} \quad (17)$$

$$\tilde{\omega}_i = \omega_i/(1 - \omega_i - r_i) \quad (18)$$

where, L denotes the number of evidence; N denotes the number of assessment grades; $\tilde{\omega}_i$ denotes the mixed weight considering the reliability and weight of evidence; $\beta_{n,i}$ denotes the initial belief degree allocated to assessment grades. $\beta_{n,e(L)}$ denotes the belief degree of the assessment result D_n. The residual support is allocated to the assessment framework, denoted by $\beta_{\Theta,e(L)}$.

The aggregated belief distribution can be expressed as follows.

$$O = \left\{(D_n, \beta_{n,e(L)}), (D_\Theta, \beta_{\Theta,e(L)}), n = 1, 2, \ldots, N\right\} \quad (19)$$

In practical application, to obtain numerical output, the belief distribution of aggregated results can be transformed into the expected utility. Assuming that the expected utility values $u(D_n)$ of all assessment grades are determined. If the aggregated belief distribution is complete ($\beta_{\Theta,e(L)} = 0$), then the expected utility of aggregated assessment result can be expressed as:

$$y = \sum_{n=1}^{N}\beta_{n,e(L)}u(D_n) \quad (20)$$

If the aggregated belief distribution is incomplete ($\beta_{\Theta,e(L)} \neq 0$), the global ignorance can be allocated to any assessment grades. The maximum, minimum, and average of the expected utility of aggregated assessment result can be expressed as follows:

$$y_{\max} = \sum_{n=1}^{N-1}\beta_{n,e(L)}u(D_n) + (\beta_{\Theta,e(L)} + \beta_{N,e(L)})u(D_N) \quad (21)$$

where $y_{\max}$ denotes maximum of the expected utility of aggregated assessment result, when $\beta_{\Theta,e(L)}$ is allocated to the most preferred assessment grades D_n.

$$y_{\min} = (\beta_{\Theta,e(L)} + \beta_{1,e(L)})u(D_1) + \sum_{n=2}^{N}\beta_{n,e(L)}u(D_n) \quad (22)$$

where, y_{min} denotes minimum of the expected utility of aggregated assessment result, when $\beta_{\Theta,\ell(L)}$ is allocated to the least preferred assessment grades D_1.

$$y_{average} = \frac{y_{max} + y_{min}}{2} \tag{23}$$

where $y_{average}$ denotes the average of the expected utility of aggregated assessment result. $\{u(D_n), |n = 1, 2, \ldots, N\}$ cannot be given accurately, which needs to be adjusted by the optimization algorithm.

4. Parameters Optimization

In this section, the optimal model is constructed to solve the second problem. The optimization of model parameters is conducted in Section 4.1. The detailed implementation process of the whole model is introduced in Section 4.2.

4.1. Optimization of Model Parameters

Due to the initial values of the evidence weight, indicator reference grades, expected utility, and transformation matrices in the assessment model are given by experts. Thus, to obtained accurate assessment results, these parameters need to be optimized based on the observation data. The optimization process is shown as Figure 3.

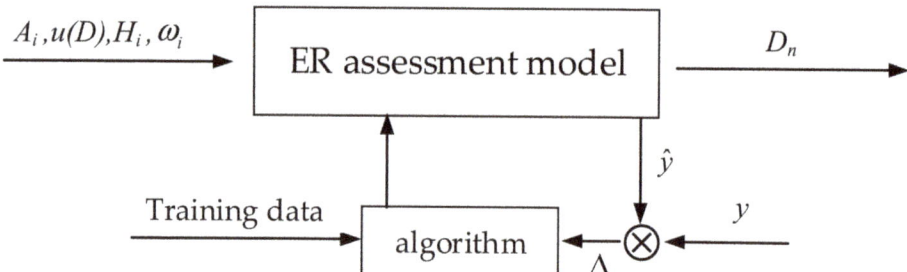

Figure 3. Optimization process of model parameters.

It should be noted that the assessment of true value of overall health is set based on experts' overall judgment in prior. According to the observation data, combined with the method of expert scoring, expert panels are set to determine the health status of the research object. The optimization objective function of the health status model is established as follows.

$$\min. \text{RMSE}(\Psi) = \sqrt{(y - \hat{y})^2} \tag{24}$$

where, y denotes the real health condition of the complex system, $\hat{y}$ denotes the assessment model output, $\Psi = \{H_{1,i}, \ldots, H_{N_i,i}, \omega_1, \ldots, \omega_l, A_1, \ldots, A_l, u(D_1), \ldots, u(D_N)\}$ is the parameter in the optimal model, and RMSE denotes the root mean square error, which is used to measure the difference between the model output and the actual output.

To ensure the accuracy of the assessment results without changing the physical meaning of the optimization parameters, the optimization range of parameters is designed according to expert knowledge, as follows.

$$b_{k,i} < H_{k,i} < c_{k,i} \tag{25}$$

$$d_{j,k} \leq a_{j,k} \leq f_{j,k}, \sum_j a_{j,k} = 1 \tag{26}$$

$$e_i < \omega_i < g_i \tag{27}$$

$$p_j < u(D_j) < q_j \quad j = 1, 2, \ldots, N, k = 1, 2, \ldots, N_i, i = 1, 2, \ldots, L \tag{28}$$

where, $H_{n,i}$ denotes the indicator reference of the ith N_i indicator; $c_{n,i}$ and $b_{n,i}$ denote the indicator reference upper and lower bounds; $a_{j,k}$ denotes the elements in row j and column k of the transformation matrix A_i; $d_{j,k}$ and $f_{j,k}$ denote respectively the lower and upper bounds; ω_i denotes evidence weight; f_i and d_i denote the weight upper and lower bounds; $u(D_n)$ denotes the utility of the nth assessment grade; e_n and g_n denote the assessment grade upper and lower bounds.

4.2. Process of Health Assessment Based on the ER Rule

The specific steps of health assessment using the ER assessment model proposed are as follows, shown in Figure 4.

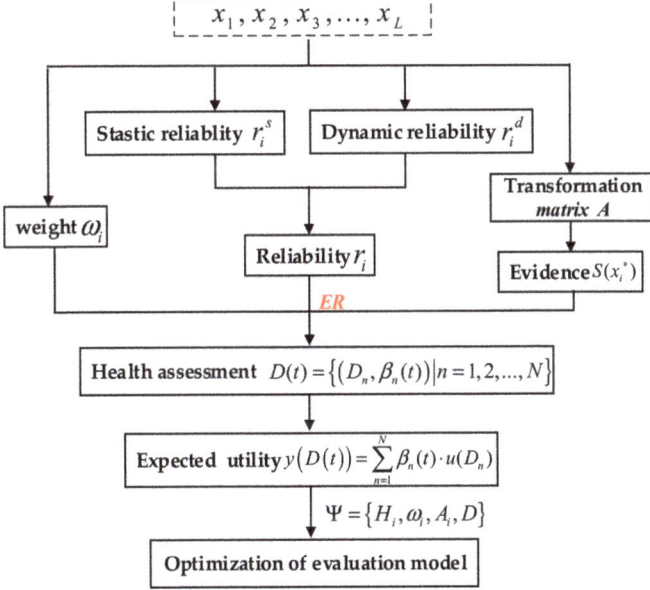

Figure 4. The implementation process of the assessment model.

Step 1: The health assessment indicator system of a complex system is established based on expert knowledge and observation data.

Step 2: Transformation matrixes are determined, then input information can be transformed into the form of initial evidence.

Step 3: The evidence weight and static reliability are given according to industry standards and expert knowledge, and the dynamic reliability is calculated based on the observation data. Then, the reliability is determined by the weighting of static reliability and dynamic reliability.

Step 4: The ER rule is employed to aggregate the initial evidence, evidence weight, and reliability, to obtain the health assessment results. The expected utility $u(D_n)$ of the assessment result grades is introduced to obtain the expected utility of the assessment result.

Step 5: The optimization of the assessment model is constituted to improve the accuracy of assessment results.

5. Experimental Research

In this section, bus of control system and engine are taken as examples to illustrate the validity of the proposed model. The health assessment of bus of control system is introduced in Section 5.1. In Section 5.2, the health assessment of engine is conducted. The result analysis is presented in Section 5.3.

5.1. Example 1—Health Assessment of the Bus of Control System

5.1.1. Background Description

The bus of control system, controlling the transmission of the test data and control instruction between the bus control (BC) and received terminal (RT), is wildly applied in rocket, missile, and other aerospace fields [24]. With the demand for rapid information transmission rate and large bandwidth, optical fiber communication technology is largely used in the bus of control system. To demonstrate the effectiveness of the proposed model, a type of the bus of control system based on a passive optical network (PON) is taken as an example. Passive optical networks are named as containing a large number of optical passive devices, such as optical fiber, optical fiber connector, and optical splitter. Because the passive optical devices in PON can be easily influenced by severe operation environment, the health status of the bus of control system can be degraded, resulting in the degradation of communication quality. Therefore, it is crucial to assess the health status of the bus of control system.

In this experiment, due to the shortage of the test data, the topology of the bus of control system is simulated based on the Optisystem software shown as Figure 5. According to the real status and fault mode analysis of the bus of control system, the different degrees of fault of the bus of control system is simulated in the simulation model. The q factor (Q) of eye diagram and received optical power (O) are selected as health status indicators [25]. The eye diagram is used to measure the signal-to-noise ratio of the signal, and q factor is one of the important parameters of the eye diagram [26]. The received optical power denotes the optical power at the optical receiver. When the received optical power is lower than the minimum received optical power of the optical receiver, the optical signal cannot be transmitted.

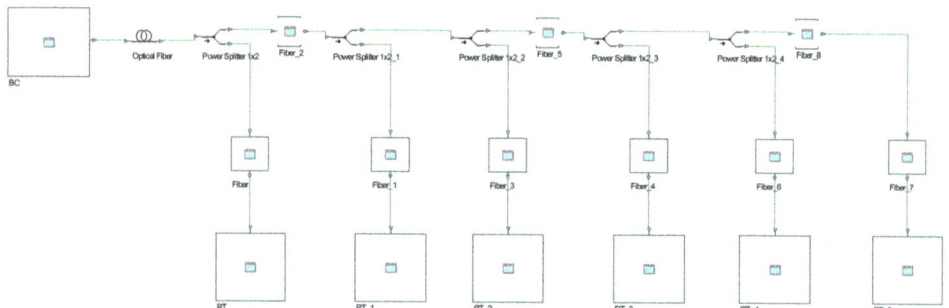

Figure 5. Simulation model of bus of control system.

As shown in Figure 6, the value of O ranges from −22.65 dBm to −17.84 dBm while Q ranges from 2.81 to 7.53. Both the curves of O and Q descended from high to the low. It can be seen in Figure 7 that the health status grades are denoted by y-axis. Meanwhile, as the failure degree of the bus of control system increases gradually, its health status can be concluded as four stage in the sequence. It is "Health" at first, followed by "Subhealth" and "Slight fault", finally "Severe fault".

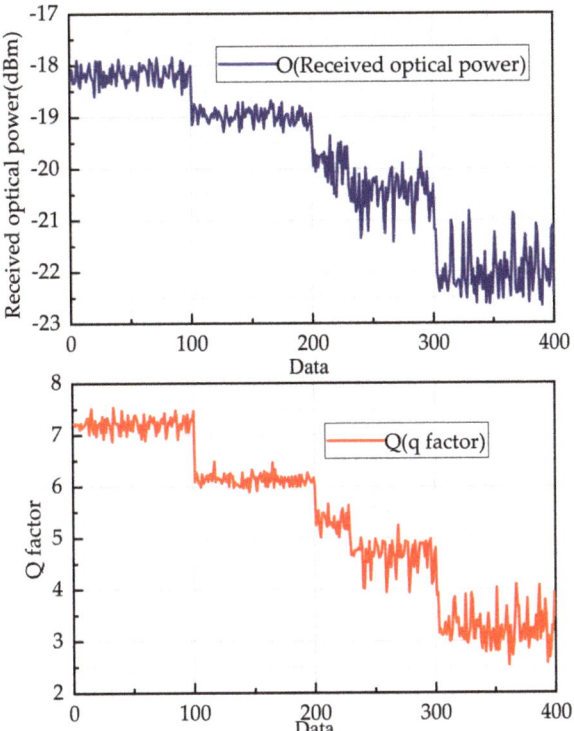

Figure 6. Observation data of O and Q.

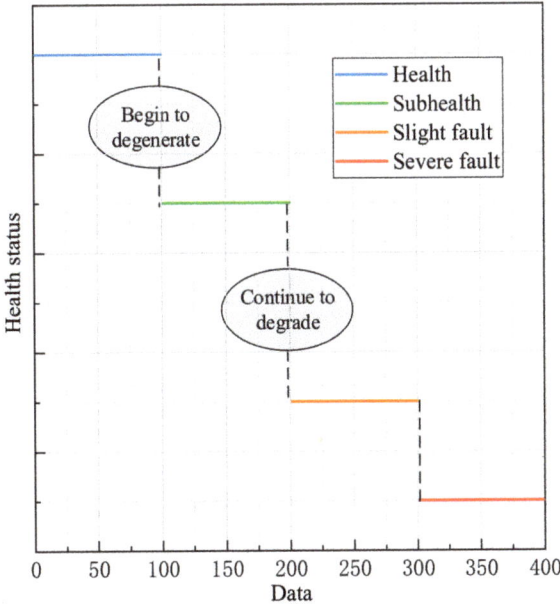

Figure 7. The health status of the bus of control system.

5.1.2. The Procedures of Health Assessment

In this subsection, the implementation process of the proposed model is conducted as the following steps:

Step1: the transformation of input information

According to the real health status of the bus of control system, assessment result grades can be defined as four parts as $D = \{D_1, D_2, D_3, D_4\} = \{Health, Subhealth, Slight\ fault, Severe\ fault\}$. However, because "subhealth" has a vague and random status, which is deduced by conjunction of multiple indicators, its reference value cannot be found in the individual indicator, resulting in the disaccord between the input and output grades. Thus, the input indicator reference grades are introduced as three parts as $H = \{H_1, H_2, H_3\} = \{High, Medium, Low\} = \{H, M, L\}$. The reference values corresponding the reference grades are determined based on experts shown in Table 1.

Table 1. The inference values of input indicators.

Indicators	H	M	L
Received optical power	[−18.5, −17]	[−21.5, −19.5]	[−26, −21.5]
Q factor	[8, 10]	[3, 6]	[0, 3]

Remark 4. *In Table 1, expert gives the intervals of reference values corresponding to reference grades, and the initial reference values are selected from the intervals. The reference values need to be optimized within intervals.*

Once the antecedents and consequent parameters of the rule are determined, the transformation matrixes can be constructed in Table 2 based on Formula (3). It should be noted that there is no ignorance in the transformation matrix.

Table 2. The parameters of transformation matrixes.

No.	Indicators	Reference Grades	$\{D_1, D_2, D_3, D_4\}$
1	Received optical power	−17.5	(0.8, 0.15, 0.05, 0)
2		−20	(0.05, 0.15, 0.5, 0.3)
3		−26	(0.05, 0.15, 0.2, 0.6)
4	Q factor	8	(0.8, 0.15, 0.05, 0)
5		4	(0.05, 0.05, 0.7, 0.2)
6		1	(0, 0, 0.2, 0.8)

Based on Table 2, the values of transformation matrixes A_1 and A_2 can be introduced as follows.

$$A_1 = \begin{bmatrix} 0.8 & 0.05 & 0.05 \\ 0.15 & 0.15 & 0.15 \\ 0.05 & 0.5 & 0.2 \\ 0 & 0.3 & 0.6 \end{bmatrix}, A_2 = \begin{bmatrix} 0.8 & 0.05 & 0 \\ 0.15 & 0.05 & 0 \\ 0.05 & 0.7 & 0.2 \\ 0 & 0.2 & 0.8 \end{bmatrix} \tag{29}$$

Based on Formulas (5)–(9), the input information can be translated into initial evidence as Figures 8 and 9.

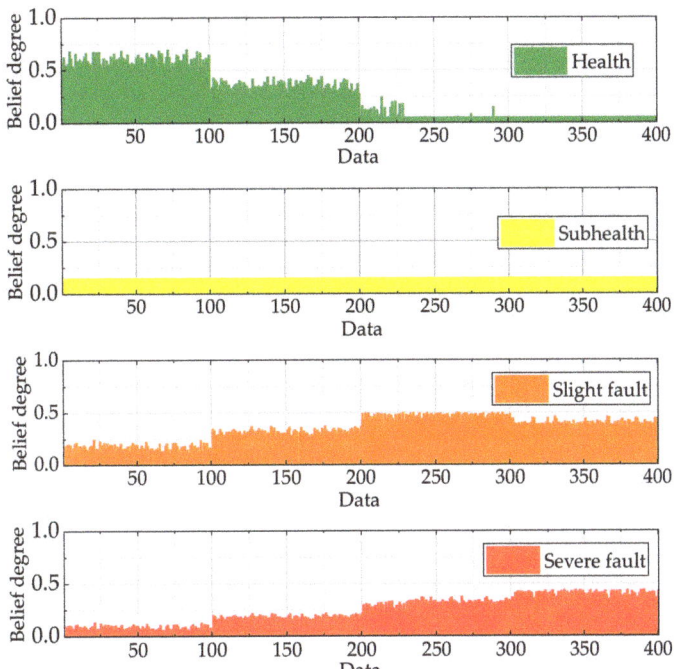

Figure 8. Belief distribution of O.

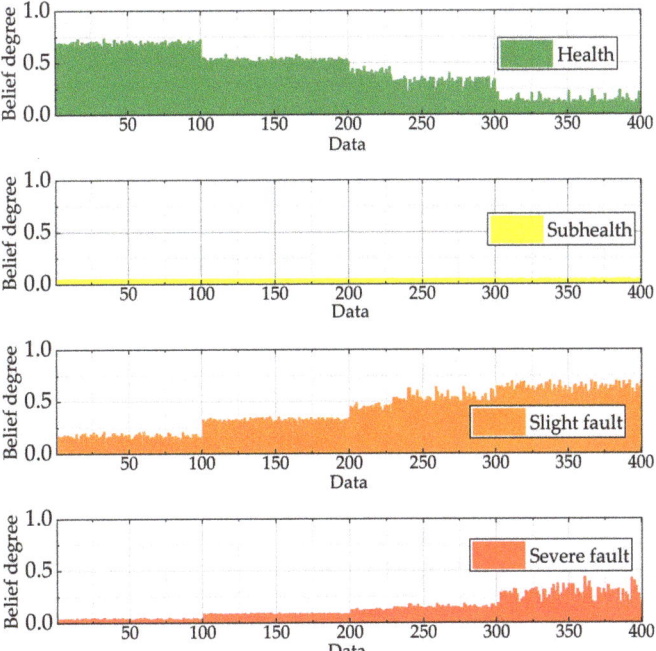

Figure 9. Belief distribution of Q.

It can be seen from Figures 8 and 9 that the belief distributions of two indicators are transformed from input information. The belief degree of two indicators of "Health" are both over 0.5 on the 0–100 sets of data, gradually decreasing to zero with the furthering of fault degree. The belief degree of "subhealth" transformed from input information is little in O and Q, as the belief degree allocated to "Subhealth" is small given by expert in Table 2. The belief degree of "slight fault" or "severe fault" is increasing as the fault continues aggravating, and reaching the greatest finally. Totally, in both figures, the declining trend of health status is conformed with real status in Figure 7.

Step2: Calculation of model parameter

The evidence weights of the two indicators are set as 0.75 and 0.95 respectively. The statistic reliabilities of the two indicators are determined as 0.7 and 0.8 respectively, based on industry standards. The dynamic reliabilities are calculated as 0.4 and 0.5 separately. Based on (11)–(14). The weighting factors δ are set to be 0.8 and 0.9 separately. Then, the reliabilities of the two indicators are 0.68 and 0.86 separately.

Step 3: Aggregation of two indicators

Based on (14)–(17), ER rule can be used to aggregate initial evidence, and the distributed health assessment results can be obtained, shown as Figure 10.

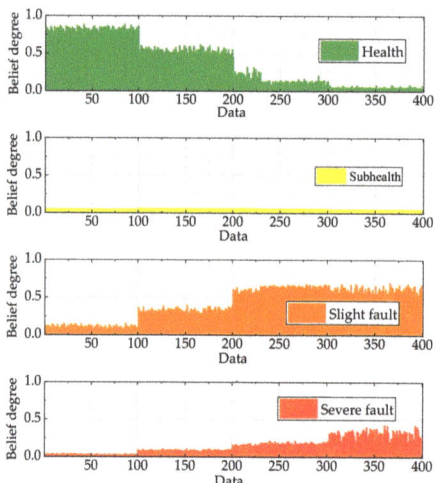

Figure 10. The aggregated health status.

It can be seen in Figure 10 that the belief degree of the "health" is clearly divided into four stages. At first, the belief degree is closed to 1, then floats around 0.5 and 0.2, and finally approaches to 0. Because the belief degree of the "subhealth" of O and Q is little in Figures 7 and 8, the aggregated belief degree is near to 0. The belief degree of "slight fault" and "severe fault" increases, which is caused by the belief distribution of O and Q.

By introducing the expected utility, the belief distribution can be transformed into numerical output. Define the utility of assessment result grades D_1, D_2, D_3, D_4 as $u(D_1) = 12$, $u(D_2) = 7, u(D_3) = 1, u(D_4) = 0$, respectively. Then, the assessment result of the initial model is shown in Figure 11. It shows that the simulated status fluctuates near the real status of the bus of control system in the first three status and deviates from the real health status in forth status. This basically matches to distributed assessment results in the Figure 10. In fact, the deviation of real status partly reflects the uncertainty of observation data and the limitation of expert's knowledge. Therefore, initial assessment model needs to be optimized based on quantitative data.

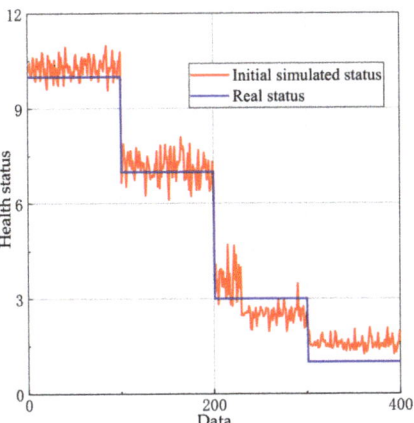

Figure 11. The comparison between initial and real status.

5.1.3. Parameters Optimization and Comparative Study

The optimization model is constructed based on Formula (24), as follows

$$\min(\text{RMSE}(\Psi)) \tag{30}$$

To ensure high accuracy, maintaining the interpretability of the assessment results, constraints of the model parameters are determined by expert as follows. The constraints of indicators reference values are given as:

$$\begin{cases} -18.5 < H_{1,Power} < -17 \\ -21.5 < H_{2,Power} < 19.5 \\ -26 < H_{3,Power} < -21.5 \\ 8 < H_{1,q} < 10 \\ 3 < H_{2,q} < 6 \\ 0 < H_{3,q} < 3 \end{cases} \tag{31}$$

The constraints of weight are given as:

$$\begin{cases} 0.5 < \omega_{Power} < 0.8 \\ 0.8 < \omega_q < 1 \end{cases} \tag{32}$$

The constraints of expected utility are given as:

$$\begin{cases} 10 < u(D_1) < 15 \\ 7 < u(D_2) < 10 \\ 3 < u(D_3) < 5 \\ 0 < u(D_4) < 1 \end{cases} \tag{33}$$

The constraints of the transformation matrixes are given as:

$$\begin{cases} 0 \leq a_{i,j} \leq 1 \\ a_{i-1,j} < a_{i,j} < a_{i+1,j} \\ \sum_{j=1}^{N} a_{i,j} = 1 \end{cases} \tag{34}$$

The above model can be optimized by the Fmincon algorithm. Fmincon algorithm is employed to find the minimum value of the objection function under nonlinear constraints. Total of 200 sets of training data are selected alternately from the 400 sets of data, and the

400 sets of data are determined as test data. The optimized parameters are obtained, as shown in Tables 3 and 4.

Table 3. The optimized transformation matrixes.

No.	Indicator	Weight	Reference Values	$\{D_1, D_2, D_3, D_4\}$
1	optical power	0.76	−17.421	{0.699, 0.209, 0.067, 0.025}
2			−21.083	{0.060, 0.060, 0.545, 0.335}
3			−26.523	{0.045, 0.121, 0.165, 0.669}
4	Q factor	0.92	7.814	{0.648, 0.266, 0.049, 0.037}
5			4.821	{0.032, 0.053, 0.586, 0.329}
6			2.790	{0.028, 0042, 0.056, 0.874}

Table 4. The optimized expected utility.

Expected Utility	$u(D_1)$	$u(D_2)$	$u(D_3)$	$u(D_4)$
Value	14.517	9.162	3.858	0.499

Remark 5. *By carefully comparing Tables 2 and 3, it can be found that reference values of indicators are not significantly changed. There are two reasons to illustrate this phenomenon:*

(1) Reference values are not quite important compared to other parameters, such as expected utility and transformation matrix.
(2) The initial values of reference values given by expert are relatively consistent with the real status of the of bus of control system, and the optimization is a mild adjustment.

The optimized model is compared with the initial model, as follows. It is shown in Figure 12 that the optimized simulated status is closer to the real status than the initial status, especially in the "Severe fault" status, in which the optimized simulated status fluctuates less than the initial simulated status. To further illustrate the effectiveness of the proposed method, the following comparative study is conducted.

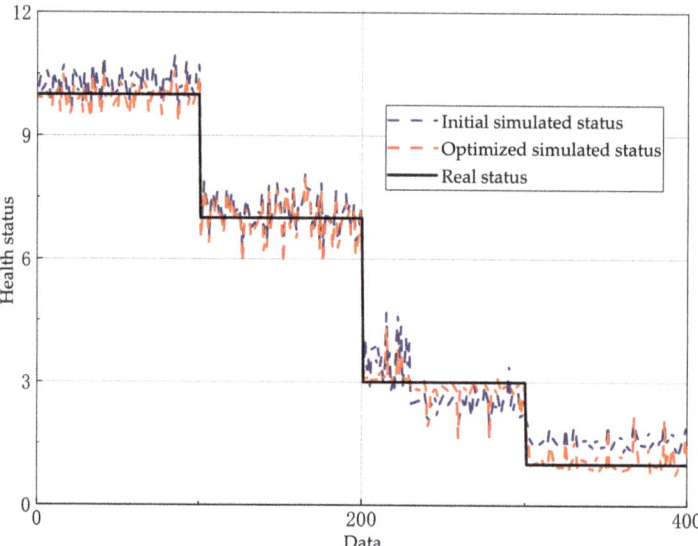

Figure 12. The comparison between the optimized model and initial model.

(1) The comparison under the ER rule framework

In this part, the traditional ER rule (model 1) and DS evidence theory (model 2) under the ER rule framework are employed to compare with the proposed model. It needs to be guaranteed the consistency between the input and output grades in the model 1 and model 2. Therefore, First, indicator reference grades are added to make it consistent with the assessment result grades. The initial reference value of "Subhealth" is given as an average value between the "Health" and "Slight fault". The reference values of both model 1 and model 2 are given in Table 5. The reliabilities are set same as the proposed model in model 1.

Table 5. The reference values of input indicators.

Reference Grades	H_1	H_2	H_3	H_4
Received optical power	−17.5	−19	−20	−26
Q factor	8	6	4	1

The constraints of indicators reference values are given as Formula (35), and the constraints of weight, expected utility are settled same as Formulas (32) and (33). Constraints are given same in model 1 and model 2, except that and reliability are set to be 1 in model 2. The same training data are used to optimize model parameters, the whole sets of data are employed as test data.

$$\begin{cases} -18.5 < H_{1,Power} < -15 \\ -20.5 < H_{2,Power} < -19.5 \\ -21.5 < H_{3,Power} < -20.5 \\ -26 < H_{4,Power} < -21.5 \\ 8 < H_{1,q} < 10 \\ 6 < H_{2,q} < 8 \\ 3 < H_{3,q} < 6 \\ 1 < H_{4,q} < 3 \end{cases} \quad (35)$$

The comparison result between the actual and simulated results are shown in Figure 13. It can be seen that the proposed model is fluctuating smaller and much closer to the real status in contrast with model 1 and model 2, especially in the "Health" status. To further compare the accuracy of different models, the root means square error can be calculated as Table 6. As can be seen from Table 6, the assessment accuracy of the proposed model is highest. Compared with the model 1, model 2 has improved 23.13%, 27.48%. In the view of the above analysis, it can be proved that the proposed model is more accurate than other methods under the ER rule framework.

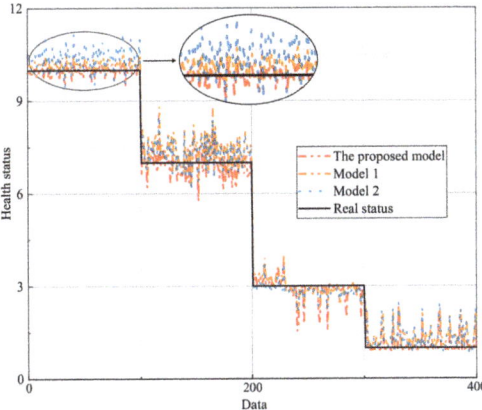

Figure 13. The comparison under ER rule framework.

Table 6. Comparison of assessment accuracy under ER framework.

Model	The Proposed Model	ER (Model 1)	DS (Model 2)
RMSE	0.3500	0.4553	0.4826

(2) The comparison with data-based models

In this part, a comparative study is implemented by adopting the data-based method, including backpropagation (BP) neural network and support vector regression (SVR). Some details of BP model parameters are shown in Table 7. The same training data and test data are utilized. The comparison results between the simulated and actual status are shown in Figure 14.

Table 7. The parameters of the BP models.

Method	Detail	
BP Neural network	Type	Feedforward neural network
	Learning rate	0.001
	The number of layers	3
	The time of training	500
	The training goal	0.0001

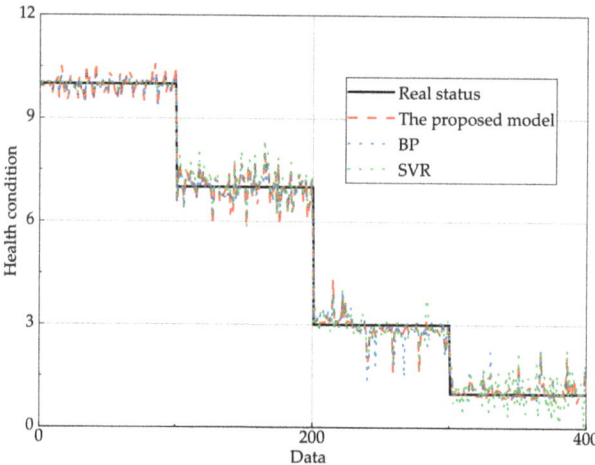

Figure 14. The comparison of data-based models.

It can be seen in Figure 14 and Table 8, the proposed model has high accuracy, which is second only to the BP model, and the accuracy of the proposed model is improved by 6.52% compared with the SVR.

Table 8. Comparison of assessment accuracy of data-based model.

Model	The Proposed Model	BPNN	SVR
RMSE	0.3500	0.3144	0.3965

At the same time, to further compare and analyze the proposed model and BP model, 10%, 25%, 50%, and 60% of the whole data sets are randomly selected as the training set, and the whole sets of data are selected as the test set. The comparative accuracy of proposed model is calculated as follows.

As shown in Table 9, when the training set randomly selects 10% and 25% of the data set, the accuracy of the proposed model is higher than that of the BP model. While the training set randomly selects 50% and 60% of the data set, the accuracy of ER model is worse than that of the BP model. It shows that the proposed model can achieve accurate health assessment by aggregating expert knowledge and observation data in the case of less observation data and sufficient prior knowledge.

Table 9. Comparative accuracy of proposed model and BP model.

Training Data	RMSE (Proposed Model)	RMSE (BP)
10%	0.4174	0.4448
25%	0.3811	0.4162
50%	0.3500	0.3144
60%	0.3170	0.2887

(3) The comparison with knowledge-based models

Belief rule base (BRB) and fuzzy reasoning (FR) are the typical qualitative knowledge-based methods. In this part, BRB and FR are implemented to compare with the proposed model. Same training data and test data are selected. The initial parameters of BRB are determined by expert' knowledge shown in Table 10, and the part parameters of fuzzy reasoning are demonstrated in Table 11.

Table 10. The initial parameters of BRB model.

No.	Rule Weight	Factors		Belief Distribution of Health Status
		O	Q	
1	1	H	H	(0.75, 0.10, 0.05, 0)
2	1	H	M	(0.55, 0.45, 0.05, 0)
3	0.1	H	L	(0.90, 0.10, 0, 0)
4	1	M	H	(0.60, 0.30, 0.10, 0)
5	1	M	M	(0.05, 0.30, 0.60, 0.05)
6	1	M	L	(0, 0.15, 0.35, 0.5)
7	0.1	L	H	(0.90, 0.10, 0, 0)
8	1	L	M	(0, 0.05, 0.45, 0.50)
9	1	L	L	(0, 0.05, 0.20, 0.75)

Table 11. The parameters of FR model.

Method		Detail
Fuzzy reasoning	Initial fuzzy matrix	[0.5, 0.3, 0.2, 0; 0, 0.6, 0.3, 0.1; 0, 0, 0.5, 0.5]
	optimized fuzzy matrix	[0.75, 0.1, 0.05, 0; 1, 0, 0, 0; 0, 0.2, 0.3, 0.5]

It is shown in Figure 15 that the assessment results of FR and BRB are relatively scattered and far from the real status. Comparing with BRB and FR, the accuracy of the proposed model is improved by 19.28% and 16.25% respectively, as shown in Table 12. It is concluded that the proposed model is most accurate compared to the qualitative knowledge-based models.

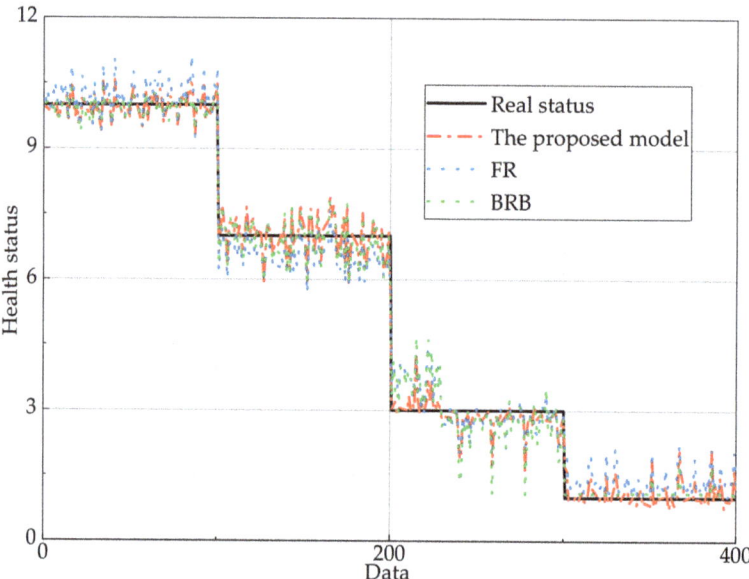

Figure 15. Comparison of qualitative knowledge-based models.

Table 12. Comparison of assessment accuracy of knowledge-based model.

Model	The Proposed Model	BRB	FR
RMSE	0.3500	0.4179	0.5058

5.2. Example 2—Health Assessment of Engine

In this subsection, the WD615 model engine is taken as a case to verify the effectiveness of the proposed model for complex system. The background description is introduced in Section 5.2.1. The implement progress of the proposed model is carried out in Section 5.2.2. In Section 5.2.3, the comparative study is conducted.

5.2.1. Background Description

In order to monitor the operation status of engine, vibration sensors are set up for amassing the vibration signal of engine [27]. Then, the vibration signal can be processed to get the time-domain characteristics, as shown as Figures 16–18. The mean, variance, and kurtosis, which reflect the center, degree of dispersion, and degree of convex of signal, are selected as the health status indicators of the engine [28]. The real status of the engine is shown in Figure 19. The assessment result grades of engine are defined as three statuses according to the different gap between the crankshaft and bearing connecting rod: First, a gap of 0.08 mm to 0.1 mm belongs to "Health"; a gap of 0.18 mm to 0.2 mm belongs to "Fault"; a gap of 0.32 mm to 0.34 mm belongs to "Failure". Thus, a frame of discernment Φ is defined as follows.

$$\Phi = \{Health, Fault, Failure\} = \{D_1, D_2, D_3\} \quad (36)$$

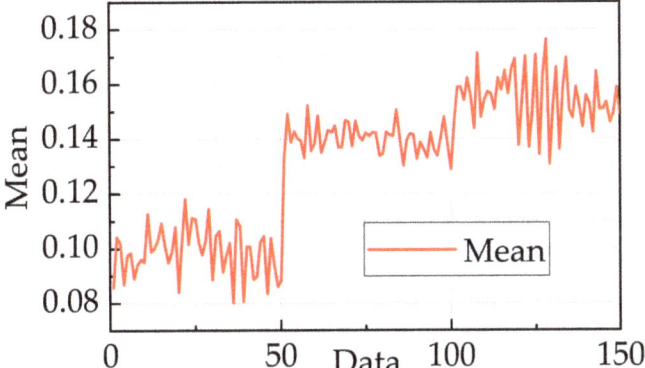

Figure 16. The mean of vibration signal.

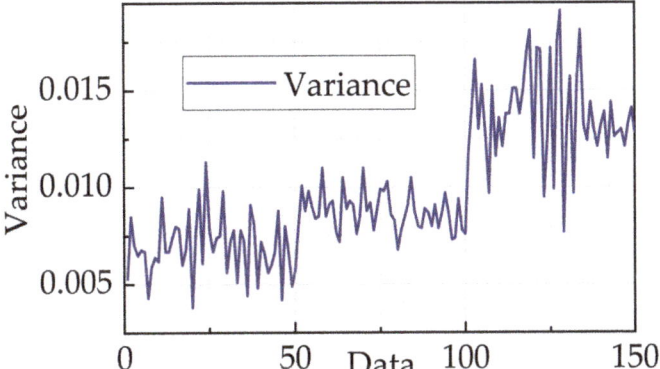

Figure 17. The variance of vibration signal.

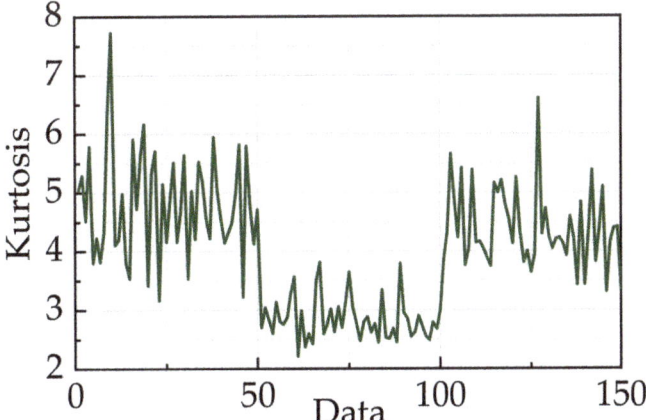

Figure 18. The kurtosis of vibration signal.

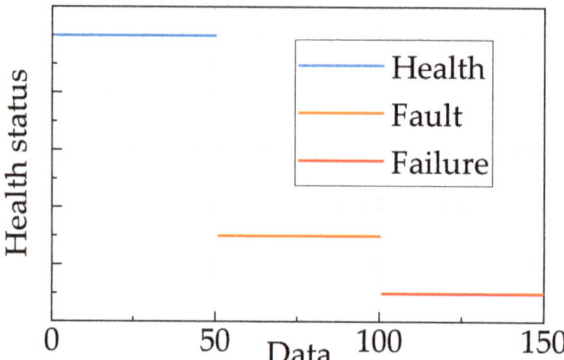

Figure 19. The health status of engine.

There are 150 sets of data, including the status of *"Health"*, *"Fault"*, and *"Failure"*. The mean of vibration signal ranges from 0.802 to 0.1761. The variance of vibration signal ranges from 0.0038 to 0.0191, and the kurtosis of vibration signal ranges from 2.2159 to 7.6801 as shown in Figures 16–18, respectively.

5.2.2. Construction and Optimization of Assessment Model

To construct a health assessment model of engine, first the indicators reference grades are determined as follows.

$$H_{mean} = H_{variance} = \{Small, Medium, Slight\ large, Large\} = \{S, M, SL, L\} \quad (37)$$

$$H_{kurtosis} = \{Average, High\} = \{A, H\} \quad (38)$$

where H_{mean}, $H_{variance}$, and $H_{kurtosis}$ denote the reference grades of mean, variance, and kurtosis.

Due to the reference grades are disaccord with assessment result grades, transformation matrixes are introduced to transform input information, and the initial values of transformation matrix and reference values are determined based on expert's knowledge in Table 13.

Table 13. The initial values of transformation matrix.

No.	Indicators	Reference Values	Belief Distribution
1	Mean	0.08	(0.90, 0.10, 0.00)
2		0.10	(0.85, 0.10, 0.05)
3		0.14	(0.15, 0.45, 0.40)
4		0.18	(0.05, 0.15, 0.75)
5	Variance	0.003	(0.70, 0.25, 0.05)
6		0.01	(0.35, 0.50, 0.15)
7		0.013	(0.05, 0.20, 0.75)
8		0.020	(0.00, 0.25, 0.75)
9	Kurtosis	2	(0.70, 0.20, 0.10)
10		8	(0.00, 0.30, 0.70)

The Table 13 can be expressed as a form of matrix as Formulas (39) and (40). Then, the initial evidence is given by using the rule-based transformation technique. According to the implement process of example 1 in Section 5.1, the optimized simulated status is introduced in Figure 18. In the process of optimization, 75 sets of data are selected alternately from 150 sets of data as training data, and the whole sets of data are taken as test data.

$$A_1 = \begin{bmatrix} 0.9 & 0.85 & 0.15 & 0.05 \\ 0.1 & 0.1 & 0.45 & 0.15 \\ 0 & 0.05 & 0.4 & 0.75 \end{bmatrix} \quad (39)$$

$$A_2 = \begin{bmatrix} 0.7 & 0.35 & 0.05 & 0 \\ 0.25 & 0.5 & 0.2 & 0.25 \\ 0.05 & 0.15 & 0.75 & 0.75 \end{bmatrix}, A_3 = \begin{bmatrix} 0.7 & 0 \\ 0.2 & 0.3 \\ 0.1 & 0.7 \end{bmatrix} \quad (40)$$

It is shown in Figure 20 that contrasting with the optimized simulated status, the error between the initial status and real status is rather large, especially in the first and third stages. By calculating the root mean square, the accuracy of the optimized status is 41.7% higher than the initial status. The optimized and calculated model parameters are given in Tables 14 and 15.

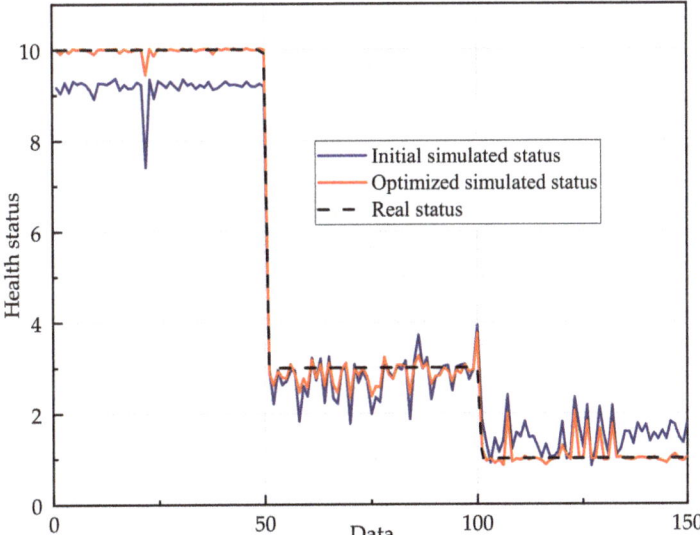

Figure 20. The comparison between the initial and optimized status.

Table 14. The parameters of optimized model parameters.

Parameters	Values		
	Mean	Variance	Kurtosis
indicators weight	0.7329	0.7950	0.922
reliability	0.8491	0.835	0.935
Health status	Health	Fault	Failure
Expected utility	10.123	4.893	0.277

Table 15. The optimized transformation matrix.

No.	Indicators	Reference Values	Belief Distribution
1	Means	0.0801	(0.959, 0.040, 0.001)
2		0.116	(0.972, 0.027, 0.001)
3		0.131	(0.106, 0.327, 0.567)
4		0.178	(0.129, 0.212, 0.659)
5	Variance	0.0031	(0.527, 0.466, 0.070)
6		0.0090	(0.503, 0.496, 0.001)
7		0.0126	(0.060, 0.275, 0.665)
8		0.0193	(0.010, 0.451, 0.539)
9	Kurtosis	1.758	(0.679, 0.224, 0.096)
10		7.720	(0.002, 0.423, 0.575)

5.2.3. Comparative Study

In this subsection, based on the same training data and test data, several kinds of qualitative knowledge-based models and data-based models are employed to compare with the proposed model.

It can be seen in the Figures 21–23 that the performance of the proposed model is better than other methods, and its assessment accuracy is improved by 10.4%, 20.6%, 8.9%, 14.4%, 15.9%, and 27.6% compared to ER, DS, BP, SVR, BRB, and FR, respectively shown in Table 16.

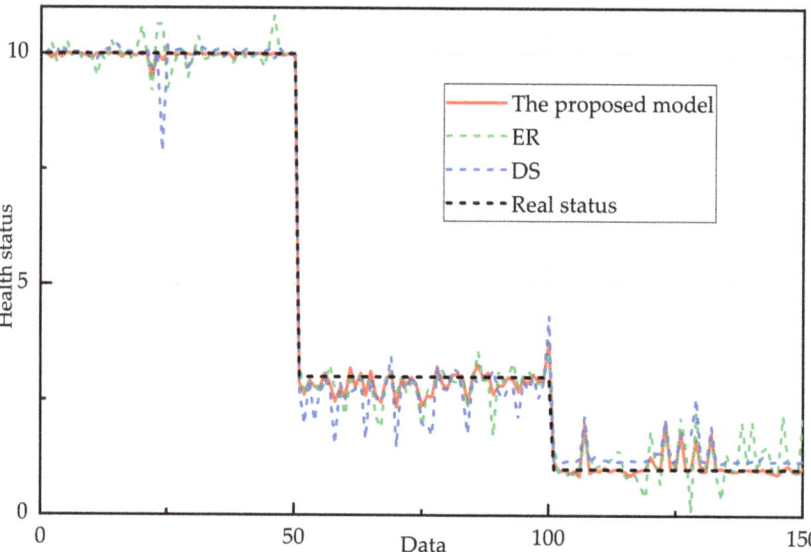

Figure 21. The comparison under ER framework.

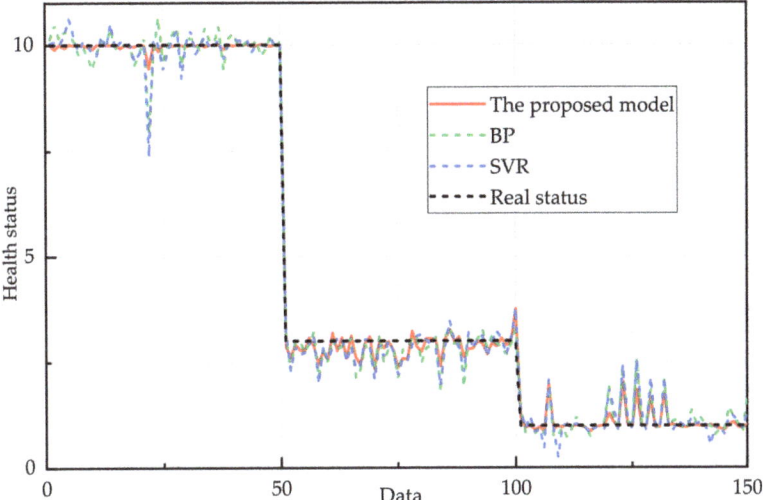

Figure 22. The comparison with data-based models.

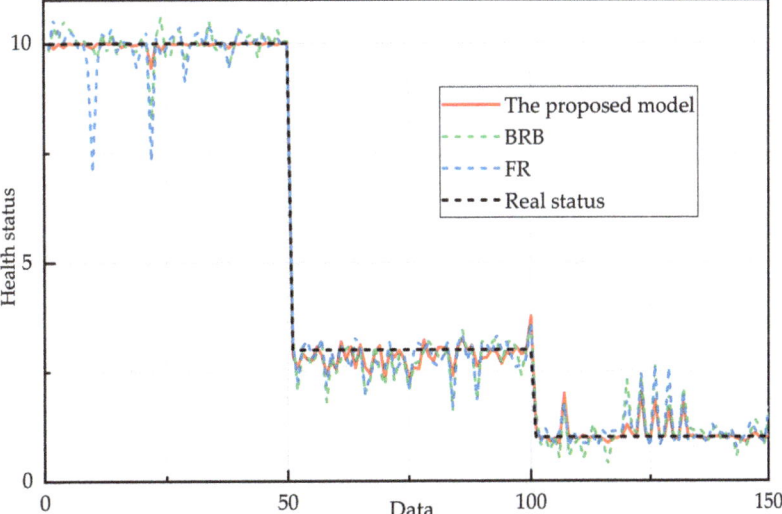

Figure 23. The comparison with knowledge-based models.

Table 16. Comparison of different models.

Model	The Proposed Model	ER	DS	BP	SVR	BRB	FR
RMSE	0.3730	0.4162	0.4695	0.4098	0.4355	0.4436	0.5154

5.3. Result Analysis

In the above two examples, the health status of the bus of control system and engine is assessed by the proposed model, where the input indicators reference grades disaccord with the assessment result grades are fully considered, and include three situations: (1) the indicators reference grades are more than the assessment result grades, (2) the input indicators reference grades are less than the assessment result grades, (3) above cases exist simultaneously. According to the above comparative research, it can be proved that the proposed method is able to combine the advantage of both data-based methods and qualitative knowledge-based method, providing an interpretable and accurate assessment result for decision-makers.

In fact, those two examples provide a general process to solve the inconsistent problem between input and output. More importantly, the proposed method can not only be applied in these two cases, but also can be extended to the dynamic assessment and other multiple indicators of health assessment.

6. Conclusions

An ER rule-based health assessment model for a complex system is proposed, where the transformation matrix is considered. In addition, case study of the bus of control system and the engine is investigated to demonstrate the validity and practicality of the proposed method.

There are mainly two contributions of this paper. First, the transformation matrix is employed to solve the disaccord problem between the input indicator reference grades and assessment result grades, which keeps the consistency and completeness of the possession of the input information transformation. Second, the calculation methods of indicator weight and reliability are conducted, where the qualitative knowledge and quantitative information are fully used. Then, the optimization method of the model is conducted, and a complete health assessment model is constructed.

According to the proposed model, the future research work can be summarized into the following two points:

(1) In engineering practice, the forms of health status threshold can be various, and the forms are not only numerical, but can also be in interval form or normal distribution form. Therefore, how to solve the disaccord problem between the indicators reference grades and assessment result grades under the different forms of threshold should be addressed.

(2) The integration model between deep learning and ER rule can be established based on the good uncertainty processing ability and interpretability of ER rule.

Author Contributions: Z.L. and Z.Z. contributed equally to this work. Conceptualization, Z.L. and Z.Z.; methodology, Z.L. and J.W.; software, X.Z.; validation, Z.L., Z.Z. and J.W.; formal analysis, Z.Z. and J.W.; investigation, W.H.; data curation, Z.L.; writing—original draft preparation, Z.L.; writing—review and editing, Z.L. and W.H.; visualization, Z.L.; supervision, W.H. and X.Z. All authors have read and agreed to the published version of the manuscript.

Funding: This work was supported in part by the Shaanxi Outstanding Youth Science Foundation under Grant 2020JC-34, in part by the Postdoctoral Science Foundation of China under Grant No. 2020M683736, in part by the Natural Science Foundation of Heilongjiang Province of China under Grant No. LH2021F038.

Institutional Review Board Statement: Not appliable.

Informed Consent Statement: Not appliable.

Data Availability Statement: Data sharing not applicable.

Conflicts of Interest: The authors declare no conflict of interest.

References

1. Duan, Z.; Sun, H.; Wu, C.; Hu, H.; Hu, H. Flow-network based dynamic modelling and simulation of the temperature control system for commercial aircraft with multiple temperature zones. *Energy* **2022**, *238*, 121874. [CrossRef]
2. Yi, T.; Jin, C.; Gao, L.; Hong, J.; Liu, Y. Nested Optimization of Oil-Circulating Hydro-Pneumatic Energy Storage System for Hybrid Mining Trucks. *Machines* **2022**, *10*, 22. [CrossRef]
3. Gao, Q. Nonlinear Adaptive Control with Asymmetric Pressure Difference Compensation of a Hydraulic Pressure Servo System Using Two High Speed on/off Valves. *Machines* **2022**, *10*, 66. [CrossRef]
4. Liu, Y.; Xu, J.; Yuan, H.; Lv, J.; Ma, Z. Health Assessment and Prediction of Overhead Line Based on Health Index. *IEEE Trans. Ind. Electron.* **2019**, *66*, 5546–5557. [CrossRef]

5. Zhou, T.; Wu, W.; Peng, L.; Zhang, M.; Li, Z.; Xiong, Y.; Bai, Y. Evaluation of urban bus service reliability on variable time horizons using a hybrid deep learning method. *Reliab. Eng. Syst. Saf.* **2022**, *217*, 108090. [CrossRef]
6. Miao, H.; Li, B.; Sun, C.; Liu, J. Joint Learning of Degradation Assessment and RUL Prediction for Aeroengines via Dual-Task Deep LSTM Networks. *IEEE Trans. Ind. Inform.* **2019**, *15*, 5023–5032. [CrossRef]
7. Li, Z.; Wu, J.; Yue, X. A Shape-Constrained Neural Data Fusion Network for Health Index Construction and Residual Life Prediction. *IEEE Trans. Neural Netw. Learn. Syst.* **2021**, *32*, 5022–5033. [CrossRef]
8. Salamai, A.; Hussain, O.; Saberi, M. Decision Support System for Risk Assessment Using Fuzzy Inference in Supply Chain Big Data. In Proceedings of the 2019 International Conference on High Performance Big Data and Intelligent Systems (HPBD&IS), Shenzhen, China, 9–11 May 2019; pp. 248–253. [CrossRef]
9. Zhu, L. An MI-BRB Based Health State Assessment Methodology for Running Gears in High-Speed Trains. In Proceedings of the 2020 3rd World Conference on Mechanical Engineering and Intelligent Manufacturing (WCMEIM), Shanghai, China, 4–6 December 2020; pp. 768–772. [CrossRef]
10. Bjornsen, K.; Selvik, J.T.; Aven, T. A semi-quantitative assessment process for improved use of the expected value of information measure in safety management. *Reliab. Eng. Syst. Saf.* **2019**, *188*, 494–502. [CrossRef]
11. Yang, J.B.; Xu, D.L. Evidential reasoning rule for evidence combination. *Artif. Intell.* **2013**, *205*, 1–29. [CrossRef]
12. Dempster, A. A Generalization of Bayesian Theory. *J. R. Stat. Soc. Ser. B Methodol.* **1968**, *30*, 205–247.
13. Xu, D.L. On the evidential reasoning algorithm for multiple attribute decision analysis under uncertainty. *IEEE Trans. Syst. Man Cybern.-Part A Syst. Hum.* **2002**, *32*, 289–304.
14. Yang, J.B.; Pratyush, S. A general multi-level evaluation process for hybrid MADM with uncertainty. *IEEE Trans. Syst. Man Cybern.* **1994**, *24*, 1458–1473. [CrossRef]
15. Zhang, Z.; Liao, H.; Tang, A. Renewable energy portfolio optimization with public participation under uncertainty: A hybrid multi-attribute multi-objective decision-making method. *Appl. Energy* **2022**, *307*, 118267. [CrossRef]
16. Ning, P.; Zhou, Z.; Cao, Y.; Tang, S.; Wang, J. A Concurrent Fault Diagnosis Model via the Evidential Reasoning Rule. *IEEE Trans. Instrum. Meas.* **2022**, *71*, 1–16. [CrossRef]
17. Zhang, C.; Zhou, Z.; Hu, G.; Yang, L.; Tang, S. Health assessment of the wharf based on evidential reasoning rule considering optimal sensor placement. *Measurement* **2021**, *186*, 110184. [CrossRef]
18. Xiong, Y.; Jiang, Z.; Fang, H.; Fan, H. Research on Health Condition Assessment Method for Spacecraft Power Control System Based on SVM and Cloud Model. In Proceedings of the 2019 Prognostics and System Health Management Conference (PHM-Paris), Paris, France, 2–5 May 2019; pp. 143–149. [CrossRef]
19. Tao, J.; Wang, X.; Liang, X. Health State Evaluation for Fuzzy Multi-state Production Systems based on MPNM. In Proceedings of the 2020 11th International Conference on Prognostics and System Health Management (PHM-2020 Jinan), Jinan, China, 23–25 October 2020; pp. 5–10. [CrossRef]
20. Yang, J.B. Rule and utility based evidential reasoning approach for multiattribute decision analysis under uncertainties. *Eur. J. Oper. Res.* **2001**, *131*, 31–61. [CrossRef]
21. Yang, J.B.; Singh, M.G. An evidential reasoning approach for multiple-attribute decision making with uncertainty. *IEEE Trans. Syst. Man Cybern.* **1994**, *24*, 1–18. [CrossRef]
22. Feng, Z.C.; Zhou, Z.; Hu, C.; Chang, L.; Hu, G.; Zhao, F. A new belief rule base model with attribute reliability. *IEEE Trans. Fuzzy Syst.* **2019**, *27*, 903–916. [CrossRef]
23. Zhao, F.; Zhou, Z.; Hu, C.; Chang, L.; Li, G. A new evidential reasoning-based method for online safety evaluation of complex systems. *IEEE Trans. Syst. Man Cybern. Syst.* **2018**, *48*, 954–966. [CrossRef]
24. Stan, O.; Cohen, A.; Elovici, Y.; Shabtai, A. Intrusion Detection System for the MIL-STD-1553 Communication Bus. *IEEE Trans. Aerosp. Electron. Syst.* **2020**, *56*, 3010–3027. [CrossRef]
25. Wang, D.; Xu, Y.; Li, J.; Zhang, M.; Li, J.; Qin, J.; Chen, X. Comprehensive Eye Diagram Analysis: A Transfer Learning Approach. *IEEE Photonics J.* **2019**, *11*, 1–19. [CrossRef]
26. Zhao, Y.; Tang, Y.; Xiao, J.; Mou, W. Eye Diagram Analysis Based on System View-Taking PCM System as an Example. In Proceedings of the 2020 5th International Conference on Mechanical, Control and Computer Engineering (ICMCCE), Harbin, China, 25–27 December 2020; pp. 2286–2289. [CrossRef]
27. Wang, M.; Qin, G.; Chen, J.; Liao, Y. Design of Vibration Monitoring and Fault Diagnosis System for Marine Diesel Engine. In Proceedings of the 2020 11th International Conference on Prognostics and System Health Management (PHM-2020 Jinan), Jinan, China, 23–25 October 2020; pp. 1–4. [CrossRef]
28. Liu, Y.; Chang, W.; Zhang, S.; Zhou, S. Fault Diagnosis and Prediction Method for Valve Clearance of Diesel Engine Based on Linear Regression. In Proceedings of the 2020 Annual Reliability and Maintainability Symposium (RAMS), Palm Springs, CA, USA, 27–30 January 2020; pp. 1–6. [CrossRef]

Article

Fault Diagnosis of Motor Vibration Signals by Fusion of Spatiotemporal Features

Lijing Wang [1,2,3], Chunda Zhang [1], Juan Zhu [3] and Fengxia Xu [1,3,*]

1. School of Mechanical and Electrical Engineering, Qiqihar University, Qiqihar 161000, China; 02926@qqhru.edu.cn (L.W.); 2019911186@qqhru.edu.cn (C.Z.)
2. School of Electrical and Electronic Engineering, Harbin University of Science and Technology, Harbin 150080, China
3. Collaborative Innovation Center of Intelligent Manufacturing Equipment Industrialization of Heilongjiang Province, Qiqihar University, Qiqihar 161006, China; 02976@qqhru.edu.cn
* Correspondence: 01541@qqhru.edu.cn

Abstract: This paper constructs a spatiotemporal feature fusion network (STNet) to enhance the influence of spatiotemporal features of signals on the diagnostic performance during motor fault diagnosis. The STNet consists of the spatial feature processing capability of convolutional neural networks (CNN) and the temporal feature processing capability of recurrent neural networks (RNN). It is used for fault diagnosis of motor vibration signals. The network uses dual-stream branching to extract the fault features of motor vibration signals by a convolutional neural network and gated recurrent unit (GRU) simultaneously. The features are also enhanced by using the attention mechanism. Then, the temporal and spatial features are fused and input into the softmax function for fault discrimination. After that, the fault diagnosis of motor vibration signals is completed. In addition, several sets of experimental evaluations are conducted. The experimental results show that the vibration signal processing method combined with spatiotemporal features can effectively improve the recognition accuracy of motor faults.

Keywords: spatiotemporal feature fusion; convolutional neural network; gated recurrent unit; attention mechanism; fault diagnosis

1. Introduction

The asynchronous motor is the most widely used mechanical drive equipment in industrial production and has become an important component in fields such as machinery manufacturing [1–3] and intelligent transportation [4,5]. Due to the harsh working environment, overload, and complex electromagnetic relationships, the motor is prone to stator winding inter-turn short circuit, broken rotor strips, air gap eccentricity, and bearing wear [6–8]. During operation, the failure of asynchronous motors may cause huge economic losses and casualties. Therefore, it is very important to evaluate the working state of the motor and detect potential faults to prevent mechanical accidents. Fault diagnosis of motors plays an important role in equipment maintenance, which can improve the quality of machines and reduce maintenance costs.

The common way of motor fault diagnosis is to use vibration signals for analysis. Vibration signals can be collected using acceleration transducers. Abnormal vibration signals can characterize equipment faults, such as asymmetry of the shaft system [9], a loose connection of components [10], and damaged rotor bearings [11]. Therefore, the acquisition and analysis of vibration signals have also become a common fault diagnosis scheme in the field of rotating machinery [12,13]. Fault diagnosis methods based on vibration signals [14,15] mainly include two stages: feature extraction and pattern recognition. The key to the asynchronous motor fault diagnosis technique is extracting feature information from non-smooth vibration signals with time-varying characteristics. In the time domain, some works [15,16] acquired amplitude, root mean square, and kurtosis for the analysis and diagnosis of vibration signals. However, it was susceptible to environmental noise and the methods have limitations. Some works [17,18] used Fourier transform to convert the signal from the time domain to the frequency domain. But the frequency characteristics of the vibration signal over time cannot be extracted effectively. The time-frequency domain analysis was performed by wavelet transform [19], short-time Fourier transform [20,21], and empirical mode decomposition [22,23], which extracted both time-domain and frequency-domain features. But the above methods are only effective for specific features and have poor adaptivity and robustness.

With the rise of deep learning, some neural networks have been introduced into the field of fault diagnosis [24–26]. The vibration characteristics of the signal can be obtained adaptively by learning the nonlinear mapping between the hidden layers in the network. Deep learning-based methods are less interpretable [27] but have high recognition accuracy. Such methods overcome the disadvantages of traditional methods that require manual feature extraction and have poor adaptability. Shi et al. [28] used a long short-term memory neural network (LSTM) to extract the temporal features of bearing vibration signals. However, the local information of the signal in the spatial dimension was ignored and the full key information could not be maintained when the data sequence is too long. Gao et al. [29] combined one-dimensional convolution and adaptive noise cancellation techniques to suppress the strong interference components in the one-dimensional time series of gearboxes. However, the time-series feature of the vibration signal was not fully utilized due to the limitation of the convolutional neural network field of perception. Zhu et al. [30] reconstructed the one-dimensional time-domain sequence into a two-dimensional data format and used two-dimensional convolution to capture the spatial features of the vibration signal. However, the dependencies between the positions of the spatial features were ignored, resulting in some important features not playing a significant role. Due to the convolutional stride and weight connection, the convolutional neural networks [31,32] cannot accurately obtain the temporal features of the vibration signal. In contrast, recurrent neural networks [33] can handle the temporal features of the signal but do not consider the information of the spatial dimension.

At present, motor fault diagnosis only uses the temporal features or spatial features of vibration signals for analysis. In this paper, spatial features and temporal features are combined to construct a spatiotemporal feature fusion network (STNet). The network solves the problem of accuracy loss caused by excessively long signal sequences and the lack of dependencies of each position. STNet is constructed for fault diagnosis of motor vibration signals. The main contributions of this paper are listed as follows.

1. The STNet utilizes the spatial feature extraction capability of a CNN and the temporal feature extraction capability of a GRU to construct a dual-stream network. The network combines temporal and spatial features for fault diagnosis of vibration signals instead of a single temporal or spatial feature.
2. The time series of vibration signals is much longer than the text in natural language processing. Recurrent neural networks do not preserve all the critical information. Therefore, a GRU with an attention mechanism is designed to extract temporal features and effectively synthesize the state and vibration features at different moments.
3. When the CNN extracts the spatial information of vibration signals, channel and position attention make the network capture the dependencies of each position. The attention mechanism obtains rich contextual features to enhance diagnostic accuracy.

The structure of this paper is as follows. Section 2 presents the attention-based mechanism for the GRU to capture the temporal features of vibration signals. Section 3 enhances the data by local mean decomposition and extracts the spatial features of vibration signals using a CNN with channel and position attention. Section 4 proposes a spatiotemporal feature fusion network. Section 5 validates the model by experiments.

2. Temporal Feature

When BP neural networks process data, there is no interrelationship between the front and back inputs of the network. However, the vibration signal of a motor is a one-dimensional time series, and the temporal relationship between each sampling point has an important impact on the performance of the diagnosis. A recurrent neural network introduces memory units to interconnect the neurons in this layer based on the ordinary neural network. The state of the hidden layer is related to the input at this moment and the state of the hidden layer at the previous moment. Therefore, the relationship of the time dimension can be extracted from the original vibration sequence by recurrent neural networks.

2.1. Gated Recurrent Unit

The spatiotemporal feature fusion network extracts the temporal features of motor vibration signals through the variant (gated recurrent unit) of the recurrent neural network. A gated recurrent unit introduces a gating mechanism to improve recurrent neural networks. A GRU can selectively forget some unimportant information while memorizing the state of the previous moment. A GRU alleviates the gradient disappearance of recurrent neural networks and solves the problem of untimely update of network parameters. The GRU controls the input, output, and state information of the hidden layer by the update gate z_t and the reset gate r_t. The internal structure is shown in Figure 1.

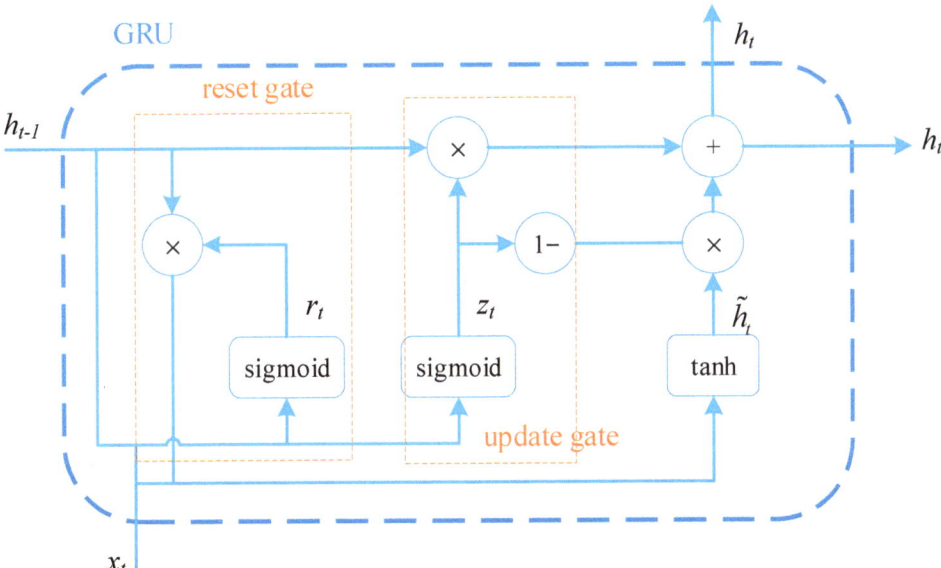

Figure 1. The GRU's internal structure.

The update gate z_t takes the current moment x_t and the previous moment information h_{t-1} by the weighting operation. Then the value between [0, 1] is obtained by the sigmoid function The value controls the effect of historical information on the state of the hidden layer at the current moment. The equation is as follows

$$z_t = \sigma(W_{tz} \cdot [h_{t-1}, x_t] + b_z) \quad (1)$$

where σ is the sigmoid function, W_{tz}, and b_z are the weights, h_{t-1} is the output at the previous moment, and x_t is the input at the current moment.

The reset gate r_t operates the current moment x_t and the previous moment information h_{t-1} with different weights, so that the model selectively forgets historical information that is irrelevant to the results. The equation is as follows

$$r_t = \sigma(W_{tr} \cdot [h_{t-1}, x_t] + b_r). \quad (2)$$

The status of the node at this moment is

$$\tilde{h}_t = \tanh(W \cdot [r_t \times h_{t-1}, x_t] + b). \quad (3)$$

The final output of the hidden layer h_t is the sum of the information to be kept at the current moment and the information to be kept at the previous moment

$$h_t = (1-z_t) \times h_{t-1} + z_t \times \tilde{h}_t. \quad (4)$$

2.2. GRU Temporal Module Based on Attention Mechanism

The length of time series of motor vibration signals is much longer than the length of text in natural language processing. Although the GRU solves the problem of gradient disappearance in long sequence learning of recurrent neural networks, it still cannot retain all the key information when the time sequence is too long. Therefore, this paper not only selects the state output of the last moment of the GRU but also combines the state features of each moment of the GRU. Moreover, the attention mechanism is introduced to assign a weight coefficient to the output of the GRU at each moment. It makes the neural network pay attention to the data features of the output at different moments adaptively. The GRU temporal module based on the attention mechanism is shown in Figure 2.

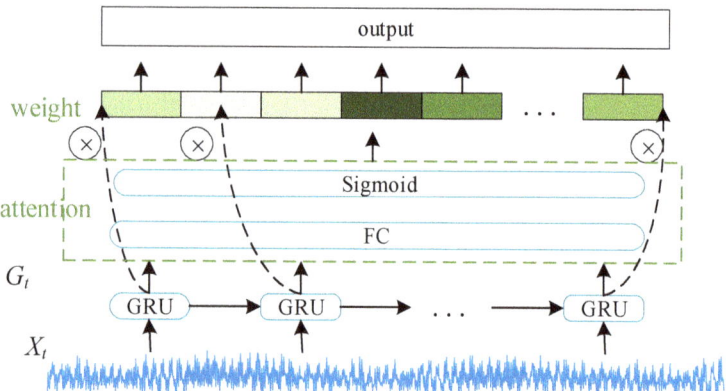

Figure 2. GRU temporal module based on an attention mechanism.

During the analysis of the vibration sequence, the output state of the GRU at the final moment determines the result of the fault diagnosis. However, the states of other moments also have many positive effects on the performance of the network. Therefore, the network not only relies on the output of the final moment but also considers the states of each moment in a comprehensive manner. The vibration signal X_t is fed into the GRU, which captures the vibration characteristics of the signal at each moment. The GRU outputs the state G_t at each moment as

$$G_t = GRU(X_t). \quad (5)$$

However, each momentary output of the GRU has a different degree of influence on the diagnosis results for different types of motor faults. Therefore, the states at each moment of the GRU are selected by the attention mechanism. The states with high relevance are kept and the states with low relevance are weakened. Then the weights of each moment state are obtained by the fully connected layer (FC) and sigmoid function. The weight parameters w_1 are

$$w_1 = \sigma(w(G_t) + b). \quad (6)$$

Finally, the output of GRU at each moment is multiplied by the weight parameter to obtain the output result O

$$O = weight \times G_t \quad (7)$$

3. Spatial Features

The GRU extracts the temporal features of vibration signals but ignores the spatially located information. This paper performs a time-frequency analysis of the vibration signal by local mean decomposition (LMD). The spatial features of the vibration signal after local mean decomposition are extracted by the convolutional neural network.

3.1. Local Mean Decomposition

The motor vibration signal is nonlinear and non-smooth. LMD adaptively decomposes the original vibration sequence into multiple instantaneous frequencies with physically meaningful product functions (PF). Each PF component is the product of a pure frequency modulation signal and an envelope signal, which can express the time-frequency distribution of the signal energy on the spatial scale. Then the vibration signal matrix is constructed and the original data is enhanced. The process of LMD for vibration signal processing is shown in Figure 3.

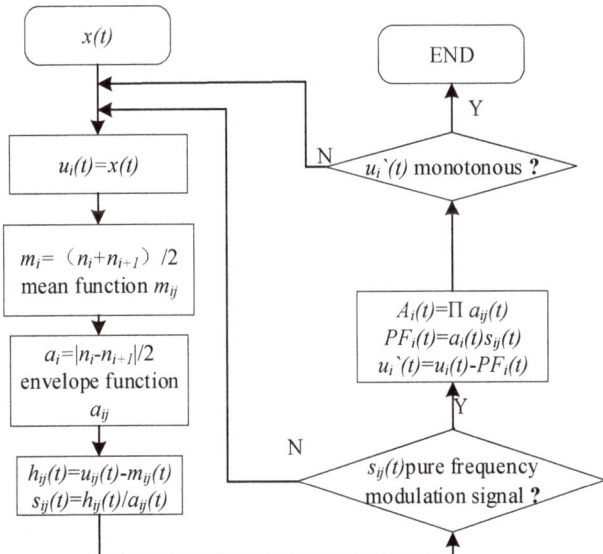

Figure 3. Local mean decomposition.

The original vibration signal $x(t)$ is decomposed by LMD and the mean value m_i of the adjacent local mean points is calculated. The curve is smoothed by the sliding average method to obtain the mean function m_{ij}. Then the envelope function a_{ij} is calculated. The mean function is separated from the original vibration signal to obtain $h_{ij}(t)$. Additionally, $h_{ij}(t)$ is demodulated to obtain $s_{ij}(t)$. If $s_{ij}(t)$ is a pure frequency modulation signal, the PF component $PF_i(t)$ and the residual signal $u_i{'}(t)$ are calculated based on the instantaneous amplitude function $a_i(t)$. If $u_i{'}(t)$ is a monotonic function, the decomposition ends and all PF components are obtained. The results of data decomposition are shown in Figure 4, where the original data $X(t)$ is decomposed into five PF components by LMD.

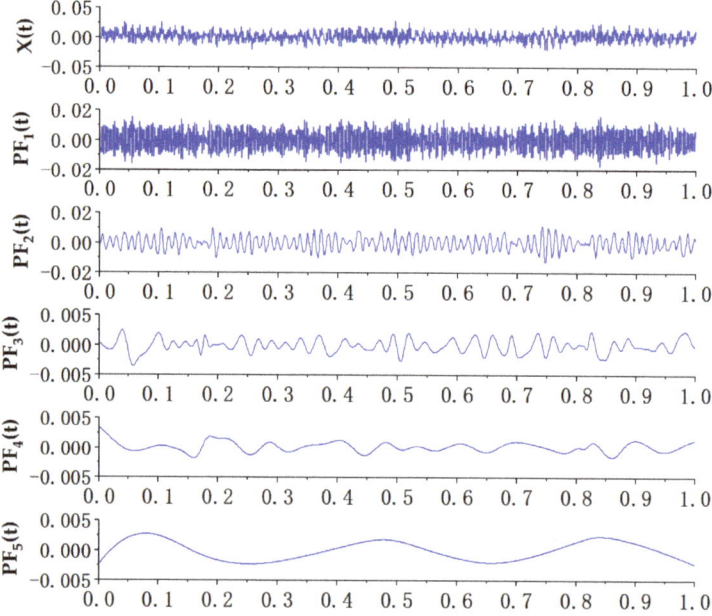

Figure 4. LMD motor vibration signal decomposition.

Convolutional neural networks are often used to process two-dimensional image signals, while the motor vibration signal $X(t)$ is a one-dimensional time-series signal, as follows

$$X(t) = [x_1, x_2, x_3, \cdots, x_t]. \tag{8}$$

Therefore, the vibration signal is converted into a two-dimensional matrix $X'(t) \in \mathrm{R}^{M \times N}$

$$X'(t) = \begin{bmatrix} x_{11} & x_{12} & \cdots & x_{1n} \\ x_{21} & x_{22} & \cdots & x_{2n} \\ \cdots & \cdots & \cdots & \cdots \\ x_{m1} & \cdots & \cdots & x_{mn} \end{bmatrix}. \tag{9}$$

Each PF component is converted into two-dimensional data as shown in Figure 5. The PF components are concatenated with the two-dimensional data $X'(t)$ of the original vibration signal in the channel dimension. The final input matrix of the convolutional neural network is obtained. The method enhances the feature representation of the vibration signal in the spatial dimension.

3.2. CNN Module Based on Attention Mechanism

The convolutional neural network takes the multidimensional matrix of the motor vibration signal as input and adaptively extracts the spatial features of the signal. The different features have different effects on the fault diagnosis results. As shown in Figure 5, the same vibration signal decomposes with different PF components. It leads to huge differences between the different channels of the input 3D matrix $X_{in} \in \mathrm{R}^{c \times M \times N}$. The different channels have different effects on the diagnosis results for different fault types. Therefore, the attention mechanism is added to the channel dimension to make the model adaptively extract different channel features.

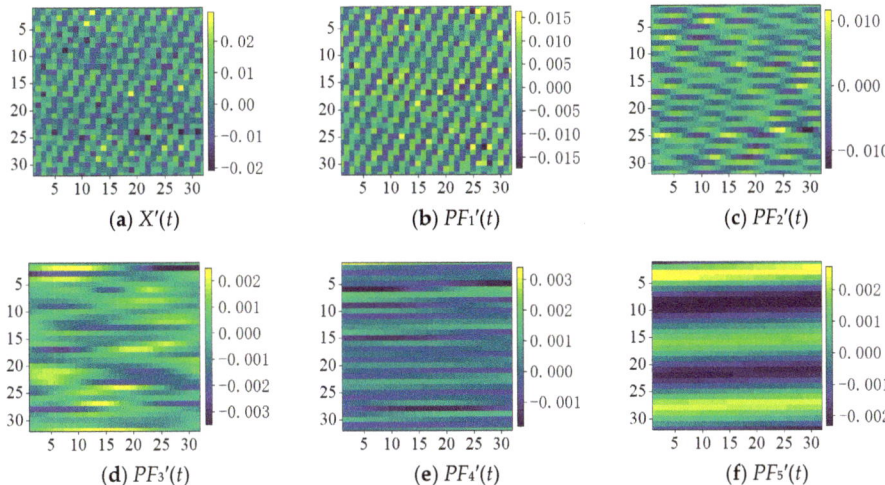

Figure 5. Two-dimensional vibration matrix visualization. (a) is the original vibration signal. (b–f) are the PF components.

The structure of the channel attention is shown in Figure 6, where the input matrix X_{in} is convolved to obtain $x \in R^{c \times m \times n}$ and $\otimes$ represents element-by-element multiplication.

$$x = w_i \otimes X_{in} + b_i \tag{10}$$

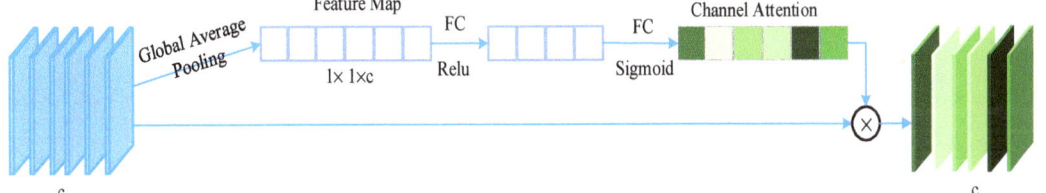

Figure 6. Channel attention module.

Then the $m \times n$ dimensions are compressed to 1×1 by global average pooling. The global feature distribution of the input matrix in the channel dimension is captured to obtain the feature map

$$map = \frac{1}{m \times n} \sum_{i=1}^{m} \sum_{j=1}^{n} x(i,j) \tag{11}$$

The feature maps are adjusted nonlinearly by the fully connected layer (FC). The module uses the sigmoid function to obtain the attentional weights of the channel dimensions C_{atte}

$$C_{atte} = \sigma(w_s \cdot (\text{Relu}(w_r \cdot map + b_r)) + b_s) \tag{12}$$

Finally, the input features X_{in} are multiplied with the channel weights to rescale the features in the channel dimension.

The channel dimension completes the rescaling of the original features, and the channel attention adjusts the different channel features. However, there are also large differences in the data of different fault types of vibration signals in the same channel, as shown in Figure 7. Convolutional neural networks also need to consider the influence of different location features on the diagnosis results when extracting features. Therefore, this paper makes the network focus on the features of vibration signals in spatial dimensional features by position attention.

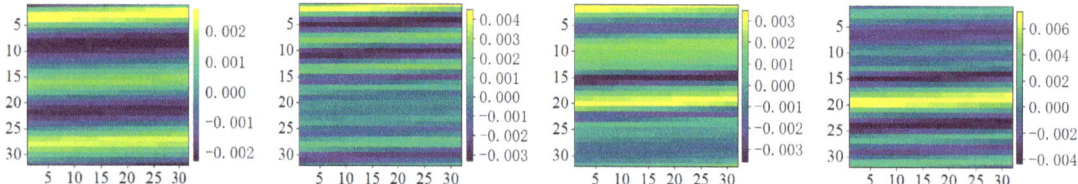

Figure 7. Data visualization of different fault types in the same channel.

The structure of the position attention is shown in Figure 8. The input features $x \in \mathrm{R}^{c \times m \times n}$ are computed separately for max pooling and average pooling to obtain feature maps $f_{max} \in \mathrm{R}^{1 \times m \times n}$ and $f_{avg} \in \mathrm{R}^{1 \times m \times n}$. Then the feature maps are concatenated in the channel dimension. Finally, the feature maps adopt convolutions and a sigmoid activation function to obtain the position attention P_{atten}

$$P_{atten} = \sigma(\mathrm{conv}(\mathrm{concat}(f_{max}, f_{avg}))). \tag{13}$$

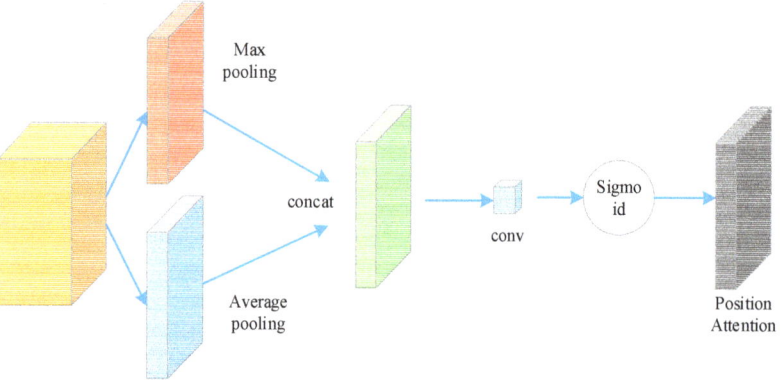

Figure 8. Position attention module.

4. Spatiotemporal Feature Fusion Network

The structure of the spatiotemporal feature fusion network is shown in Figure 9. The STNet uses a GRU to extract the temporal features of one-dimensional vibration signals. The GRU branch introduces the attention mechanism to synthesize the effect of each moment state on the performance in the long sequence signal. Meanwhile, the original vibration sequence is decomposed by LMD for time-frequency analysis. The original vibration data and each PF component are converted into multidimensional matrices as the input of the CNN. The CNN branch adaptively extracts the spatial features of the input matrix by convolutions. Meanwhile, considering the channel features and the influence of different fault features, the CNN branch adds channel attention and position attention to selectively enhance the spatial features of the signal. The attention mechanism acquires rich contextual information. Finally, the spatial and temporal features of the vibration signal are fused, and the softmax layer classifies the fused features.

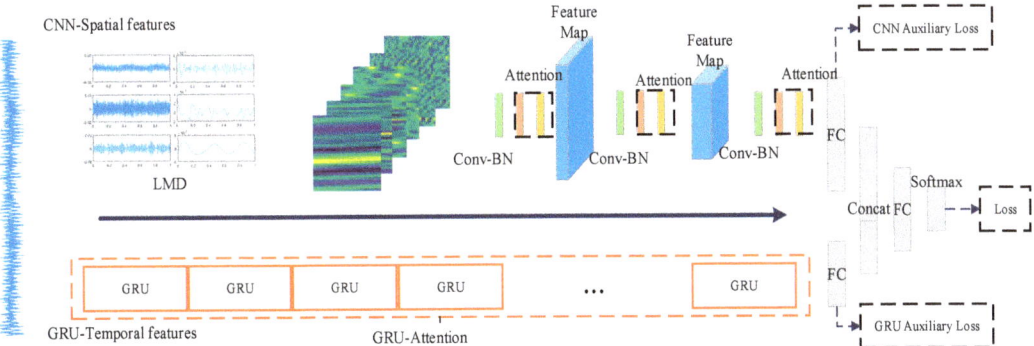

Figure 9. Spatiotemporal feature fusion network.

STNet is a dual-stream network consisting of a GRU branch and CNN branch. The specific network layers are shown in Table 1, where Conv-BN denotes the convolution layer and batch normalization layer, and FC is the fully connected layer. The input of the CNN branch is the vibrational signal matrix with the size of 6 × 32 × 32. The network uses the convolution kernel with the size of 3 × 3 to extract features. The padding type of the convolution kernel is "SAME". Then, the kernel is normalized by the BN layer with a Relu activation function. The CNN branch recalibrates the original features by channel attention and position attention. The spatial resolution of the feature map at each stage becomes half that of the previous stage, and the number of channels becomes twice that of the previous stage. The network obtains a feature map with the size of 128 × 8 × 8 by three stages of feature extraction. The captured features are then fed into the fully connected layer with 1024 neurons. The input of the GRU branch is the original vibration signal with 1024 sampling points. The network obtains the temporal features through the 2-layer GRU attention unit, and the features are fed into the fully connected layer with 128 neurons. The fully connected layers of the CNN branch and GRU branch are concatenated, and the number of neurons is 1152. The network is nonlinearly adjusted by two fully connected layers. Finally, the diagnosis results of eight faults are output by the softmax function.

When the STNet extracts features, there are significant differences between the spatial features extracted by the CNN and the temporal features extracted by the GRU. Therefore, the CNN auxiliary loss function and GRU auxiliary loss function are added respectively during the training process. The auxiliary loss function supervises the temporal features and spatial features extracted by the network separately to reduce the generation of invalid information. The auxiliary loss function not only promotes the backpropagation of the network but also enhances the canonical representation of temporal and spatial features. The final loss function (L_{total}) of the network is shown as follows

$$L = \frac{1}{N}\sum_i L_i = -\frac{1}{N}\sum_i \sum_{c=1}^{M} y_{ic} \log(p_{ic}) \tag{14}$$

$$L_{\text{total}} = \alpha L_{\text{CNN}} + \beta L_{\text{GRU}} + L_{\text{loss}} \tag{15}$$

where M is the number of categories; y_{ic} is the symbolic function; p_{ic} is the probability that sample i belongs to c; α and β are the weights of the auxiliary loss function.

Table 1. The STNet's structure.

Layer	Node	Stride	Output Size	Layer	Node	Stride	Output Size
CNN Branch				GRU Branch			
			6 × 32 × 32				1024
Conv-BN	32	2	32 × 16 × 16	FC	990	-	990
Channel-Position Attention	-	1	32 × 16 × 16	GRU	330	-	330
Conv-BN	64	2	64 × 8 × 8	Attention	-	-	330
Channel-Position Attention	-	1	64 × 8 × 8	GRU	110	-	110
Conv-BN	128	2	128 × 8 × 8	Attention	-	-	110
Channel-Position Attention	-	1	128 × 8 × 8	FC	128	-	128
FC	1024	-	1024				
			Concat (1152)				
			FC (512)-FC (128)				
			Softmax (8)				

5. Experiments

5.1. Data

The main types of faults in the experimental motor vibration data are inter-turn short circuit, air gap eccentricity, rotor broken strips, bearing seat damage, bearing wear, etc. There are 8 kinds of samples, the number of samples is 8000, and the number of sampling points per second is 1024, as shown in Table 2. The deep learning framework is PaddlePaddle 1.8.4. The CPU of the training platform is Intel Xeon Gold 6171C. The GPU is Nvidia Tesla V100 (16G). GPU acceleration is performed by CUDA 10.1, and the experimental dataset is divided into training and test sets (7:3).

Table 2. Fault types.

Label	Types	Numbers
0	Normal	1000
1	2 turns short circuit	1000
2	4 turns short circuit	1000
3	8 turns short circuit	1000
4	Air gap eccentricity	1000
5	Broken rotor strip	1000
6	Bearing seat damage	1000
7	Bearing wear	1000

5.2. Experiment Analysis

When the network extracts features using the GRU, only the features in the time domain of the vibration signal are captured. However, the vibration signal also contains rich features in the frequency domain. Therefore, the original vibration data is decomposed by LMD. The decomposition results of each fault type are shown in Figure 10. When abnormal vibration occurs in the accelerometer, each PF component can show the amplitude modulation and frequency modulation signals of the abnormal vibration. The vibration signal is enhanced so that the CNN extracts the vibration features by the original vibration sequence and each PF component.

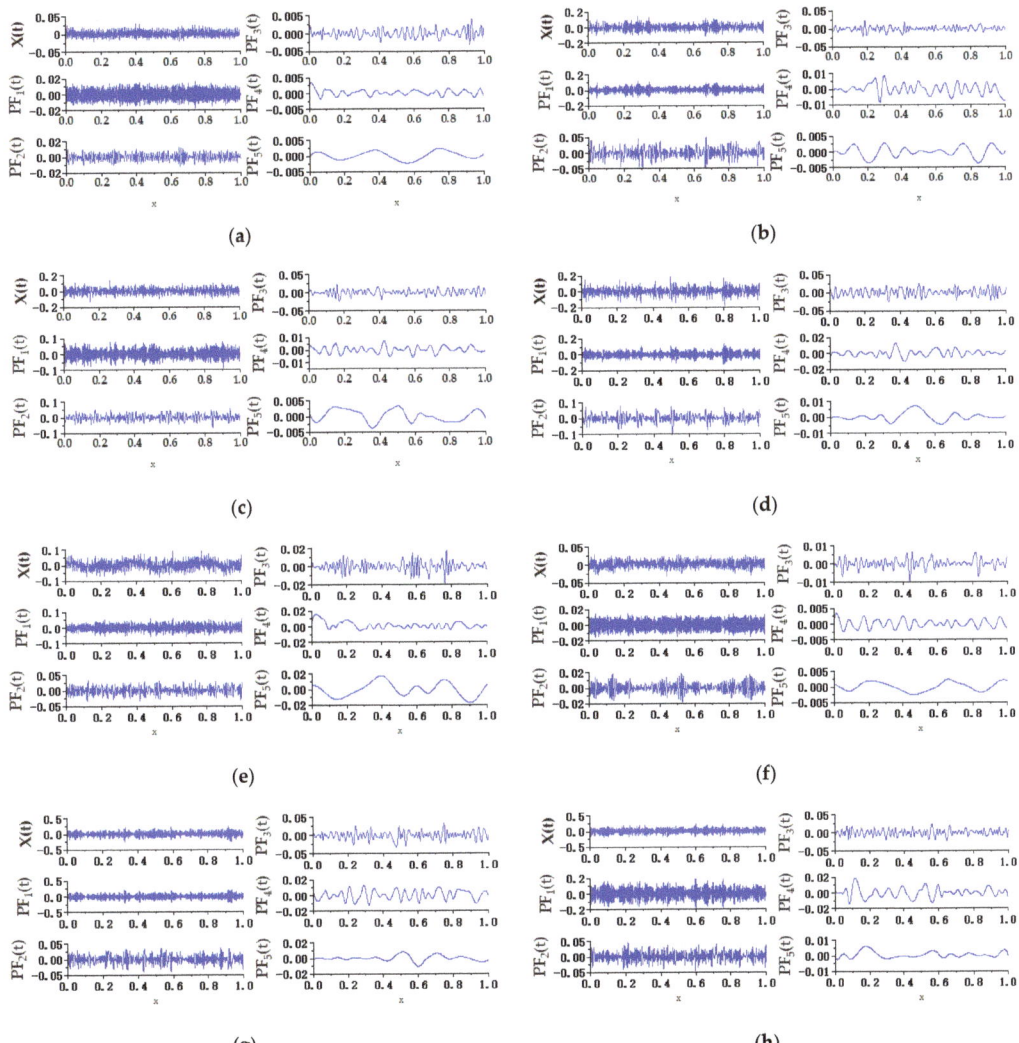

Figure 10. Visualization of local mean decomposition of fault signals. (**a**) normal; (**b**) 2 turns short circuit; (**c**) 4 turns short circuit; (**d**) 8 turns short circuit; (**e**) air gap eccentricity; (**f**) broken rotor strip; (**g**) bearing seat damage; (**h**) bearing wear.

A convolutional neural network has unique superiority in two-dimensional image recognition due to the special structure of local weight sharing and the presence of the local perceptual field. The visualization results of each fault signal transformed into the two-dimensional matrix are shown in Figure 11. The original vibration signal is 1024 sampling points, and the size of the transformed 2D matrix is 32 × 32. Similarly, each PF component is also transformed into a two-dimensional matrix and connected to the two-dimensional matrix of the original vibration signal in the channel dimension. Finally, the input size of the CNN branch is 6 × 32 × 32. The visualization results of the two-dimensional matrix show that the PF component matrices of different faults have large differences in different dimensions, and the fault features extracted by the CNN would have a positive effect on the performance of diagnosis.

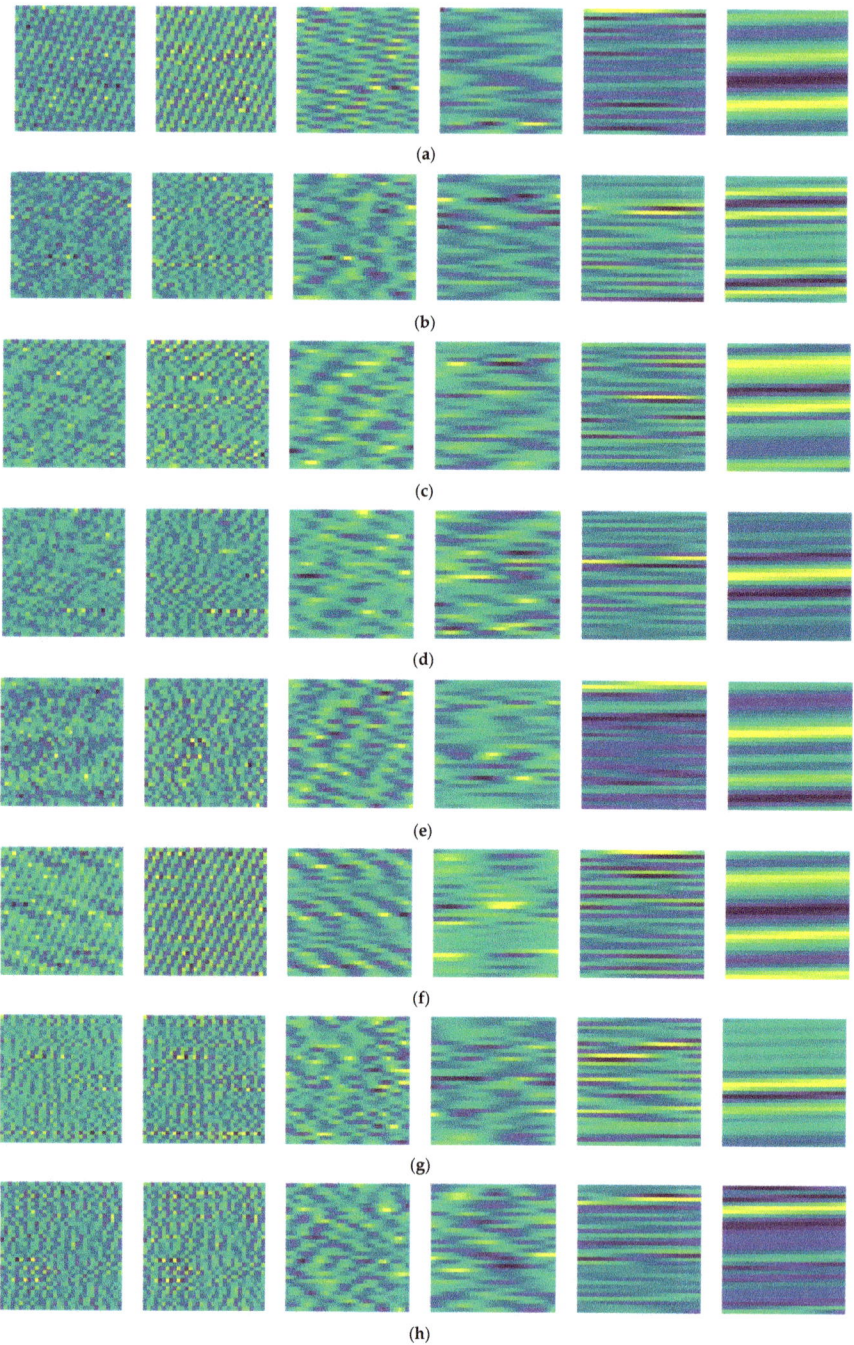

Figure 11. Two-dimensional matrix visualization of fault data. (**a**) normal; (**b**) 2 turns short circuit; (**c**) 4 turns short circuit; (**d**) 8 turns short circuit; (**e**) air gap eccentricity; (**f**) broken rotor strip; (**g**) bearing seat damage; (**h**) bearing wear.

The input size of the CNN branch is 6 × 32 × 32, and the sequence length of the GRU branch input is 1024. The number of network training epochs is 100. The batch size is 600. The model parameters are updated using the Adam optimization algorithm. The learning rate adjustment strategy is "Poly", with an initial learning rate of 0.001 and a power of 0.9. The loss function is the cross-entropy loss function. The weight of the CNN network auxiliary loss function is 0.1. The weight of the GRU network auxiliary loss function is 0.9. The evaluation index is the accuracy rate. The loss and accuracy curves of the training set and test set with the number of epochs are shown in Figure 12.

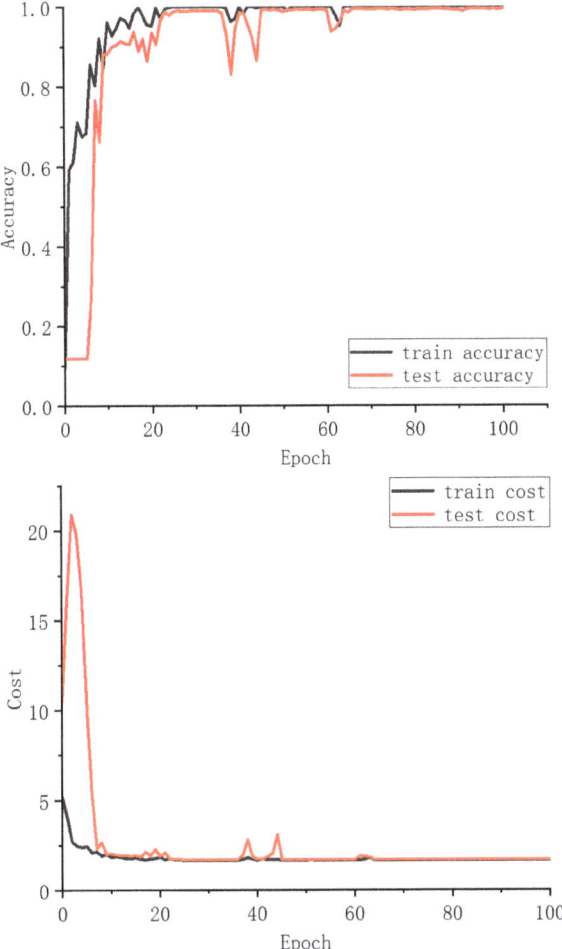

Figure 12. Training process loss and accuracy variation.

The test set loss increases sharply in the first 10 rounds of training, but the training set and test set losses gradually decrease with the increase of iterations. It indicates that the model is converging and approaching 0. After 60 epochs, the training set loss and test set loss are close to overlapping. The waveforms do not have large fluctuations and there are no overfitting problems.

The model is validated for each type of fault after training, and the results are shown in Table 3. The number of error samples for inter-turn short circuit fault is three, and the number of error samples for bearing seat damage is three. The recognition accuracy of each type of fault is above 99%. The model has high recognition accuracy.

Table 3. The result of each category of fault identification.

Label	Types	Accuracy
0	Normal	100%
1	2 turns short circuit	99.67%
2	4 turns short circuit	99.33%
3	8 turns short circuit	100%
4	Air gap eccentricity	100%
5	Broken rotor strip	100%
6	Bearing seat damage	99%
7	Bearing wear	100%

To verify the performance of each module in the STNet, five ablation experiments are set up. The results are shown in Table 4. The accuracy of the temporal features extracted from the vibration signal using the GRU is 98.58%, while the accuracy of the spatial features captured from the vibration signal using the CNN is 98.83%. The CNN + GRU model with the fusion of temporal and spatial features improves the accuracy by 0.39% and 0.04%, respectively. Compared with the single branch, it indicates that both temporal and spatial features of the vibration signal are indispensable parts for fault diagnosis. The CNN + GRU + attention model with the attention module on the CNN branch and GRU branch improves the accuracy by 0.59% compared to the model without attention. The attention mechanism considers the importance of different features and makes the important features play a significant role in the network. The final accuracy of the STNet with auxiliary loss function is 99.75%. The auxiliary loss function facilitates the network backpropagation to update the parameters and enhances the feature representation of each branch.

Table 4. Ablation experiments.

Model	Accuracy
GRU	98.58%
CNN	98.83%
CNN + GRU	98.97%
CNN + GRU + Attention	99.56%
CNN + GRU + Attention + Auxiliary Loss	99.75%

To further investigate the effect of the attention module on the network performance, the attention matrices of the GRU branch and the CNN branch are visualized. Figure 13a represents the channel attention for the three-stage feature extraction in the CNN branch with channel dimensions of 32, 64, and 128. The shallow layer of the CNN branch requires sufficient feature extraction of the vibration signal to preserve all feature information as much as possible. Therefore, the attention varies from 0.48 to 0.51, which is not a large range. Due to the number of network layers increasing and the number of channels increasing, the redundant features are increased. The network needs to suppress the redundant channels, while the effective channel features are enhanced. So, the range of variation of channel attention increases. Figure 13b represents the position attention of the three-stage feature extraction in the CNN branch with dimensions of 16×16, 8×8, and 8×8. The position attention becomes more and more focused because the local features of the convolutional neural network are extracted. Figure 13c represents the attention of the output features of the second GRU in the GRU branch. The GRU module outputs the prediction results of multiple time series. The output represents the impact of each moment on the diagnostic results. It retains the results with high relevance by the attention mechanism, so the GRU attention does not fluctuate greatly.

To further verify the fault diagnosis capability of the STNet, it is compared with BP, 1D-CNN, multichannel-CNN, and inception-LSTM models. The experimental results are shown in Table 5. The BP network diagnoses the fault types by nonlinear mapping without considering the temporal and spatial features of the signal. Therefore, the recognition accuracy is only 96.12%. The 1D-CNN model uses 1D convolution to obtain the abstract features and local features of the vibration signal. The 1D-CNN model improves the accuracy by 2.12% compared to the BP network. The multichannel-CNN model weights different receptive fields and captures contextual information at different scales. The inception-LSTM model extracts temporal information under several different receptive fields with an accuracy of 99.34%. Compared with BP, 1D-CNN, multichannel-CNN, and inception-LSTM models, the STNet obtains the highest accuracy of 99.75%. The STNet combines spatial features and temporal features instead of single features, compared with BP, 1D-CNN, and multichannel-CNN models. Compared with the inception-LSTM model, STNet uses the attention mechanism to select features adaptively. Therefore, both temporal and spatial features have a positive impact on the performance of diagnosis during the analysis of vibration signals. The number of parameters of STNet is 9.2876 M and the number of floating-point operations (FLOPs) is 0.02 G.

Table 5. Model comparison experiments.

Model	Accuracy
BP	96.12%
1D-CNN	98.24%
Multichannel-CNN	99.17%
Inception-LSTM	99.34%
STNet	99.75%

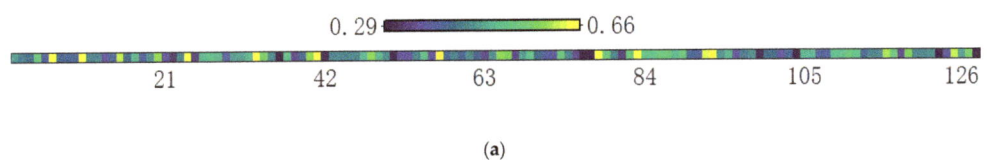

(a)

Figure 13. *Cont.*

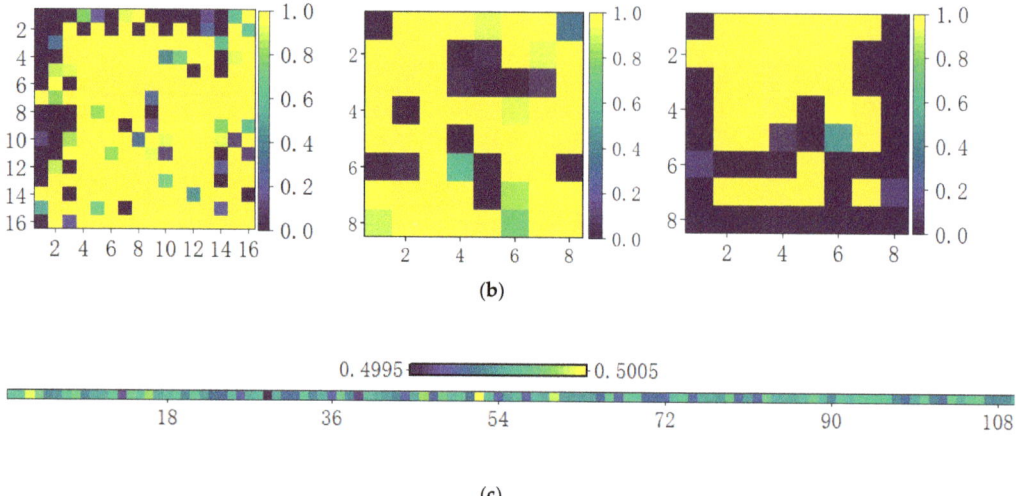

Figure 13. Attention visualization. (**a**) CNN branching channel attention visualization; (**b**) CNN branching position attention visualization; (**c**) GRU branch attention visualization.

6. Conclusions

In the paper, the fault diagnosis for motor vibration signals has been investigated based on spatiotemporal feature fusion. The method has used gated recurrent units and convolutional neural networks to extract the temporal and spatial features of vibration signals. Since the time series of vibration signals were too long to retain all the key information, a GRU has extracted the temporal features by an attention mechanism to effectively synthesize the states of different time series and the vibration features at different moments. When extracting spatial features, the one-dimensional time-domain signal has been converted into a two-dimensional matrix using local mean decomposition and matrix transformation to extend the data dimensionality. The CNN model based on the attention mechanism adaptively has extracted the channel and location features of the signal. In the experimental evaluation of eight different vibration signals, the vibration signal processing method combined with spatiotemporal feature fusion has obtained 99.75% recognition accuracy. The method has improved the diagnostic performance effectively, which is important for the safe detection and stable operation of the system.

Author Contributions: Conceptualization, L.W. and F.X.; methodology, L.W.; software, C.Z.; validation, L.W., C.Z., J.Z. and F.X.; writing—review and editing, L.W., C.Z., J.Z. and F.X.; visualization, F.X.; supervision, F.X.; project administration, F.X.; funding acquisition, L.W. and F.X. All authors have read and agreed to the published version of the manuscript.

Funding: This work was supported in part by the Heilongjiang Province Key R&D Program (GA21A304), the Fundamental Research Funds in Heilongjiang Provincial Universities (135509404 and 135409102), and the Open project of Agricultural multidimensional sensor information Perception of Engineering and Technology Center in Heilongjiang Province (DWCGQKF202105).

Institutional Review Board Statement: Not applicable.

Informed Consent Statement: Not applicable.

Data Availability Statement: Not applicable.

Conflicts of Interest: The authors declare no conflict of interest.

References

1. Yu, J.; Liu, X. One-dimensional residual convolutional auto-encoder for fault detection in complex industrial processes. *Int. J. Prod. Res.* **2021**, *196*, 1–20. [CrossRef]
2. Han, T.; Liu, C.; Yang, W.; Jiang, D. A novel adversarial learning framework in deep convolutional neural network for intelligent diagnosis of mechanical faults. *Knowl. Based Syst.* **2019**, *165*, 474–487. [CrossRef]
3. Chen, Z.; Gryllias, K.; Li, W. Intelligent fault diagnosis for rotary machinery using transferable convolutional neural network. *IEEE Trans. Ind. Inform.* **2019**, *16*, 339–349. [CrossRef]
4. Chen, H.; Jiang, B. A review of fault detection and diagnosis for the traction system in high-speed trains. *IEEE Trans. Intell. Transp. Syst.* **2020**, *21*, 450–465. [CrossRef]
5. Chen, H.; Jiang, B.; Ding, S.; Huang, B.S. Data-driven fault diagnosis for traction systems in high-speed trains: A survey, challenges, and perspectives. *IEEE Trans. Intell. Transp. Syst.* **2020**, *23*, 1–17. [CrossRef]
6. Hsueh, Y.-M.; Ittangihal, V.R.; Wu, W.-B.; Chang, H.-C.; Kuo, C.-C. Fault diagnosis system for induction motors by CNN using empirical wavelet transform. *Symmetry* **2019**, *11*, 1212. [CrossRef]
7. Kao, I.H.; Wang, W.J.; Lai, Y.H.; Perng, J.W. Analysis of permanent magnet synchronous motor fault diagnosis based on learning. *IEEE Trans. Instrum. Meas.* **2019**, *68*, 310–324. [CrossRef]
8. Namdar, A.; Samet, H.; Allahbakhshi, M.; Tajdinian, M.; Ghanbari, T. A robust stator inter-turn fault detection in induction motor utilizing kalman filter-based algorithm. *Measurement* **2022**, *187*, 110181. [CrossRef]
9. Vinayak, B.; Anand, K.; Jagadanand, G. Wavelet-based real-time stator fault detection of inverter-fed induction motor. *IET Electr. Power Appl.* **2020**, *14*, 82–90. [CrossRef]
10. Ben Abid, F.; Sallem, M.; Braham, A. Robust interpretable deep learning for intelligent fault diagnosis of induction motors. *IEEE Trans. Instrum. Meas.* **2020**, *69*, 3506–3515. [CrossRef]
11. Hasan, M.J.; Islam, M.M.M.; Kim, J.M. Bearing fault diagnosis using multidomain fusion-based vibration imaging and multitask learning. *Sensors* **2022**, *22*, 56. [CrossRef]
12. Karabacak, Y.E.; Özmen, N.G.; Gümüşel, L. Intelligent worm gearbox fault diagnosis under various working conditions using vibration, sound and thermal features. *Appl. Acoust.* **2022**, *186*, 108463. [CrossRef]
13. Mao, W.; Feng, W.; Liu, Y.; Zhang, D.; Liang, X. A new deep auto-encoder method with fusing discriminant information for bearing fault diagnosis. *Mech. Syst. Signal Process.* **2021**, *150*, 107233. [CrossRef]
14. Yan, X.; Liu, Y.; Jia, M.; Zhu, Y. A multi-stage hybrid fault diagnosis approach for rolling element bearing under various working conditions. *IEEE Access* **2019**, *7*, 138426–138441. [CrossRef]
15. Jan, S.; Lee, Y.-D.; Shin, J. Sensor fault classification based on support vector machine and statistical time-domain features. *IEEE Access* **2017**, *5*, 8682–8690. [CrossRef]
16. Cui, L.; Huang, J.; Zhang, F. Quantitative and localization diagnosis of a defective ball bearing based on vertical horizontal synchronization signal analysis. *IEEE Trans. Ind. Electron.* **2017**, *66*, 8695–8706. [CrossRef]
17. Shao, S.-Y.; Sun, W.-J.; Yan, R.-Q.; Wang, P.; Gao, R.X. A deep learning approach for fault diagnosis of induction motors in manufacturing. *Chin. J. Mech. Eng.* **2017**, *30*, 1347–1356. [CrossRef]
18. Lin, H.; Ye, Y.; Huang, B.; Su, J. Bearing vibration detection and analysis using enhanced fast Fourier transform algorithm. *Adv. Mech. Eng.* **2016**, *8*, 1687814016675080. [CrossRef]
19. Qu, J.; Zhang, Z.; Gong, T. A novel intelligent method for mechanical fault diagnosis based on dual-tree complex wavelet packet transform and multiple classifier fusion. *Neurocomputing* **2016**, *171*, 837–853. [CrossRef]
20. Li, L.; Han, N.N.; Jiang, Q.T. A chirplet transform-based mode retrieval method for multicomponent signals with crossover instantaneous frequencies. *Digit. Signal Process.* **2022**, *120*, 103262. [CrossRef]
21. Li, H.; Zhang, Q.; Qin, X.; Sun, Y.T. Fault diagnosis method for rolling bearings based on short-time Fourier transform and convolution neural network. *Shock Vib.* **2018**, *37*, 124–131.
22. Ali, J.B.; Fnaiech, N.; Saidi, L.; Chebel-Morello, B.; Fnaiech, F. Application of empirical mode decomposition and artificial neural network for automatic bearing fault diagnosis based on vibration signals. *Appl. Acoust.* **2015**, *89*, 16–27.
23. Yu, X.; Dong, F.; Ding, E.; Wu, S.; Fan, C. Rolling bearing fault diagnosis using modified LFDA and EMD with sensitive feature selection. *IEEE Access* **2017**, *6*, 3715–3730. [CrossRef]
24. Zhang, W.; Li, C.; Peng, G.; Chen, Y.; Zhang, Z. A deep convolutional neural network with new training methods for bearing fault diagnosis under noisy environment and different working load. *Mech. Syst. Signal Process.* **2018**, *100*, 439–453. [CrossRef]
25. Zhu, J.; Chen, N.; Peng, W. Estimation of bearing remaining useful life based on multiscale convolutional neural network. *IEEE Trans. Ind. Electron.* **2019**, *66*, 3208–3216. [CrossRef]
26. Li, D.; Zhang, M.; Kang, T.; Li, B.; Xiang, H.; Wang, K.; Pei, Z.; Tang, X.; Wang, P. Fault diagnosis of rotating machinery based on dual convolutional-capsule network (DC-CN). *Measurement* **2022**, *187*, 110258. [CrossRef]
27. Chen, H.; Liu, Z.; Alippi, C.; Huang, B.; Liu, D. Explainable Intelligent Fault Diagnosis for Nonlinear Dynamic Systems: From Unsupervised to Supervised Learning. *TechRxiv* **2022**. Preprint. [CrossRef]
28. Shi, H.; Guo, L.; Tan, S.; Bai, X.; Sun, J. Rolling bearing initial fault detection using long short-term memory recurrent network. *IEEE Access* **2019**, *7*, 171559–171569. [CrossRef]
29. Gao, J.; Guo, Y.; Wu, X. Gearbox bearing fault diagnosis based on SANC and 1-D CNN. *Shock Vib.* **2020**, *39*, 204–209.

30. Zhu, X.; Hou, D.; Zhou, P.; Han, Z.; Yuan, Y.; Zhou, W.; Yin, Q. Rotor fault diagnosis using a convolutional neural network with symmetrized dot pattern images. *Measurement* **2019**, *138*, 526–535. [CrossRef]
31. Guo, S.; Zhang, B.; Yang, T.; Lyu, D.; Gao, W. Multitask Convolutional Neural Network with Information Fusion for Bearing Fault Diagnosis and Localization. *IEEE Trans. Ind. Electron.* **2019**, *67*, 8005–8015. [CrossRef]
32. Cao, P.; Zhang, S.; Tang, J. Preprocessing-free gear fault diagnosis using small datasets with deep convolutional neural network-based transfer learning. *IEEE Access* **2018**, *6*, 26241–26253. [CrossRef]
33. Liu, H.; Zhou, J.; Zheng, Y.; Jiang, W.; Zhang, Y. Fault diagnosis of rolling bearings with recurrent neural network-based autoencoders. *ISA Trans.* **2018**, *77*, 167–178. [CrossRef] [PubMed]

Article

Multipoint Feeding Strategy of Aluminum Reduction Cell Based on Distributed Subspace Predictive Control

Jiarui Cui [1,2], Peining Wang [1], Xiangquan Li [3], Ruoyu Huang [2], Qing Li [1,*], Bin Cao [2,4] and Hui Lu [2]

1. School of Automation and Electrical Engineering, University of Science and Technology Beijing, Beijing 100083, China; cuijiarui@ustb.edu.cn (J.C.); g20198738@xs.ustb.edu.cn (P.W.)
2. Guiyang Aluminum Magnesium Design and Research Institute Co., Ltd., Guiyang 550081, China; ry_huang@chalieco.com.cn (R.H.); caobinh@yeah.net (B.C.); hui_lu@chalieco.com.cn (H.L.)
3. School of Information Engineering, Jingdezhen University, Jingdezhen 333000, China; b20180295@xs.ustb.edu.cn
4. Chinalco Intelligent Technology Development Co., Ltd., Hangzhou 311199, China
* Correspondence: liqing@ies.ustb.edu.cn

Abstract: With the continuous development of large-scale aluminum reduction cells, the problem of the uniform distribution of alumina concentration in the cell has become more and more serious for the reduction process. In order to achieve the uniform distribution of the alumina concentration, a data-driven distributed subspace predictive control feeding strategy is proposed in this paper. Firstly, the aluminum reduction cell is divided into multiple sub-systems that affect each other according to the position of the feeding port. Based on the subspace method, the prediction model of the whole cell is identified, and the prediction output expression of each sub-system is deduced by decomposition. Secondly, the feeding controller is designed for each aluminum reduction cell subsystem, and the input and output information can be exchanged between each controller through the network. Thirdly, under consideration of the influence of other subsystems, each subsystem solves the Nash-optimal control feeding quantity, so that each subsystem realizes distributed feeding. Finally, the simulation results show that, compared with the traditional control method, the proposed distributed feeding control strategy can significantly improve the problem of the uniform distribution of alumina concentration and improve the current efficiency of the aluminum reduction cell.

Keywords: aluminum reduction process; alumina concentration; subspace identification; distributed predictive control

1. Introduction

In order to improve labor productivity and reduce investment costs, large-capacity pre-baked reduction cells are currently used in various enterprises. Due to their high efficiency and low energy consumption, pre-baked cells of 400–600 kA have gradually become the mainstream cell type in China's aluminum electrolytic industry [1]. The capacity of the reduction cell is continuously increasing, while the auxiliary facilities and the intelligent control technology of aluminum electrolytic are relatively backward. Thus, the problems of the local anode effect and local precipitation in the large reduction cell have increasingly become the main factors of instability in the production process [2], which has caused serious economic losses and some casualties. The main cause of these problems is the uneven distribution of alumina concentration in the anode bottom surface of the large aluminum reduction cell [3]. Through some studies and experiments, it is known that the concentration of alumina is generally controlled between 1.5% and 3.5%. Currently, the change of groove resistance and the concentration of alumina is basically linear and easy to identify, and the current efficiency is also the highest [4]. If the concentration of alumina is excessively high, there will be problems such as increased energy consumption, fluctuation of aluminum liquid layer, etc., and even induced precipitation at the bottom of the cell and crusting on the cell side which reduce the service life of the reduction cell [5]. An excessively low concentration of alumina will lead to frequent anode effects in the reduction cell. Once the anode effect occurs, the cell voltage rises sharply. Meanwhile, Moxnes et al. [6] found that when the alumina concentration distribution is more uniform, the current efficiency of the reduction cell is higher, and the probability of abnormal cell conditions such as the anode effect is lower. Therefore, adjusting the feeding interval of each feeder of the large-scale aluminum reduction cell and accurately controlling the amount of alumina feeding to make the alumina concentration evenly distributed has become a general concern and urgent issue that has extremely important practical significance for the further development of large and super large aluminum electrolytic technology [5].

The uniform distribution of alumina concentration plays a vital role in the stable operation and efficient production of large aluminum electrolysis. In order to achieve this goal, many scholars have carried out in-depth research on alumina. In [7], a method combining fuzzy control theory and expert experience was proposed to control the alumina concentration by changing the feeding interval. In [4], by improving the crust breaking and feeding device and control system of the reduction cell, the feeding interval was set separately for each feeding point, and the single-point precise feeding control is achieved. However, for large aluminum reduction cells with multiple feeders, single-point feeding control method cannot effectively control the uniform distribution of the alumina concentration. With the development of soft measurement technology [8,9] and distributed data measurement technology, more and more scholars have begun to try to apply soft-sensing the aluminum electrolysis industry. The least squares support vector machine method for the alumina concentration soft measurement model is established in [10]. In [11], in order to obtain more accurate results, a soft-sensor model of alumina concentration was proposed that introduces time series to optimize the input parameters of a deep belief network (DBN). In [12], a KPI was developed through a probabilistic soft sensing based on maximizing the coefficient of determination to estimate the alumina concentration. An improved Kalman filter for the soft sensing of alumina concentration is presented in [13]. Therefore, intelligent control methods based on the soft measurement model of the alumina concentration emerge endlessly. The generalized regression neural network (GRNN) was adopted to identify the alumina concentration model and a fuzzy cerebellar model neural network (FCMAC) controller was proposed for the feeding equipment to control the alumina concentration in [14]. A data-driven intelligent control system based on the least squares support vector machine alumina concentration soft sensing model was proposed in [15]. An extended Kalman filter (EKF) was used to estimate the local alumina concentration to design a multivariable blanking control strategy in [16]. However, these control methods ignore the influence of each feeding port on the alumina concentration near other feeding ports due to the flow of electrolytes during the aluminum reduction process, and only consider the overall parameters such as cell voltage and cell resistance, and do not fully utilize important distribution parameters. In order to fully understand the distribution of an alumina concentration in aluminum reduction cells, the dissolution and diffusion of alumina were studied in [17–20]. In [21], a multi-coupling distributed alumina simulation model was constructed by ANSYS, and the uniform distribution of alumina concentration was achieved through the simulation model. However, it is still very difficult to establish an accurate model through mechanism analysis. Some scholars have studied the production process using a large amount of data. At present, the most advanced technology was that the relationship between the distributed current and the feeding rate was established using random forest, and the stability of the aluminum reduction cell was maintained by controlling the distributed current to be consistent in [22]. However, the final simulation results did not verify the uniform distribution of alumina concentration.

In order to solve the problem of the uneven distribution of alumina concentration and adapt to the aluminum electrolysis process with difficulty in mechanism modeling, a distributed subspace predictive control data-driven method is applied to large aluminum reduction cells, and the subsystem model is established by fully using distributed data.

The distributed control algorithm is used to realize the distributed feeding of multiple feeders in large electrolytic cell considering the influence of each feeder, and the simulation is carried out in MATLAB. The simulation results show that the application of this method in a large aluminum reduction cell is feasible, and it has important guiding significance for realizing the uniform distribution of alumina concentration in a large aluminum reduction cell in industrial processes.

Compared with the existing research on alumina concentration control strategy, the contributions of this paper are as follows:

(1) The large aluminum reduction cell is divided into several subsystems according to the position of the feeder. Compared with the work in [14,15], the difference is that this paper considers the influence of each feeding port caused by the flow of the electrolyte between subsystems on the alumina concentration near other feeding ports.

(2) Inspired by the work of [22], this paper designs the controller by establishing a prediction model between the feed rate and alumina concentration in each subsystem, and the input and output information can be exchanged between each subsystem through the network.

(3) Compared with the traditional timing grouping feeding strategy, a new distributed control feeding strategy is designed in this paper, so that each feeding device is controlled by an independent controller. Each feeder works in coordination with the influence of other subsystems' feeding, realizing on-demand distributed feeding, and improving the control performance of each subsystem [23].

The rest of this paper is organized as follows. In Section 2, a distributed feeding control strategy for aluminum electrolysis is proposed. Section 3 introduces the distributed subspace predictive control algorithm and discusses the implementation of the proposed algorithm in aluminum electrolysis in detail. In Section 4, the feasibility of the algorithm is verified by MATLAB simulation. In Section 5, relevant conclusions are given.

2. Design of Distributed Feeding Control Scheme for Aluminum Electrolysis

The top view of a 400 kA large aluminum reduction cell in an aluminum plant is shown in Figure 1. FD1, FD2, FD3, FD4, FD5, and FD6 are the six feed ports of the cell. Twenty-four anode guides are on the B side. Due to the flow of electrolytes caused by carbon dioxide and carbon monoxide gas produced by anode, electromagnetic field, temperature and concentration difference, when feeding at any feeding port, the alumina concentration in other regions will be affected to some extent. Taking the six feeding ports as centers, the aluminum reduction cell is divided into six subsystems. For aluminum electrolysis, which is a complex large system composed of multiple mutually influencing subsystems, distributed control cannot only better consider the impact of feeding between subsystems but the computation complexity is greatly reduced compared to centralized control. As predictive control has been widely used in engineering applications and has high control accuracy, distributed model predictive control has received more attention from scholars.

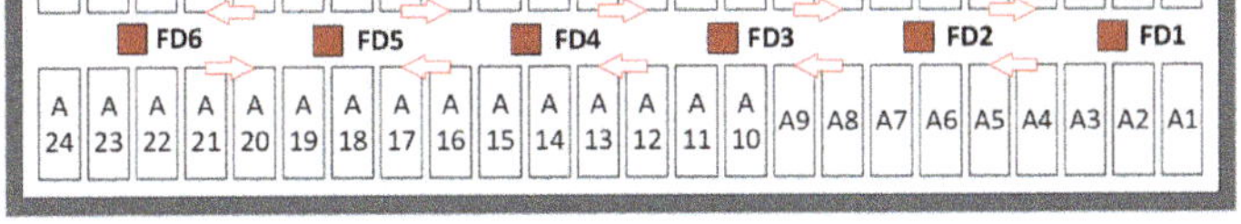

Figure 1. Top view of 400 kA large aluminum reduction cell.

The main idea of the distributed predictive control algorithm is to transform a large-scale online optimization problem into a small-scale distributed optimization of each subsystem, and at the same time, each subsystem communicates and shares information through the network, thereby improving the control performance of the system. For the general distributed model predictive control method, the design of the controller must be based on accurate modeling. Due to the complexity of the aluminum reduction process, there are many difficulties in the research on the multi-point distributed feeding control of the aluminum electrolysis process. Firstly, the aluminum reduction process is a dynamic system with complex physical and chemical reactions, multi-coupling and large delay [24]. Secondly, it is difficult to establish a precise distributed multi-point feeding mechanism model due to the extremely complex environment such as high temperature and strong corrosion in industrial aluminum reduction cells. Finally, it is difficult to obtain the coupling relationship between each subsystem. The subspace identification method is not limited to the prior structure information and mechanism model of the system but directly uses the historical input and output data to solve the prediction model [25,26], and can obtain the prediction model of each subsystem through parameter decomposition. This method is more suitable for complex large systems composed of multiple subsystems and it is difficult to establish an accurate mechanism model [27,28]. Therefore, the data-driven subspace identification method and predictive control are designed in a control system design framework, which is applied to the aluminum electrolysis system with model uncertainty and has better control performance [29–31].

For large aluminum reduction cells, since the distributed alumina concentration data cannot be obtained in real-time, the development of distributed current measurement technology provides the basis for its soft measurement. According to the relevant mechanism of an aluminum reduction cell, there is a close relationship between the distributed alumina concentration and distributed current. This research group has also conducted in-depth research on the soft measurement of distributed alumina concentration [32,33]. Therefore, the distributed alumina concentration data required in this paper can be obtained by using the soft sensing model, to carry out the follow-up work.

For the 400 kA aluminum reduction cell, the structure of the predictive control principle of one of the subsystems is shown in Figure 2. It is mainly composed of a subspace prediction model and distributed controller.

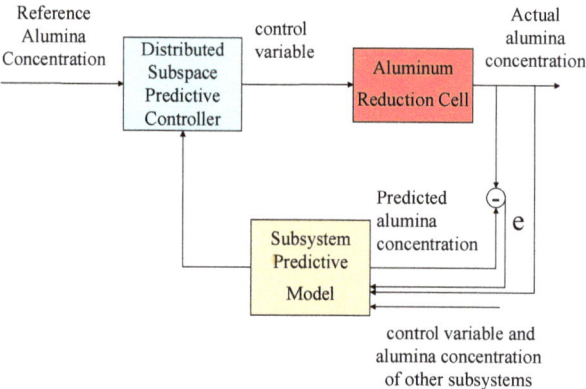

Figure 2. Predictive control diagram of aluminum electrolysis subsystem.

The subspace prediction model is a model of the entire system identified by the input and output data. After parameter decomposition, the prediction model of the subsystem can be obtained. The prediction model of each subsystem includes the influence of other subsystems on itself. The aluminum electrolysis system shown in Figure 1 can be described as due to the flow of electrolyte, whilst other subsystems cause changes in alumina concentration to a subsystem. Each subsystem has a separate controller to control the feeder responsible for supplying the alumina powder, and a distributed control algorithm is designed under the condition that the subsystems can communicate with each other. At each moment, each subsystem solves the optimal control signal of its system when the optimal control signals of other subsystems are known, and the global optimality is also guaranteed in the case of achieving local optimality. For the aluminum electrolysis system, the advantage of this method is that the six feeding ports can change the original group feeding or timing feeding strategy so that the six feeding ports can consider the influence of other feeding ports. The purpose of distributed feeding is to make the alumina concentration distribution more uniform, reduce the occurrence of local precipitation and local anode effect, ensure the stable and efficient operation of the entire cell, and improve the production efficiency of the aluminum plant.

3. Distributed Subspace Predictive Control

The basic idea of the data-driven distributed subspace predictive controller design is to first obtain the input and output data of length n, then use these data to solve the distributed prediction model, and finally use the obtained distributed prediction model to design the controller [24]. According to the actual data collection situation on-site, the input variable $u_1, u_2, \cdots, u_m (m = 6)$ is determined as the alumina feeding amount of the six subsystems, and the output variable $y_1, y_2, \cdots, y_m (m = 6)$ is determined as the alumina concentration for the six subsystems. Among them, the alumina feeding amount data are obtained by combining the dissolution and consumption mechanism of alumina and the feeding interval. According to the relevant mechanism of an aluminum reduction cell, there is a close relationship between the distributed alumina concentration and distributed current [22]. The soft sensor model of the distributed current and distributed alumina concentration is established by using the current data of a single anode guide rod and the alumina concentration data of different areas collected by field test. The alumina concentration data and alumina discharge data of six subsystems with n = 1000 can be obtained.

3.1. Data-Driven Distributed Prediction Model

For an aluminum reduction cell with six subsystems, the output prediction model can be described as: at time k, the relationship between the output prediction vector $\hat{y}_f(k) = [\hat{y}_{f\_1}(k)^T, \cdots, \hat{y}_{f\_m}(k)^T]^T$ composed of the alumina concentration prediction values of each subsystem and the input and output data is [30]:

$$\hat{y}_f(k) = L_w \cdot w_p(k) + L_u \cdot u_f(k) \tag{1}$$

where $L_w \in R^{mN \times 2mN}$ and $L_u \in R^{mN \times mN}$ are the unknown parameter matrix, which is obtained by subspace identification, N is the length of the prediction window, $u_f(k)$ is the input vector composed of the future alumina feeding amount of each subsystem, and $w_p(k)$ is the vector composed of the past input and output data, respectively, defined as follows:

$$w_p(k) \triangleq \begin{bmatrix} w_{p\_1}(k) \\ \vdots \\ w_{p\_m}(k) \end{bmatrix}; u_f(k) \triangleq \begin{bmatrix} u_{f\_1}(k) \\ \vdots \\ u_{f\_m}(k) \end{bmatrix}; w_{pi}(k) \triangleq \begin{bmatrix} u_{pi}(k)^T & y_{pi}(k)^T \end{bmatrix}^T;$$

$$u_{p\_i}(k) \triangleq \begin{bmatrix} u_i(k-N) \\ \vdots \\ u_i(k-1) \end{bmatrix}; \hat{y}_{p\_i}(k) \triangleq \begin{bmatrix} y_i(k-N) \\ \vdots \\ y_i(k-1) \end{bmatrix}; u_{f\_i}(k) \triangleq \begin{bmatrix} u_i(k) \\ \vdots \\ u_i(k+N-1) \end{bmatrix}; \hat{y}_{f\_i}(k) \triangleq \begin{bmatrix} \hat{y}_i(k) \\ \vdots \\ \hat{y}_i(k+N-1) \end{bmatrix}.$$

where $i = 1, 2, \ldots, 6$, the subscripts "p" and "f" represent the past and the future, respectively.

In order to realize distributed feeding control, a distributed prediction model needs to be established. Equation (1) is decomposed into the following form:

$$\begin{pmatrix} \hat{y}_1(k) \\ \hat{y}_2(k) \\ \vdots \\ \hat{y}_m(k) \end{pmatrix} = \begin{bmatrix} L_w(1) \\ L_w(2) \\ \vdots \\ L_w(m) \end{bmatrix} \cdot w_p(k) + \begin{bmatrix} L_{u(1,1)} & L_{u(1,2)} & \cdots & L_{u(1,m)} \\ L_{u(2,1)} & L_{u(2,2)} & \cdots & L_{u(2,m)} \\ \vdots & \vdots & \vdots & \vdots \\ L_{u(m,1)} & L_{u(m,2)} & \cdots & L_{u(m,m)} \end{bmatrix} \cdot \begin{pmatrix} u_1(k) \\ u_2(k) \\ \vdots \\ u_m(k) \end{pmatrix} \quad (2)$$

Based on this decomposition, the alumina concentration prediction model of each subsystem can be obtained:

$$\hat{y}_i(k) = L_{u(i,i)} u_i(k) + \underbrace{\sum_{j=1, j \neq i}^{m} L_{u(i,j)} u_j(k)}_{Effects\ of\ other\ subsystems, i = 1, 2, \ldots, 6} + L_{w(i)} \cdot w_p(k) \quad (3)$$

Each prediction model includes the influence of the feeding number of other subsystems on itself.

3.2. Design of Distributed Predictive Controller for Aluminum Reduction Cell System

The control target of the distributed aluminum reduction cell system is to control the alumina concentration to track the reference value. After certain research and experiments, the reference alumina concentration value of each subsystem can be obtained:

$$R_{ref} = \begin{bmatrix} r_1 & r_2 & \cdots & r_m \end{bmatrix}^T \quad (4)$$

The control strategy of the distributed aluminum reduction cell system is shown in Figure 2. According to Equation (3), predicting the output of any subsystem requires knowing the input and output data of all subsystems at the previous time [28]. Therefore, the controller of each subsystem must have a communication function to transmit its information to other subsystems. The difference between distributed control and centralized control is that the global performance index can be expressed as the sum of the performance indexes of all subsystems [34]:

$$J = \sum_{i=1}^{m} J_i \quad (5)$$

then the performance index of the ith subsystem can be expressed as

$$\min_{u_{f\_i}(k)} J_i = \left[y_{\hat{f}\_i}(k) - r_i(k) \right]^T Q_i \left[y_{\hat{f}\_i}(k) - r_i(k) \right] + u_{f\_i}(k)^T R_i u_{f\_i}(k) \quad (6)$$

where $r_i(k)$ is the reference input signal vector of the subsystem; and Q_i and R_i are the positive definite weighting matrixes. Substituting Equation (3) into Equation (6), we can obtain:

$$J_i = [L_{w(i)} w_p - r_i(k) + \sum_{j=1}^{m} L_{u(i,j)} \cdot u_j(k)]^T Q_i$$
$$[L_{w(i)} w_p - r_i(k) + \sum_{j=1}^{m} L_{u(i,j)} \cdot u_j(k)] + u_{f\_i}(k)^T R_i u_{f\_i}(k) \quad (7)$$

Differentiate the objective function to find the extremum:

$$\frac{\partial J_i}{\partial u_{f\_i}(k)} = 0 \quad (8)$$

The controller for each subsystem can be obtained

$$\vdots$$

$$u_{f\_i}(k) = -\left[R_i + L_{u(i,i)}^T Q_i L_{u(i,i)}\right]^{-1} L_{u(i,i)}^T Q_i \left[L_{w(i)} w_p(k) - r_i(k) + \sum_{j=1, j\neq i}^{m} L_{u(i,j)} \cdot u_{f\_j}(k)\right] \quad (9)$$

In actual control, we only put u_{f_i} the first component of the controlled input into the future input data matrix and pass this matrix to other subsystems. When performing predictive control iteration and rolling optimization in the above steps, the optimal control quantity is always calculated. Therefore, the actual output alumina concentration of each subsystem of aluminum electrolysis can be synchronized with the reference alumina concentration to achieve the control target.

3.3. Determination of Parameters of Aluminum Reduction Cell Prediction Model and Design of Data-Driven Distributed Predictive Control Algorithm

In order to obtain the prediction model of Equation (3), the above input and output data of n = 1000 are used to solve the parameter matrix L_w and L_u in Equation (1). The prediction step N is set to 5. Input data at time $(0, 1, \cdots, N-1)$ and output data at time $(0, 1, \cdots, 2N-1)$ are used to predict the output at time $(N, N+1, \cdots, 2N-1)$ according to Equation (1), as shown in Figure 3. After that, we move the time window and the input data at time $(1, 2, \cdots, 2N)$ and output data at time $(1, 2, \cdots, N)$ are used to predict the output at time $(N+1, N+2, \cdots, 2N)$ according to Equation (1), as shown in Figure 4.

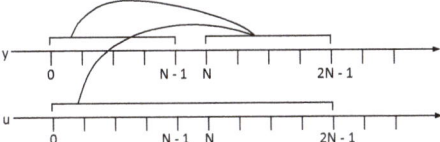

Figure 3. Output predictions ($N = 5$).

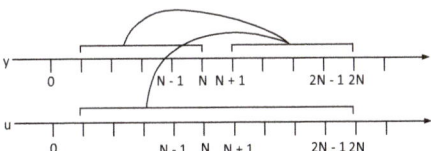

Figure 4. Step 2: Output prediction.

In order to solve the parameter matrix L_w and L_u, Equation (1) is rewritten into the Hankel matrix:

$$Y_f = L_w \cdot W_p + L_u \cdot U_f = \begin{bmatrix} L_w & L_u \end{bmatrix} \begin{bmatrix} W_p \\ U_f \end{bmatrix} \quad (10)$$

where Y_f, W_p and U_f are the Hankel matrix composed of input and output data, defined as follows:

$$\hat{Y}_f = \begin{bmatrix} \hat{Y}_{f\_i} \\ \vdots \\ \hat{Y}_{f\_m} \end{bmatrix}; U_f = \begin{bmatrix} U_{f\_i} \\ \vdots \\ U_{f\_m} \end{bmatrix}; Y_p = \begin{bmatrix} Y_{p\_i} \\ \vdots \\ Y_{p\_m} \end{bmatrix}; U_p = \begin{bmatrix} U_{p\_i} \\ \vdots \\ U_{p\_m} \end{bmatrix}; W_p = \begin{bmatrix} W_{p\_i} \\ \vdots \\ W_{p\_m} \end{bmatrix}$$

$$W_{p\_i} \triangleq \begin{bmatrix} U_{p\_i}^T, Y_{p\_i}^T \end{bmatrix}^T \in R^{2N \times M}$$

$$U_{p\_i} \triangleq \begin{bmatrix} u_i(0) & u_i(1) & \cdots & u_i(M-1) \\ u_i(1) & u_i(2) & \cdots & u_i(M) \\ \vdots & \vdots & \cdots & \vdots \\ u_i(N-1) & u_i(N) & \cdots & u_i(N+M-2) \end{bmatrix} ; U_{f\_i} \triangleq \begin{bmatrix} u_i(N) & u_i(N+1) & \cdots & u_i(N+M-1) \\ u_i(N+1) & u_i(N+2) & \cdots & u_i(N+M) \\ \vdots & \vdots & \cdots & \vdots \\ u_i(2N-1) & u_i(2N) & \cdots & u_i(2N+M-2) \end{bmatrix} ;$$

$$Y_{p\_i} \triangleq \begin{bmatrix} y_i(0) & y_i(1) & \cdots & y_i(M-1) \\ y_i(1) & y_i(2) & \cdots & y_i(M) \\ \vdots & \vdots & \cdots & \vdots \\ y_i(N-1) & y_i(N) & \cdots & y_i(N+M-2) \end{bmatrix} ; \hat{Y}_{f\_i} \triangleq \begin{bmatrix} \hat{y}_i(N) & \hat{y}_i(N+1) & \cdots & \hat{y}_i(N+M-1) \\ \hat{y}_i(N+1) & \hat{y}_i(N+2) & \cdots & \hat{y}_i(N+M) \\ \vdots & \vdots & \cdots & \vdots \\ \hat{y}_i(2N-1) & \hat{y}_i(2N) & \cdots & \hat{y}_i(2N+M-2) \end{bmatrix}$$

The problem is solved by least squares method:

$$\min_{L_w, L_u} \left\| Y_f - \begin{bmatrix} L_w & L_u \end{bmatrix} \begin{bmatrix} W_p \\ U_f \end{bmatrix} \right\|_F^2 \quad (11)$$

This problem can be solved by orthogonal projection. According to the subspace projection theorem, Y_f projects to the column space of W_p and U_f:

$$\hat{Y}_f = Y_f / \begin{bmatrix} W_p \\ U_f \end{bmatrix} = Y_f \begin{bmatrix} W_p \\ U_f \end{bmatrix}^T \left(\begin{bmatrix} W_p \\ U_f \end{bmatrix} \begin{bmatrix} W_p \\ U_f \end{bmatrix}^T \right)^+ \begin{bmatrix} W_p \\ U_f \end{bmatrix} \quad (12)$$

where the symbol "+" stands for pseudo-inverse; "/" stands for data space projection, then:

$$\begin{bmatrix} L_w & L_u \end{bmatrix} = Y_f \begin{bmatrix} W_p \\ U_f \end{bmatrix}^T \left(\begin{bmatrix} W_p \\ U_f \end{bmatrix} \begin{bmatrix} W_p \\ U_f \end{bmatrix}^T \right)^+ \quad (13)$$

substituting Equation (13) into Equation (2), we obtain:

$$\hat{Y}_f = Y_f \begin{bmatrix} W_p \\ U_f \end{bmatrix}^+ \begin{bmatrix} W_p \\ U_f \end{bmatrix} \quad (14)$$

QR decomposition of matrix $\begin{bmatrix} W_p \\ U_f \\ Y_f \end{bmatrix}$ is shown as

$$\begin{bmatrix} W_p \\ U_f \\ Y_f \end{bmatrix} = R^T Q^T = \begin{bmatrix} R_{11} & 0 & 0 \\ R_{21} & R_{22} & 0 \\ R_{31} & R_{32} & R_{33} \end{bmatrix} Q^T \quad (15)$$

then, Equation (12) can be rewritten as

$$\hat{Y}_f = \begin{bmatrix} R_{31} & R_{32} & R_{33} \end{bmatrix} Q^T \left(\begin{bmatrix} R_{11} & 0 & 0 \\ R_{21} & R_{22} & 0 \end{bmatrix} Q^T \right)^+ \cdot \begin{bmatrix} W_p \\ U_f \end{bmatrix} =$$
$$\begin{bmatrix} R_{31} & R_{32} & R_{33} \end{bmatrix} Q^T Q^{T+} \begin{bmatrix} R_{11} & 0 & 0 \\ R_{21} & R_{22} & 0 \end{bmatrix}^+ \cdot \begin{bmatrix} W_p \\ U_f \end{bmatrix} = \begin{bmatrix} R_{31} & R_{32} \end{bmatrix} \begin{bmatrix} R_{11} & 0 \\ R_{21} & R_{22} \end{bmatrix}^+ \begin{bmatrix} W_p \\ U_f \end{bmatrix} \quad (16)$$

comparing Equations (2) and (16), we obtain the solution of L_w and L_u:

$$\begin{bmatrix} L_w & L_u \end{bmatrix} = \begin{bmatrix} R_{31} & R_{32} \end{bmatrix} \begin{bmatrix} R_{11} & 0 \\ R_{21} & R_{22} \end{bmatrix}^+ \quad (17)$$

furthermore, we obtain:

$$L_w = \begin{bmatrix} L_w(1) \\ L_w(2) \\ \vdots \\ L_w(m) \end{bmatrix}; L_u = \begin{bmatrix} L_{u(1,1)} & L_{u(1,2)} & \cdots & L_{u(1,m)} \\ L_{u(z,1)} & L_{u(2,z)} & \cdots & L_{u(z,m)} \\ \vdots & \vdots & \ddots & \vdots \\ L_{u(m,1)} & L_{u(m,2)} & \cdots & L_{u(m,m)} \end{bmatrix} \quad (18)$$

where L_w is a 30 × 60 matrix, L_u is a 30 × 30 matrix, then after decomposition, we obtain $L_{w(i)}$ and $L_{u(i,j)}$:

$$L_{w(i)} = L_w(i:m:mN-m+i,:)$$

$$L_{u(ij)} = L_u(i:m:mN-m+i,j:m:mN-m+j)$$

where $i = 1, 2, \cdots, m$, $j = 1, 2, \cdots, m$, $m = 6$, $N = 5$. The prediction model of six subsystems of an aluminum reduction cell is further obtained.

The Nash optimal method is used to design the distributed control algorithm. The definition of Nash optimal is as follows.

For a complex large system with m subsystems, if there is a vector solution that $u^N = (u_1^N, \cdots, u_i^N, \cdots, u_m^N)$ satisfies the following inequalities for all subsystems $u_i(i = 1, 2, \cdots, m)$ [35]:

$$\begin{aligned} J_i(u_1^N, \cdots, u_i^N, \cdots, u_m^N) &\leqslant \\ J_i(u_1^N, \cdots, u_{i-1}^N, u_i, u_{i+1}^N, \cdots, u_m^N) & \end{aligned} \quad (19)$$

then, the vector is $u^N = (u_1^N, \cdots, u_i^N, \cdots, u_m^N)$ called the optimal Nash solution of the system. This solution optimizes the control performance of the entire large system, and all subsystems will not change this control decision.

As mentioned in the algorithm introduction above, in distributed predictive control, each subsystem must know the optimal control signal of other subsystems before solving its own optimal control signal, but each controller solves the optimal control signal at every moment simultaneously. In order to make all subsystems optimal at the same time, the iterative method is usually adopted. At each sampling moment, an iterative calculation is carried out to obtain the optimal control input signal of each subsystem at the sampling moment and determine whether it satisfies the Nash optimal solution. At the same time, the control signal is transmitted to other subsystems. When the iterative values of all subsystems meet the conditions, the iteration ends, so the global optimization of a complex large system is realized. The detailed steps of the iterative algorithm are as follows:

- Step 1 At the k sampling time, take the initial value of the control input variable of each subsystem $(u_1^0, u_2^0, \ldots, u_m^0)$ and pass the initial value to other subsystems, so that the iteration ordinal $l = 0$;
- Step 2 Use the last iteration value $\{u_1(k)^l, u_2(k)^l, \cdots, u_m(k)^l\}$ calculate the value of iteration $u_i(k)^{l+1}$ $l+1$ for the ith subsystem;
- Step 3 Pass the calculation result $u_i(k)^{l+1}$ to other subsystems through the network;
- Step 4 If the Nash optimality is satisfied for all subsystems $\|u_i(k)^{l+1} - u_i(k)^l\| \leqslant \varepsilon_i$ or the maximum number of iterations is reached, the iteration is ended, otherwise, return to the second step;
- Step 5 Each subsystem executes the optimal control signal $\{u_1^N, u_2^N, \cdots, u_m^N\}$ and uses it as the initial value at the next moment;
- Step 6 End the calculation of this sampling time, and wait for the next sampling time $k + 1$.

Flowchart is shown in Figure 5:

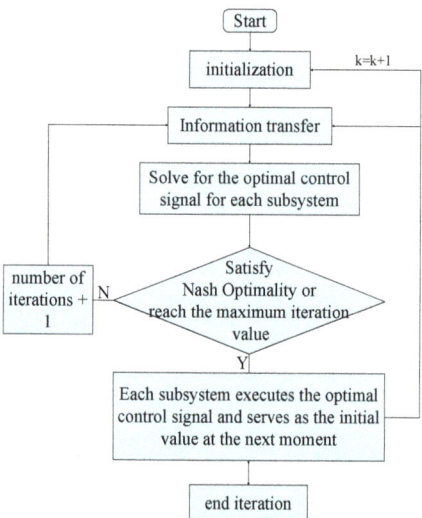

Figure 5. Flow chart of distributed predictive control algorithm.

4. Simulation Experiments

Based on the actual data of an aluminum plant, this section compares the control effect of the traditional feeding control method and the distributed subspace predictive control method through simulation results. The simulation includes the control effect under the condition of no interference and inaccurate feed quantity.

4.1. Data Acquisition

The on-site collection situation of an aluminum plant is shown in Figure 6. The actual working area of modern aluminum electrolysis is shown in Figure 6a. The data collected in the field include a feeding interval, distributed alumina concentration and distributed current. In Figure 6b, the data of the feeding interval were obtained using a stopwatch recording each feeding time. In Figure 6c, the distributed alumina concentration is scooped out by the field workers, cooled, bagged, and sent to the laboratory for analysis. The distributed current was obtained by a data collector installed on the anode guide rod, as shown in Figure 6d. Using the data collected in the field, 1000 sets of data for simulation mentioned in Section 2 can be obtained. The simulation parameters are: the input constraints of the six subsystems $U = [0\ 0.1]$, and the error accuracy $\varepsilon = 0.05$, and each subsystem expects an output setpoint $r(k)$ of 2.5.

4.2. Control Effect without Any Interference

Under the premise that there is no model mismatch and external interference in the aluminum reduction cell, as shown in Figure 7, the first 1000 s is the control effect of the traditional control strategy, and the control effect of the control strategy in this paper is after 1000 s. The traditional control strategy is based on the relationship between the cell resistance and the concentration to drive the six feeding devices' group timing feeding: FD1, FD3, FD5 are a group of simultaneous feeding, FD2, FD4, and FD6 are a group of simultaneous feeding, and each group of feeding is staggered by half the feeding cycle.

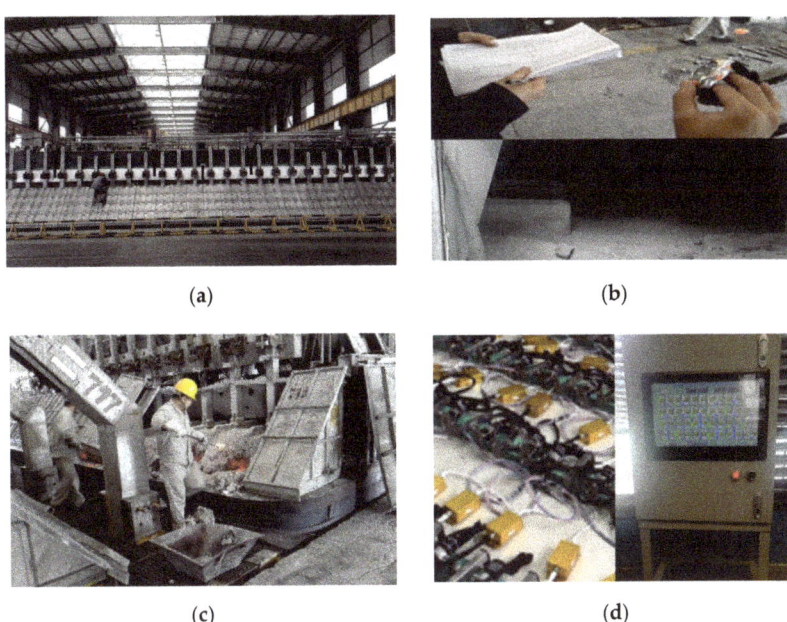

Figure 6. Field collection diagram: (**a**) actual working area of modern aluminum electrolysis; (**b**) data acquisition diagram of feeding interval; (**c**) scoop out the electrolyte diagram; and (**d**) distributed current acquisition diagram.

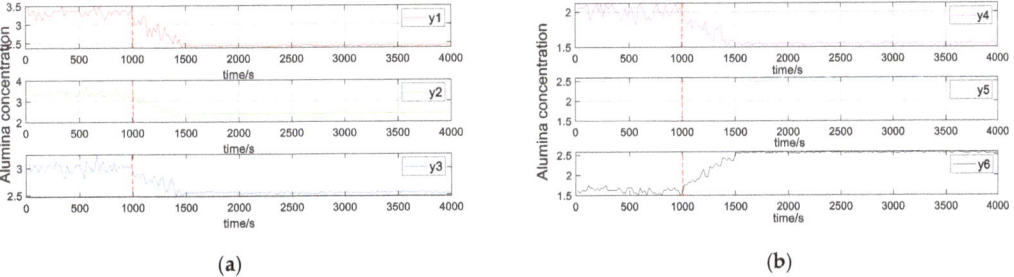

Figure 7. Variation of alumina concentration: (**a**) subsystems 1–3; and (**b**) subsystems 4–6.

In the first 1000 s of Figure 7, the concentration of the six subsystems is distributed very unevenly, although it is roughly in the appropriate range after feeding for a period using the traditional control strategy. After 1000 s, the distributed subspace predictive control method proposed in this paper is used to control the aluminum reduction cell. Each feeder is distributed as needed under the influence of other feeders, so that the variation of the alumina concentration is greatly reduced in space and time, and the concentration of the six areas is well controlled near the set value, the alumina concentration distribution in the entire cell is more uniform. The continuous feeding amount of each feeder in Figure 8. Since alumina is dumped in discrete 1.8 kg batches each time during the actual operation on site, the actual feeding interval is calculated according to the theoretical consumption rate of alumina in Figure 9. It can be seen from Figure 9 that the control method proposed in this paper can make the six feeders of the aluminum electrolysis cell distribute according to the demand, considering the influence of other feeders. The distribution of alumina concentration throughout the cell is more uniform and can be effectively controlled within the set value.

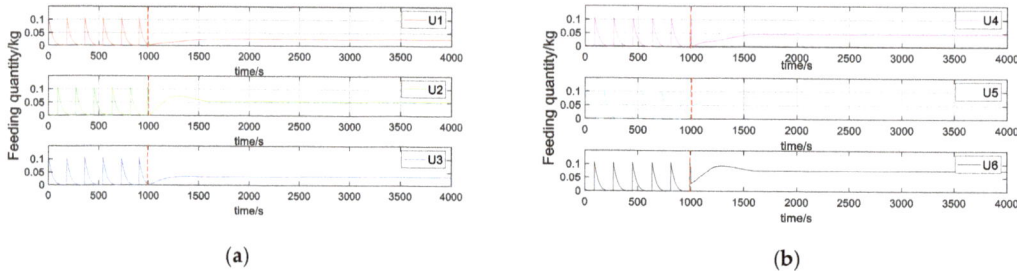

Figure 8. Distributed feeding quantity control: (**a**) subsystems 1–3; (**b**) subsystems 4–6.

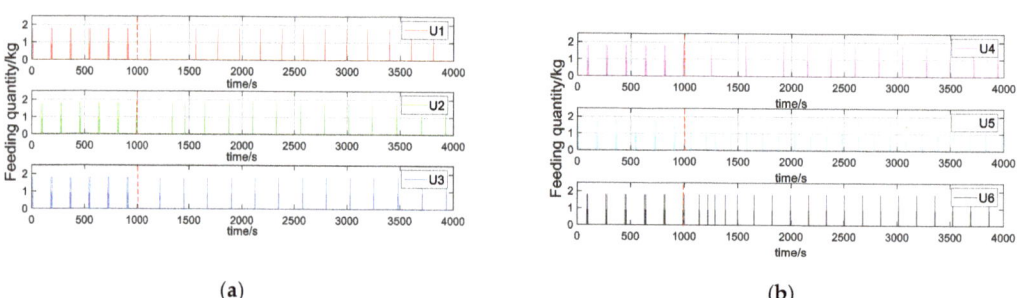

Figure 9. Distributed feeding interval control: (**a**) subsystems 1–3; (**b**) subsystems 4–6.

4.3. The Control Effect when the Feeding Amount of the Feeder Is Inconsistent with the Actual Set Value

In practice, the feeder may be blocked or overloaded. Therefore, the disturbance of inaccurate feed quantity is introduced to test the stability of the proposed control method. As shown in Figure 10, after the second 500 s, the inaccurate feeding amount was simulated for the feeding ports of subsystem 2 and subsystem 6, and the control effects were increased by 15% and decreased by 15%, respectively.

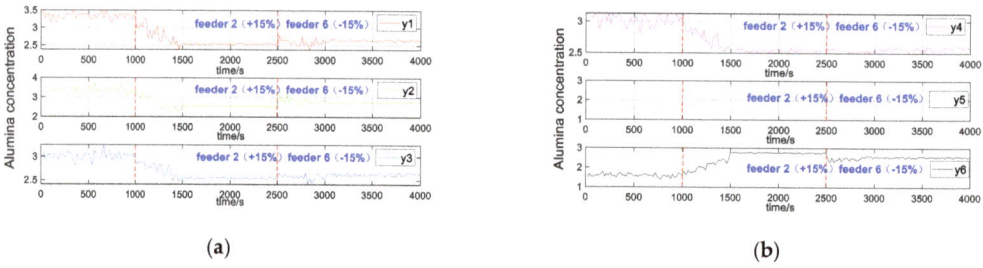

Figure 10. Variation of the alumina concentration: (**a**) subsystems 1–3; and (**b**) subsystems 4–6.

As can be seen from Figure 10, after the simulation of a 15% increase and 15% decrease in feeding port 2 and feeding port 6, the concentration of subsystem 2 will increase for a short time, and the alumina concentration of subsystem 6 will decrease for a short time. Due to the flow of electrolyte in the reduction cell, the alumina concentration of other subsystems will also be affected, but the controller can quickly stabilize the alumina concentration of each subsystem, indicating that the controller designed in this paper has a good stability. The continuous feeding amount of each feeder is shown in Figure 11. Since 1.8 kg alumina is discretely dumped each time during the actual operation on site, the actual feeding interval is calculated according to the theoretical consumption rate of alumina in Figure 12.

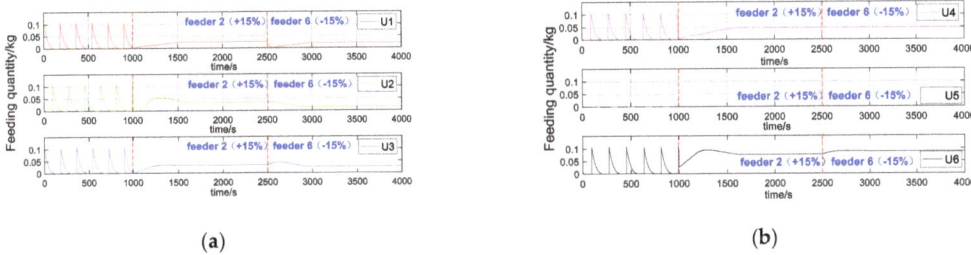

Figure 11. Distributed feeding quantity control: (**a**) subsystems 1–3; (**b**) subsystems 4–6.

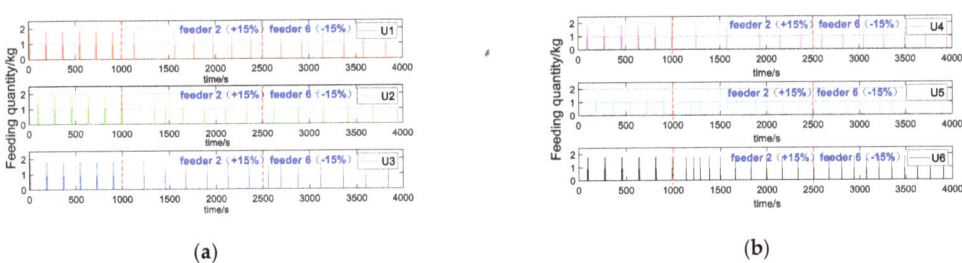

Figure 12. Distributed feeding interval control: (**a**) subsystems 1–3; (**b**) subsystems 4–6.

It can be seen from Table 1 that the control method in this paper can still maintain a small error in the presence of interference. The main reason is that the method proposed in this paper considers the influence of adjacent subsystems on itself so that each feeder can act independently to control the local alumina concentration to maintain the setpoint while making the concentration of the entire cell uniformly distributed, which is conducive to the stable operation of the cell.

Table 1. Mean squared error (MSE) of the actual concentration and set concentration when disturbance occurs.

Subsystem	MSE without Interference	MSE with Interference
Subsystem 1	0.0309	0.0387
Subsystem 2	0.0306	0.0667
Subsystem 3	0.0140	0.0203
Subsystem 4	0.0156	0.0161
Subsystem 5	0.0414	0.0475
Subsystem 6	0.0421	0.0633
Subsystem 1	0.0309	0.0387
Average	0.0291	0.0421

5. Conclusions

This paper proposes a multi-point feeding strategy for aluminum reduction cell based on distributed subspace predictive control. This method combines the subspace method with the idea of distributed model predictive control using process data and designs a distributed controller through the input and output data. Therefore, it overcomes the shortcomings of centralized control and decentralized control and achieves the performance optimization of the entire complex large system at a lower cost. Compared with traditional methods, the proposed control strategy has the following advantages:

(1) Each feeding device is controlled by an independent controller, and the distributed control method which combines the advantages of centralized and decentralized control is adopted, overcoming their shortcomings.

(2) The mutual influence between the various subsystems and the influence of sudden interference are considered. For example, when the feeding amount is inaccurate, the controller can also control the concentration of alumina well to ensure the stability of the reduction cell.

Compared with traditional control strategies, the method developed in this paper can control the uniform distribution of alumina concentration more effectively, improve the production efficiency of aluminum plants and save production costs. However, in the actual production process, with the passage of time, the change of aluminum reduction cell health status will affect the accuracy of the prediction model and further affect the control accuracy. Therefore, combining the distributed subspace predictive control with the adaptive idea and improving the adaptability of the method by updating the parameters of the predictive model are the key avenues of research future.

Author Contributions: Conceptualization, J.C. and Q.L.; methodology, P.W. and X.L.; software, P.W.; validation, R.H. and H.L.; formal analysis, J.C.; investigation, P.W.; resources, Q.L. and B.C.; data curation, J.C.; writing—original draft preparation, P.W.; writing—review and editing, J.C. and X.L.; visualization, P.W. and J.C.; supervision, Q.L. and R.H.; project administration, H.L. and B.C.; funding acquisition, J.C. All authors have read and agreed to the published version of the manuscript.

Funding: This research was funded by the China Postdoctoral Science Foundation, grant number 2021M690798; Guizhou Province Science and Technology Plan Project, grant number [2021] General 085; National Natural Science Foundation of China, grant number 61603034; The Fundamental Research Funds for the Central Universities, grant number FRF-DF-20-14.

Institutional Review Board Statement: Not applicable.

Informed Consent Statement: Not applicable.

Data Availability Statement: Not applicable.

Conflicts of Interest: The authors declare no conflict of interest.

References

1. Li, Z.Y.; Yang, S.; Zou, Z.; Li, J. Research progress of on-line detection for spatial distribution information in large-amperage aluminum reduction cells. *Light Met.* **2019**, *9*, 22–30.
2. Bai, W.B. Discussion on the control of alumina concentration during aluminum electrolysis under complex electrolyte system. *Sci. Technol. Innov.* **2018**, *36*, 53–54.
3. Wang, Z.W.; Gao, B.L.; Hu, X.W.; Lu, Y.; Li, Y.; Shi, Z.N.; Yu, J.Y. Some Issues in Scale Aluminum Electrolysis Cell. In Proceedings of the 15th session of the 15th China Association for Science and Technology Annual Conference: National Seminar on Aluminum Metallurgy Technology Proceedings of the Conference, Guiyang, China, 25–27 May 2013.
4. Wu, Z.W.; Ouyang, X.Y. Production practice of accuracy feeding control for aluminum reduction pots. *Light Met.* **2017**, *8*, 26–29.
5. Li, X.; Liu, M.Z. Effect of adjusting feeding Interval on technical parameters of aluminum electrolysis. *Light Met.* **2011**, *S1*, 228–230.
6. Lv, Z.M. Basic characteristics and control requirements of stable production mode of large aluminum reduction cell. *Alum. Magnes. Commun.* **2012**, *3*, 24–26.
7. Zeng, S.P.; Zhang, Q.P.; Zhao, G.X. Fuzzy control for the alumina concentration in aluminum cells. *Metall. Autom.* **2001**, *5*, 9–11.
8. Kong, J.Y.; Li, G.F.; Xiong, H.G.; Jiang, G.Z.; Yang, J.T.; Wang, X.D.; Hou, Y. Research on Soft-sensing Modeling Methods and its Application in Industrial Production. *Mach. Tool Hydraul.* **2007**, *6*, 149–151.
9. Yin, H.M.; Wang, M.L.; Fan, J.J. High Speed Milling Cutting Temperature Soft Measurement Modeling and Algorithm Implementation Based on PSO Algorithm. *Mach. Des. Res.* **2016**, *32*, 128–131.
10. Cui, J.R.; Li, W.H.; Su, G.C.; Cao, B.; Huang, R.Y.; Yang, X.; Li, Q. Research progress of distributed all-element model of large-amperage aluminum pots for intelligent manufacturing. *Light Met.* **2021**, *11*, 30–38.
11. Cui, J.R.; Zhang, N.; Yang, X. Soft sensing of alumina concentration in aluminum electrolysis industry based on deep belief network. In Proceedings of the 2020 Chinese Automation Congress (CAC), Shanghai, China, 6–8 November 2020.
12. Zhang, Y.; Yang, X.; Shardt, Y.A.W.; Cui, J.; Tong, C. A KPI-based probabilistic soft sensor development approach that maximizes the coefficient of determination. *Sensors* **2018**, *18*, 3058. [CrossRef]
13. Yang, X.; Zhang, Y.; Shardt, Y.A.W.; Li, X.; Cui, J.; Tong, C. A KPI-based soft sensor development approach incorporating infrequent, variable time delayed measurements. *IEEE Trans. Control Syst. Technol.* **2019**, *28*, 2523–2531.
14. Li, J.J.; Feng, D.D. Intelligent feeding control strategy based on aluminum concentration identification in aluminum electrolysis. *Light Met.* **2019**, *2*, 31–36.
15. Huang, H.; Wei, Y. Research on Intelligent Control of aluminum reduction cell Based on Data Drive. *Electron. Manuf.* **2015**, *1*, 32.

16. Shi, J.; Yao, Y.; Skyllas-Kazacos, M.; Welch, B.J. Multivariable Feeding Control of Aluminum Reduction Process Using Individual Anode Current Measurement. *IFAC Pap. Online* **2020**, *53*, 11907–11912. [CrossRef]
17. Kaszás, C.; Kiss, L.; Poncsák, S.; Guérard, S.; Bilodeau, J.-F. Spreading of Alumina and Raft Formation on the Surface of Cryolitic Bath. In Proceedings of the 146th TMS Annual Meeting and Exhibition/Conference on Light Metals, San Diego, CA, USA, 26 February–2 March 2017.
18. Zhan, S.Q.; Li, M.; Zhou, J.-M.; Yang, J.-H.; Zhou, Y.-W. CFD simulation of effect of anode configuration on gas-liquid flow and alumina transport process in an aluminum reduction cell. *J. Cent. South Univ.* **2015**, *22*, 2482–2492. [CrossRef]
19. Kovács, A.; Breward, C.; Einarsrud, K.; Halvorsen, S.A.; Nordgård-Hansen, E.; Manger, E.; Münch, A.; Oliver, J.M. A heat and mass transfer problem for the dissolution of an alumina particle in a cryolite bath. *Int. J. Heat Mass Transf.* **2020**, *162*, 120232. [CrossRef]
20. Gylver, S.E.; Omdahl, N.H.; Prytz, A.K.; Meyer, A.J.; Lossius, L.P.; Einarsrud, K.E. Alumina feeding and raft formation: Raft collection and process parameters. In Proceedings of the Light Metals Symposium at the 148th TMS Annual Meeting, San Antonio, TX, USA, 10–12 March 2019.
21. Einarsrud, K.E.; Eick, I.; Wei, B.; Feng, Y.; Hua, J.; Witt, P.J. Towards a coupled multi-scale, multi-physics simulation framework for aluminium electrolysis. *Appl. Math. Model.* **2015**, *44*, 3–24. [CrossRef]
22. Wang, R.G.; Bao, J.; Yao, Y.C. A data-centric predictive control approach for nonlinear chemical processes. *Chem. Eng. Res. Des.* **2018**, *142*, 154–164. [CrossRef]
23. Xi, L.; Sun, M.M.; Chen, S.S.; Zhu, J.Z.; Sun, Q.Y.; Liu, Z.J. Multi-region cooperative control method for distributed grid. *Electr. Mach. Control* **2021**, *25*, 75–86.
24. Wang, Z.B.; Li, C.M.; He, W.Y. Control of alumina concentration in aluminum electrolysis production. *Des. Nonferrous Met.* **2018**, *45*, 101–103.
25. Yang, X.; Gao, J.J.; Huang, B. Data-driven design of fault detection and isolation method for distributed homogeneous systems. *J. Frankl. Inst.* **2021**, *358*, 4929–4949. [CrossRef]
26. Yang, X.; Gao, J.J.; Li, L.L.; Luo, H.; Ding, S.X.; Peng, K.X. Data-driven design of fault-tolerant control systems based on recursive stable image representation. *Automatica* **2020**, *122*, 109246. [CrossRef]
27. Han, P.; Liu, M.; Jia, H. Data Driven Pre Tuning Adaptive Subspace Model Predictive Control. *J. Syst. Simul.* **2018**, *30*, 332–340.
28. Wu, X.; Shen, J. Subspace identification and predictive control of boiler-turbine coordination system. *J. Southeast Univ. Nat. Sci. Ed.* **2012**, *42*, 281–286.
29. Dong, T.T.; Li, L.J.; Xiong, L.; Xu, O.G. Distributed Predictive Control Based on Associated Subsystems. *Control Eng.* **2015**, *22*, 1201–1206.
30. Chen, J.M.; Yang, F.W. Communication-Based Data-Driven Distributed Predictive Control. *J. East China Univ. Sci. Technol. Nat. Sci. Ed.* **2014**, *40*, 113–119.
31. Chen, Q.; Li, S.Y.; Xi, Y.G. Distributed Predictive Control Based on Plant-Wide Optimality. *J. Shanghai Jiao Tong Univ.* **2005**, *03*, 349–352.
32. Yao, Y.; Cheung, C.-Y.; Bao, J.; Skyllas-Kazacos, M.; Welch, B.J.; Akhmetov, S. Estimation of spatial alumina concentration in an aluminum reduction cell using a multilevel state observer. *AIChE J.* **2017**, *63*, 2806–2818. [CrossRef]
33. Yao, Y.; Cheung, C.Y.; Bao, J.; Skyllas-Kazacos, M. Monitoring Local Alumina Dissolution in Aluminum Reduction Cells Using State Estimation. *Light Met.* **2015**, *2015*, 577–581.
34. Wahab, N.A.; Katebi, R.; Balderud, J.; Rahmat, M.F. Data-driven adaptive model-based predictive control with application in wastewater systems. *IET Control Theory Appl.* **2011**, *5*, 803–812. [CrossRef]
35. Alexander, M.K. Why are enzymes less active in organic solvents than in water? *Trends Biotechnol.* **1997**, *15*, 97–101.

Article

A Model for Flywheel Fault Diagnosis Based on Fuzzy Fault Tree Analysis and Belief Rule Base

Xiaoyu Cheng [1,†], Shanshan Liu [2,†], Wei He [1,3,*], Peng Zhang [3,4], Bing Xu [1], Yawen Xie [1] and Jiayuan Song [1]

[1] School of Computer Science and Information Engineering, Harbin Normal University, Harbin 150025, China; chengxiaoyu1104@163.com (X.C.); bingxv_0227@163.com (B.X.); xie_yw@foxmail.com (Y.X.); mintpepper2216@gmail.com (J.S.)
[2] School of Computer and Information Security, Guilin University of Electronic Technology, Guilin 541004, China; wxci339701@163.com
[3] High-Tech Institute of Xi'an, Xi'an 710025, China; zpdyxdz@126.com
[4] State Key Laboratory of Astronautic Dynamics, Xi'an Satellite Control Center, Xi'an 710043, China
* Correspondence: he_w_1980@163.com; Tel.: +189-4567-2266
† These authors contributed equally to this work.

Abstract: In the fault diagnosis of the flywheel system, the input information of the system is uncertain. This uncertainty is mainly caused by the interference of environmental factors and the limited cognitive ability of experts. The BRB (belief rule base) shows a good ability for dealing with problems of information uncertainty and small sample data. However, the initialization of the BRB relies on expert knowledge, and it is difficult to obtain the accurate knowledge of flywheel faults when constructing BRB models. Therefore, this paper proposes a new BRB model, called the FFBRB (fuzzy fault tree analysis and belief rule base), which can effectively solve the problems existing in the BRB. The FFBRB uses the Bayesian network as a bridge, uses an FFTA (fuzzy fault tree analysis) mechanism to build the BRB's expert knowledge, uses ER (evidential reasoning) as its reasoning tool, and uses P-CMA-ES (projection covariance matrix adaptation evolutionary strategies) as its optimization model algorithm. The feasibility and superiority of the proposed method are verified by an example of a flywheel friction torque fault tree.

Keywords: flywheel fault diagnosis; belief rule base; fuzzy fault tree analysis; Bayesian network; evidential reasoning

1. Introduction

The flywheel [1] system is a key actuator for spacecraft attitude control, which is widely used in the aerospace field. The normal operation of a flywheel system is very important for spacecraft. However, the spacecraft environment where the flywheel system is located has a harsh operating environment and complex structure. Once a failure occurs, it will pose a great threat to space safety. Therefore, to ensure the reliability and orderly operation of the flywheel system, it is of great significance to diagnose the faults of the flywheel system quickly and accurately.

Many scholars have carried out a lot of research on the fault diagnosis of flywheel systems. Changrui Chen et al. [2] proposed a 3D associated dimension diagnosis method, it is improved by K-Medoids clustering technology for different typical states of satellite flywheel bearings and verified the feasibility of the method through experiments. Xinchang Zhang et al. [3] developed a set of methods for inputting correct premises, and based on consistency test results, presented a fault diagnosis model based on finite state machines, which could locate and diagnose some faults. Junweir Lin et al. [4] proposed a new fault diagnosis scheme for linear analog circuits. The author constructs a diagnostic evaluator, which can diagnose faults through digital signals and diagnose media after analyzing and modeling the components. Bo Chen et al. [5] studied the distributed fault diagnosis technology and combined it with software technology, computer network, artificial intelligence and fault diagnosis to improve the self-fault diagnosis function of an expert system. Zijian Qiao et al. [6] proposed a second-order stochastic resonance method based on fractional derivative enhancement, which uses strong background noise to enhance the weak fault characteristics. It is used for mechanical fault diagnosis. Wenjun Sun et al. [7] studied a deep neural network based on a sparse self-code device for induction motor fault diagnosis. This method is used in the sparse automatic process to add noise encoding using the sparse automatic learning feature, which is the unsupervised feature learning that is required to measure the data without marking. Yao Cheng et al. [8] studied a set of combined fault diagnoses based on observer redundancy in the background of a satellite attitude control system. The modified scheme can solve actuator and sensor faults that are difficult to solve by traditional methods.

It can be seen from the above, most of the existing flywheel fault diagnosis schemes are designed on the basis of the data-driven method [9]. However, the current flywheel fault diagnosis still lacks an effective diagnosis scheme for the following two problems: First, the model accuracy cannot be guaranteed under small sample data. It is difficult to obtain accurate diagnosis results by using small sample data in actual fault diagnosis. This is because in the system life cycle, it is difficult to obtain a large number of flywheel fault samples, and more difficult to obtain fault samples under different fault modes; second, the black box model has the disadvantage of unexplainable diagnostic processes.

BRB (belief rule base) is a general rule-based reasoning method proposed by Yang Jianbo et al. [10] on the basis of evidentiary reasoning, which has important applications in mechanism analysis [11], health status assessment [12,13] and fault diagnosis [14]. BRB is suitable for flywheel systems, mainly reflected in three aspects: First, BRB can effectively describe the uncertainty of flywheel systems; second, the BRB modeling method is suitable for flywheel systems. It uses expert knowledge for modeling and data for model training; third, BRB has shown to be a good treatment effect for small sample problems. However, applying BRB to the actual fault diagnosis of the flywheel system cannot solve problems such as the difficulty in constructing an expert knowledge base, the unclear logical relationship between the flywheel fault events and the unclear fault index. FFTA (fuzzy fault tree analysis) [15,16] enables the logical relationship between different events to be clearly expressed. This is because FFTA can present the cause of failure and events caused by this cause in the form of a fault tree from the perspective of the fault mechanism. At the same time, FFTA makes the occurrence probability of each event in the fault tree better describe the uncertainty, because it introduces the theory of fuzzy mathematics. The combination of FFTA and BRB not only enables the fault index to be clearly established and the event fuzziness to be better described, but also enables the advantages of BRB to be applied in the fault diagnosis of the flywheel system, which makes comprehensive use of the advantages of the two. Therefore, this paper establishes the FFBRB (fuzzy fault tree analysis and belief rule base) model, which makes full use of the FFTA and BRB's advantages.

The main contributions of the FFBRB model proposed in this paper are as follows: (1) The way FFTA is used to build the initial BRB model. In this paper, the FFTA mechanism is used to expand the BRB knowledge base and solve the problem of constructing an expert knowledge base of complex flywheel system; (2) A new flywheel fault diagnosis model based on BRB is proposed. This model can obtain relatively accurate data even with a small number of samples and has higher applicability. It uses expert knowledge to construct the initial parameters of the model and uses training samples to optimize the model parameters.

The main structure of this paper is as follows: In the first part, the fault diagnosis model of the original flywheel system is analyzed and discussed. On the basis of revealing the shortcomings of the original model, the fault diagnosis model of the FFBRB flywheel system is proposed; In the second part, it describes the problems that need to be solved in the process of flywheel system modeling and gives the general solution diagram; In the third part, it defines and describes the fault diagnosis model of FFBRB flywheel system, and describes its transformation mechanism and inference optimization process in detail; In the fourth part, this paper uses a concrete example to verify the method in this paper and gives the experimental conclusion; In the fifth part, it gives the summary of this thesis.

2. Problem Description

This section describes the problems and solutions encountered in the fault diagnosis of the flywheel system, and puts forward and introduces the FFBRB model.

2.1. Clarifying Questions

Constructing the FFBRB flywheel system fault diagnosis model needed a solution to the following problems:

Problem 1. *How to use the FFTA mechanism and integrate it into the BRB knowledge base was the first problem to be solved. In the BRB, the relationship between the input and output is described by a series of belief rules, and belief rules are built based on expert knowledge. However, when the BRB is applied to the practical flywheel system, expert knowledge is difficult to embed into the fault diagnosis model of the flywheel system (see Section 3.2.).*

To realize the FFTA to BRB conversion, it is necessary to describe the correspondence between FFTA logic gates and BRB belief rules, and the correspondence between FFTA events and BRB input and output. The function to solve this problem is denoted as CovBridge(*) and ϱ is the set of parameters in this process, then the process can be described by the following expression:

$$\text{BRB}(\text{BeliefRule}, \text{input/output}) = \text{CovBridge}\,(\text{FFTA}(\text{LogicGate}, \text{event}),\ \varrho) \quad (1)$$

This is a nonlinear mapping. It is not executed in a specific software language. With CovBridge(*), logic gates in the FFTA were converted into belief rules in the BRB, and events in the FFTA were converted into inputs and outputs in the BRB. The inputs of the CovBridge(*) function were logic gates, events, and parameter sets in the FFTA, and the outputs were belief rules and their inputs and outputs in the BRB.

Problem 2. *How to build a reasonable and complete FFBRB model was the second problem to be solved. In order to solve the problem of how to diagnose various faults in the actual flywheel system, it is necessary to design the reasoning process and optimization process of the FFBRB model reasonably and establish a reasonable and accurate model (See Section 3.3).*

The function to solve this problem is denoted as FFBRB(*). ζ is the set of parameters in this process, y then the process can be described by the following expression:

$$y = \text{FFBRB}\,(x, \zeta) \quad (2)$$

This is a nonlinear mapping. x is the failure probability of the bottom event in the FFTA, and y is the output utility value of the BRB, corresponding to the occurrence probability of the top event. ζ is the set of parameters in this process.

Remark 1. *In order to solve the problem of small sample size, it could usually take two solutions. First, sample data with similar characteristics to the research question should be sought to expand the sample data volume, such as transfer learning [17,18]. Second, through the analysis of the model mechanism to expand the amount of information input. The BRB belongs to the second type of method, which can expand the model information input through expert knowledge, so as to realize model training under small samples.*

2.2. Overview of FFBRB Fault Diagnosis Model Principle

To solve the above problems, the FFBRB flywheel fault diagnosis model is proposed in this paper. In this model, the existing FFTA is used to construct the initial belief rules of BRB, and the transformation rules from FFTA to BRB are given. The model used the ER (evidential reasoning) algorithm to give the reasoning process of the model. In this model, the P-CMA-ES (projection covariance matrix adaptation evolutionary strategies) algorithm was used to optimize the parameters of the model, which improved the accuracy of the model. Figure 1 shows the overall transformation process of the model.

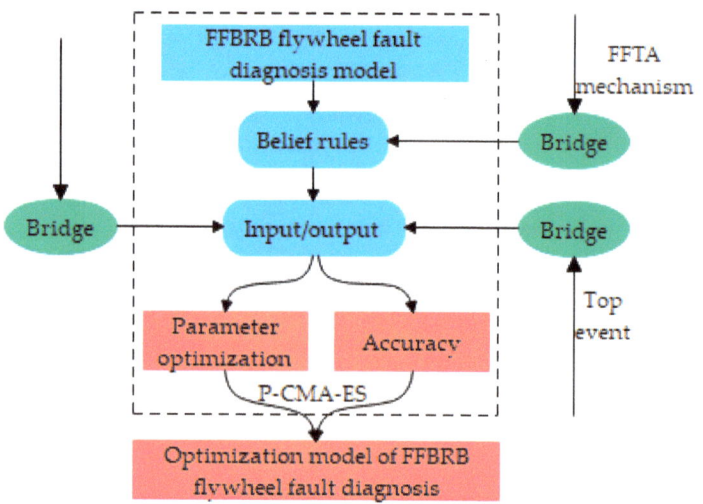

Figure 1. Fault diagnosis schematic diagram of FFBRB model.

Remark 2. *The similar learning ability of the BRB and neural networks was noted in the literature [19]. Therefore, the fault diagnosis of complex systems could be achieved through constructing deep BRB or hierarchical BRB models [20].*

3. Construction and Inference of the FFBRB Model

This section mainly introduces three parts:

- The basic structure of the FFTA flywheel system. In this part, fuzzy fault tree analysis is carried out for the flywheel system (see Section 3.1);
- The process of constructing the BRB model is based on FFTA. This part mainly describes the conversion process from FFTA to BRB (see Section 3.2);
- Reasoning and optimization process of the FFBRB model. This part is actually the reasoning and optimization process of BRB (see Section 3.3).

3.1. Basic Structure of the FFTA Flywheel System

In a practical flywheel system, FFTA analysis mainly depends on how the probability of each event in a fuzzy fault tree is calculated and expressed, and how to apply them to BRB. The overall fuzzy fault tree analysis structure is shown in Figure 2.

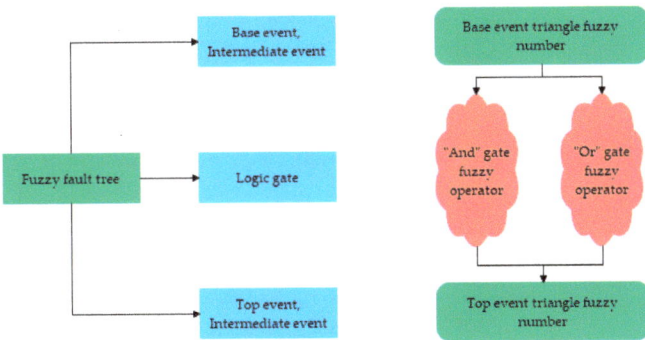

Figure 2. FFTA structure diagram of flywheel.

The fuzzy fault tree graph of the flywheel system is mainly composed of logic gates and related events, and its faults include sensor faults and system faults. The complete flywheel system fault tree [21] is shown in Figure 3 below:

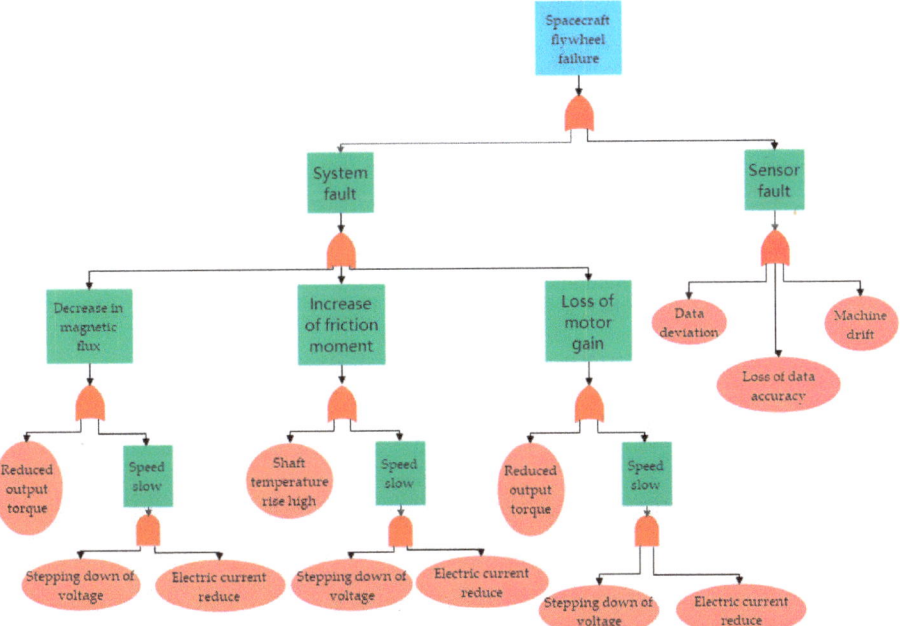

Figure 3. Graph of flywheel system fault tree.

3.2. The Process of Constructing the BRB Model Based on the FFTA

3.2.1. Analysis of Conversion Mechanism between FFTA and BRB

FFTA and BRB have differences in inputs and outputs. The input and output in BRB are mainly described by a series of belief rules, whereas the input and output in the FFTA are mainly described by logic gates and events. Therefore, it needed a bridge to enable the transition and transformation between the FFTA and BRB. The fault tree established in FFTA can sort out the relationship between fault events and clarify the context of different events. Bayesian networks describe the state of a part of the modeled thing and are associated with probability, also known as reliability networks. There is a certain mapping relationship between fuzzy fault tree and Bayesian network, which is expressed as follows [22]:

- Nodes in Bayesian networks correspond to events in FFTA. Specifically, all the top events of FFTA correspond to all the leaf nodes in the Bayesian network, and all the basic events of FFTA correspond to all the root nodes in the Bayesian network.
- Conditional probability distribution of nodes in Bayesian networks is represented by logic gates in FFTA.
- The direction of node arrows in the Bayesian network also represents the logical relationship of events in the FFTA, that is, the relationship between input and output of logic gates.

In order to describe the correspondence between FFTA and Bayesian networks, an example is listed in Figure 4 for reference.

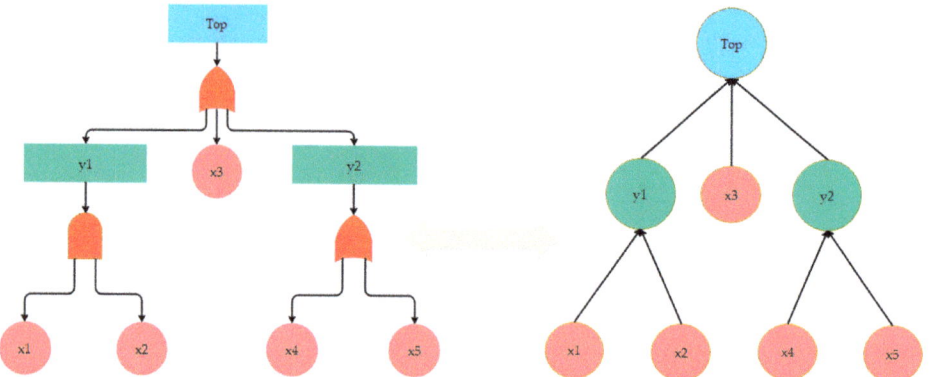

Figure 4. The corresponding expression graph between FFTA and Bayesian network graph.

BRB consists of three important parts: knowledge base, inference machine and optimization method. BRB's knowledge base is composed of a series of belief rules, which represent the relationship between input and output. ER, as the reasoning machine of BRB, is an evidential reasoning method [23]. The literature proves that the Bayesian inference can be extended to ER, where ER has weighted reliable inaccurate information, and the relationship between Bayes rules and ER rules can be revealed. The literature comes to the following conclusion: when each event is independent of the other, conditional probability is equivalent to belief degree. Therefore, it can be concluded that the Bayesian inference can be transformed into ER inference. ER [24], as the inference machine of BRB, is a part of BRB. Therefore, Bayesian inference can be transformed into BRB inference. The corresponding relationship between BRB and Bayesian network [25–27] is as follows:

- The input of the BRB corresponds to the parent node in the Bayesian network;
- The belief of the BRB can be transformed from conditional probability in the Bayesian network;
- Bayesian inference can be transformed from the ER to BRB inference.

Thus, as can be seen from the above analysis, it can conclude the complete FFTA to BRB conversion process, and the schematic conversion diagram from the FFTA to BRB is shown in Figure 5:

- The three numbers in the triangular fuzzy number of FFTA's base event failure probability are divided into three groups corresponding to the root node of the Bayesian network, respectively, which are used as the input of BRB;

- The three numbers in the triangle fuzzy number of FFTA intermediate event occurrence probability are divided into three groups corresponding to the root leaf nodes of the Bayesian network, respectively, which serve as the input and output of BRB;
- The three numbers in the triangular fuzzy number of FFTA top event occurrence probability are divided into three groups of night nodes corresponding to the Bayesian network, respectively, which are used as the output of BRB.

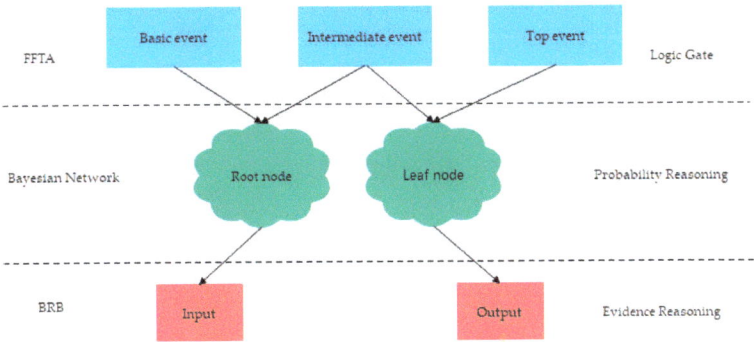

Figure 5. Schematic conversion diagram from FFTA to BRB.

3.2.2. Conversion Rules from FFTA to BRB

It can be seen from the above that the logic gate in FFTA corresponds to the conditional probability distribution of the corresponding node in the Bayesian network. Different logic gate pairs should have different transformation rules, and this section defines the transformation process.

Probability Representation of Transformation Space Condition Corresponding to Different Logic Gates

x_i is used to represent the i-th base event in FFTA, then the conditional probability rule in the Bayesian network corresponding to the logic gate of type "and" in FFTA can be described as expression 3, and the conditional probability rule in the Bayesian network corresponding to the logic gate of type "or" can be described as expression 4.

$$p(Top|x_1, x_2, \ldots, x_n) = \prod_{i=1}^{n} x_i \qquad (3)$$

$$p(Top|x_1, x_2, \ldots, x_n) = \sum_{i=1}^{n} x_i \qquad (4)$$

The Belief Rule and Rule Activation Weight Representation of the BRB Corresponded to the Logic Gate

Attribute importance withdrawal in BRB is the weight of attribute, and the importance of rules is the weight of rules. In this section, this paper defined different transformation rules for different logic gates, which also correspond to different rule activation weights.

The set of input reference values in FFTA below, that is, the set of reference values of the base event is represented by A_i. $Top_1, Top_2, \ldots, Top_n$ represents n results; under the k belief rule, the corresponding belief degree of each result is determined by $\beta_i (i = 1 \cdots N)$, N indicates the number of results; this paper used $\delta_i (i = 1 \ldots M)$ which represents the attribute weight of each premise attribute, M represents the number of attributes, and θ_k represents the rule weight of the belief rule in the article k, K is the number of belief rules.

- Under the condition of "and" logic gates, the BRB's belief rules [28] can be described as follows:

$$\text{BeliefRule}_k:$$
If x_1 is $A_1 \wedge x_2$ is $A_2 \wedge \ldots \wedge x_n$ is A_n

Then result is $\{(Top_1, \beta_1), (Top_2, \beta_2), \ldots, (Top_n, \beta_N)\}$ (5)

with rule weight $\theta_1, \theta_2, \ldots, \theta_K$

and attribute weight $\delta_1, \delta_2, \ldots, \delta_M$

where a_i^k represents the rule matching degree under rule k (the adaptability of input sample and belief rule), l indicates two adjacent activation rules, two rules are activated when the input falls between them, and the rule activation weight calculation under the "and" gate condition is as follows:

$$\omega_k = \frac{\theta_k \prod_{i=1}^{M} (a_i^k)^{\delta_i}}{\sum_{i=1}^{K} \theta_l \prod_{i=1}^{M} (a_i^l)^{\delta_i}} \qquad (6)$$

$$a_i^k = \begin{cases} \frac{A_i^{l+1} - x_i}{A_i^{l+1} - A_i^l} & k = l, A_i^l \leq x_i \leq A_i^{l+1} \\ 1 - a_i^k & k = l+1 \\ 0 & k = 1 \cdots K, k \neq l, l+1 \end{cases} \qquad (7)$$

- Under the condition of "or" logic gates, the BRB's belief rules could be described as follows:

If x_1 is $A_1 \vee x_2$ is $A_2 \vee \ldots \vee x_n$ is A_n

Then result is $\{(Top_1, \beta_1), (Top_2, \beta_2), \ldots, (Top_n, \beta_N)\}$ (8)

with rule weight $\theta_1, \theta_2, \ldots, \theta_K$

and attribute weight $\delta_1, \delta_2, \ldots, \delta_M$

where a_i^k represents the rule matching degree (the adaptability of input sample and belief rule), the rule activation weight calculation under the "and" gate condition is as follows:

$$\omega_k = \frac{\theta_k \sum_{i=1}^{M} (a_i^k)^{\delta_i}}{\sum_{l=1}^{K} \theta_l \sum_{i=1}^{M} (a_i^k)^{\delta_i}} \qquad (9)$$

The calculation of the rule matching degree is the same as the above "and" logic gate condition.

3.3. Establishment of the FFBRB Model and Inference Optimization

The FFBRB flywheel system fault diagnosis model established in this paper is shown in Figure 6.

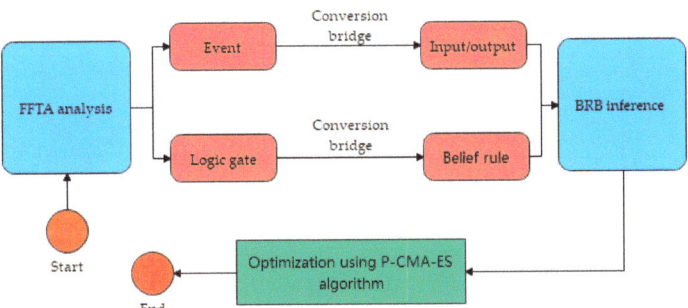

Figure 6. FFBRB flywheel system fault diagnosis model diagram.

3.3.1. Analysis of Reasoning Process from FFTA to BRB

The reasoning process of the FFBRB model, which is actually the reasoning process of the BRB, is shown in Figure 7.

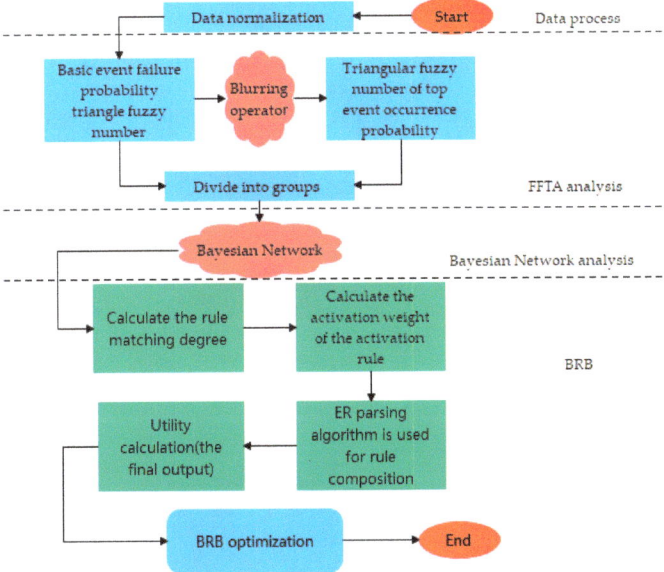

Figure 7. Diagram of FFBRB model inference process.

In particular, this model uses the triangle fuzzy number FFTA in the probability of events, from the upper and lower bounds of the triangular fuzzy number representation and event probability values are divided into three groups, respectively, after dealing with the BRB, can go through BRB to optimize the processing of the top event probability triangle fuzzy number, see FFTA analysis of the fitting effect of the result of the probability of the top event.

FFBRB model makes the FFTA knowledge mechanism embedded in the BRB expert knowledge base, which solves the problem that it is difficult to embed BRB expert knowledge. The FFBRB model uses BRB to train a series of sample data, which further improves the accuracy of the data and solves a considerable part of the uncertainty problems of the flywheel model. This section mainly introduces the reasoning process of FFBRB model fault diagnosis, that is, the reasoning process of BRB.

The specific fault diagnosis process of the FFBRB model is as follows:

Step 1: Data preprocessing. This paper first normalized the data samples and limited the data within the range of 0–1 to characterize the probability, so as to better describe the problem.

Step 2: Fuzzy fault tree analysis. Firstly, the logical relationship between events is sorted out and the fault tree graph of the fault diagnosis model is drawn. Then, this paper used a triangle fuzzy number to represent the failure probability of the FFTA basic event, introduce a fuzzy interval operator, calculate the triangle fuzzy number of occurrence probability of the middle event and top event and divide the data into three groups. For example, a triangle fuzzy number is used to represent the failure probability of a base event $x1(a1, m1, b1)$ and base event $x2(a2, m2, b2)$, and interval fuzzy operator formula is used to obtain the occurrence probability of an intermediate event or top event (a, m, b). In order to facilitate subsequent data processing, this paper divided these data into three groups $(a1, a2, a)$, $(m1, m2, m)$, $(b1, b2, b)$.

Step 3: Taking the Bayesian network as a bridge, FFTA is mapped to BRB. The equivalence of FFTA logic gate input and output and BRB input and output was explained through the bridge of the Bayesian network. According to the mapping rules mentioned above, fault tree graphs are mapped to the Bayesian network graphs and then BRB analysis is carried out, respectively, according to the graphs.

Step 4: Input the sample data integrating FFTA fault mechanism knowledge into BRB and use BRB for fault diagnosis. There are four steps to achieve concrete reasoning:

- Rule matching is calculated, that is, the degree of adaptation between input sample and belief rule. The calculation formula is shown in Formula (7).
- According to the activation weight formulas of different rules corresponding to different logic gates above (Formulas (6) and (9)), the activation weight of activation rules is calculated.
- ER analytic algorithm is used to synthesize rules and obtain the belief degree output of BRB. L indicates the number of activation rules. The calculation process is as follows:

$$\beta_n = \frac{\mu \times \left[\prod_{l=1}^{L}\left(\omega_l \beta_{n,l} + 1 - \omega_l \sum_{i=1}^{N} \beta_{i,l}\right) - \prod_{l=1}^{L}\left(1 - \omega_l \sum_{i=1}^{N} \beta_{i,l}\right)\right]}{1 - \mu \times \left[\prod_{l=1}^{L}(1 - \omega_l)\right]} \quad (10)$$

$$\mu = \frac{1}{\sum_{n=1}^{N}\prod_{l=1}^{L}\left(\omega_l \beta_{n,l} + 1 - \omega_l \sum_{i=1}^{N}\beta_{i,l}\right) - (N-1)\prod_{l=1}^{L}\left(1 - \omega_l \sum_{i=1}^{N}\beta_{i,l}\right)} \quad (11)$$

- Utility calculation, the final output.

$$y = \sum_{n=1}^{N} u(Top_n)\beta_n \quad (12)$$

Step 5: BRB optimization. In this step, the optimization algorithm is used to process the parameters to make the BRB output more accurate.

3.3.2. Optimization of the FFBRB Fault Diagnosis Model

This section describes the optimization process of the FFBRB model, as shown in Figure 8 below:

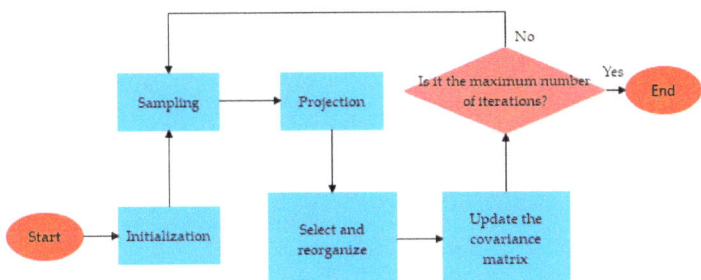

Figure 8. Optimization process flow chart.

In this model, the data generated by fuzzy fault tree analysis are still uncertain after BRB processing. In order to reduce the error between the parameters processed by the initial BRB and the real data and complete the optimization of parameters, an optimization mechanism is introduced in this model. P-CMA-ES [29] algorithm is used. The optimization function can be described as follows:

$$min\ MSE(\varsigma)$$
$$s.t. \sum_{n=1}^{N} \beta_{n,k} = 1, k = 1 \cdots K \tag{13}$$
$$0 \leq \beta_{n,k} \leq 1$$
$$0 \leq \theta_k \leq 1$$

In the upper form, the actual output of the square error is used by the $MSE(\varsigma)$, ς is the parameter that appears in the process and this paper used the lower formula to represent the average error of the output of the prediction:

$$MSE(\varsigma) = \frac{1}{K} \sum_{k=1}^{K} (y^* - y)^2 \tag{14}$$

In the above expression, y represents the actual output, y^* represents the predicted output, and the number of training samples is expressed by K. The realization process of the P-CMA-ES algorithm is described in detail below:

- Set initial parameters. The number of solutions is defined as Num in the population, Pn in the optimal subgroup, the dimension of the problem is defined as D, the optimal subgroup is defined as μ, the weight of the optimal subgroup is defined as ω_i;

$$\sum_{i=1}^{\mu} \omega_i = 1, \quad \omega_1 \geq \omega_2 \geq \cdots \geq \omega_\mu \geq 0 \tag{15}$$

- Sampling. The mean value of the optimal subgroup solution is the desired output value, and the population is normally distributed. The calculation process is as follows:

$$\varsigma_i^{h+1} = average^h + \eta^h H(0, To^h) \tag{16}$$

In the population of generation $h + 1$, the $i(0 < i < Num)$ solution is represented to ς_i^{h+1}; $average^h$ is the average of optimal subgroup solutions in the population; η^h is the h the generation of evolutionary steps; $H(*)$ is the normal distribution function representation of data; population h generation covariance matrix is represented by To^h;

- Projection. The process of performing a projection operation for each equality constraint can be described as follows:

$$\varsigma_i^{h+1}(1 + m \times (\tau - 1) : m \times \tau)$$
$$= \varsigma_i^{h+1}(1 + m \times (\tau - 1) : m \times \tau) - Q^T \times (Q \times Q^T)^{-1} \quad (17)$$
$$\times \varsigma_i^{h+1}(1 + m \times (\tau - 1) : m \times \tau) \times Q$$

The $m = (1 \ldots M)$, expression of the number of variables can be expressed as m in the equality constraint, $m = (1 \ldots M)$, M represents the solutions in each equality constraint, and $\tau = (1 \ldots M + 1)$, when the constraints are equal, its quantity can be expressed by τ. In addition, $Q = [1, 1, \ldots, 1]_{1 \times N}$ is the way to represent parameter vectors;

- Select and reorganize. Select the optimal subgroup and calculate the solution set of the mean. In the optimal subgroup, the weight of the $i - th(i=1 \ldots P_n)$ solution can be expressed as h_i, which is calculated as follows:

$$average^{h+1} = \sum_{i=1}^{P_n} h_i \varsigma_i^{h+i}, \sum_{i=1}^{P_n} h_i = 1 \quad (18)$$

- Update the covariance matrix. The specific calculation process is as follows:

$$To^{h+1} = (1 - e_1 - e_{Pn})T^h + e_1 s_c^{h+1}(s_c^{h+1})^T + e_{Pn} \sum_{i=1}^{P_n} h_i \left(\frac{\varsigma_i^{h+1} - average^h}{\eta^g} \right) \times \left(\frac{\varsigma_i^{h+1} - average^h}{\eta^g} \right)^T \quad (19)$$

$$s_c^{h+1} = (1 - e_c)s_c^h + \sqrt{e_c(2 - e_c)(\sum_{i=1}^{P_n} h_i^2)^{-1}} \times \frac{average^h average^{h+1}}{\eta^g} \quad (20)$$

$$\eta^{h+1} = \eta^h \exp\left(\frac{e_\eta}{o_\eta}\left(\frac{\|s_\varsigma^{h+1}\|}{\|H(0,J)\|} - 1\right)\right) \quad (21)$$

$$s_\eta^{h+1} = (1 - e_\eta)s_\eta^h + \sqrt{e_c(2 - e_c)(\sum_{i=1}^{P_n} h_i^2)^{-1}} \times To^{h-\frac{1}{2}} \times \frac{average^{h+1} - average^h}{\eta^h} \quad (22)$$

In the above calculation expression, the learning rate is expressed as e_1, e_{Pn}, e_c, e_η; The hth evolutionary step is expressed as $s_\eta^h, s_\eta^h = 0$; The evolution path of the hth covariance matrix is expressed as $s_c^h, s_c^h = 0$. In addition, J is used to represent the identity matrix, and the damping coefficient is denoted by o_η, Normal distribution of mathematical expectation $H(o, To^h)$ use $F \parallel N(o, I) \parallel$.

The above steps describe the specific calculation process of the P-CMA-ES algorithm. This algorithm was an improvement of the CMA-ES (projection covariance matrix adaptation evolutionary strategies) algorithm, which successfully solved the equality constraint problem in the BRB and was suitable for the fault diagnosis model proposed in this paper.

4. Case Study

The sub-tree of friction torque fault was the research object selected in this paper. The drop of voltage and current would slow down the speed of the flywheel, which would lead to a friction torque fault. The friction torque fault is also directly related to the shaft temperature (source used in this article from NASA). There were voltage, current, speed, shaft and friction moment data in this. One group of them could be chosen for the experiment. After selecting the data, they needed to be preprocessed. After the normalization of the data, fuzzy operator formula and ER fusion were used to obtain the data as the real value.

The fault diagnosis principle of the FFBRB flywheel system proposed in this paper included four parts: First, this paper normalized the collected data to make the data more accurate in practical application. Second, the normalized data were input into the fuzzy fault tree of the flywheel system, and the fuzzy probability of the intermediate event and the top event is calculated according to the corresponding formula. Third, this paper mapped the fuzzy fault tree to the BRB through the transformation space of the Bayesian network, so that the analysis process of the fuzzy fault tree corresponded to the inference process of BRB, and the input and output of the fuzzy fault tree correspond to the input and output of BRB, respectively. Finally, the data were handed over to the BRB for processing to realize the one-to-one correspondence between the BRB optimized value and the real value.

4.1. Construction of the FFBRB Fault Diagnosis Model

4.1.1. The Fault Tree of the Friction Torque Fault of the Flywheel System Is Constructed

In the following description, the fault tree of the flywheel friction torque fault is preliminarily constructed to sort out the logical relationship between each fault event and determine the cause of the fault. The friction torque fault tree is shown in Figure 9:

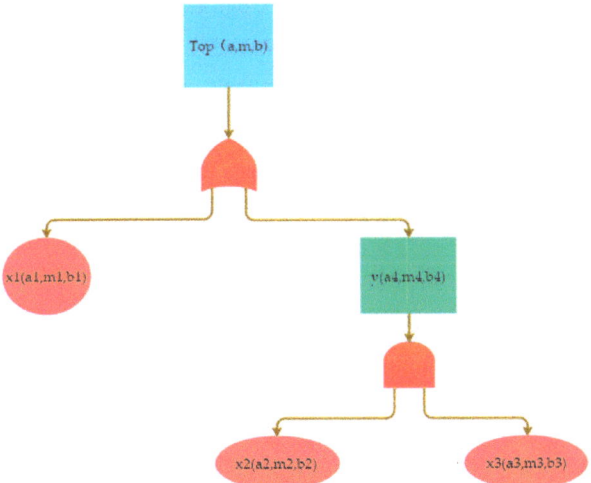

Figure 9. Friction torque fault tree.

In the fuzzy fault tree graph of the case, the triangle fuzzy number is marked to limit the probability of each event within a range. This paper marked the meanings of each symbol in the fault tree below in advance to better describe the problem. The meanings of specific symbols are shown in Table 1.

Table 1. FFTA indicates the letters in the fault tree.

Id	Letters	Meaning
1	X1	Shaft temperature rise high
2	X2	Stepping down of voltage
3	X3	Electric current reduce
4	y	Speed slow
5	Top	Increase in friction moment

4.1.2. FFTA Is Mapped to the BRB Using the Bayesian Network as a Bridge

After the establishment of the fault tree, this paper used the bridge of the Bayesian network to map the fault tree of FFTA to several different BRBS, so that the transformation from FFTA to BRB is perfectly realized, and the FFBRB model can be initially established. The relationship between the transformed Bayesian network graph and BRB is shown in Figure 10.

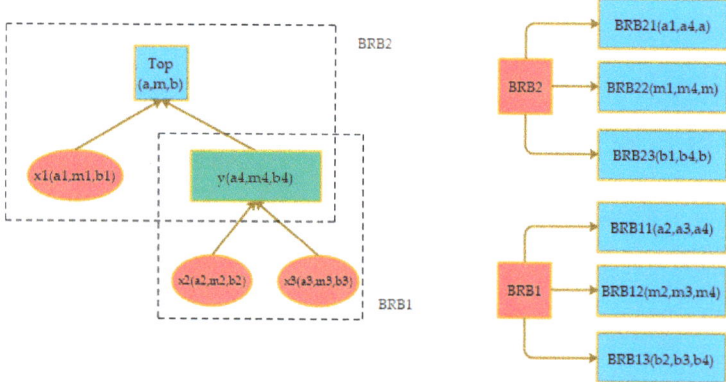

Figure 10. FFTA to BRB Bayesian network transformation diagram.

4.1.3. Determining the Fuzzy Number of Occurrence Probability of Bottom Event and Top Event

This step first needed to determine the trigonometric fuzzy number of the occurrence probability of the bottom event, and then calculate the trigonometric fuzzy number of the occurrence probability of the top event by using the formulas of fuzzy operators under different logic gates. The failure probability of the bottom event corresponds to the input of the BRB, and the occurrence probability of the top event corresponds to the output of the BRB, which is ready for the subsequent processing of the BRB program.

According to the previous introduction, corresponding data are divided into three groups (a1, a2, a), (m1, m2, m) and (b1, b2, b) according to the rules before. The data of the three groups are carried into the subsequent BRB, respectively, for fault diagnosis.

Triangulation fuzzy numbers of event probability in the BRB2 experiment are listed in Table 2 for reference.

Remark 3. *Each event in the above table only captures the data listed in article 10, from the data in the floating range there is a probability value of 10% of the incident left and if the interval data value is less than zero, the table is down to zero, if the data interval right value is greater than 1, the table down to 1, so the data that are limited to 0 to 1 can better describe probability.*

4.1.4. Built Initial Belief Rules

$$\text{If } x_1 \text{ is } A_1 \wedge x_2 \text{ is } A_2$$
$$\text{Then result is } \{(Top_1, \beta_1), (Top_2, \beta_2), (Top_3, \beta_3), (Top_4, \beta_4)\} \tag{23}$$
$$\text{with rule weight } \theta_1, \theta_2, \ldots, \theta_K$$
$$\text{and attribute weight } \delta_1, \delta_2$$

The initialization of BRB requires belief rule construction. In this case, the belief rule construction of BRB is as above.

Table 2. Trigonometric fuzzy number of event probability in FFTA.

Event	Ai	Mi	Bi
Base Event 1	0.3000	0.3333	0.3667
	0.0000	0.0000	0.0000
	0.9000	1.0000	1.0000
	0.0000	0.0000	0.0000
	0.0600	0.0667	0.0733
	0.7200	0.8000	0.8800
	0.4800	0.5333	0.5867
	0.0000	0.0000	0.0000
	0.0000	0.0000	0.0000
	0.0000	0.0000	0.0000
Base Event 2	0.9000	1.0000	1.0000
	0.0600	0.0667	0.0733
	0.1200	0.1333	0.1467
	0.0600	0.0667	0.0733
	0.6000	0.6667	0.7333
	0.4200	0.4667	0.5133
	0.3000	0.3333	0.3667
	0.6600	0.7333	0.8067
	0.8400	0.9333	1.0000
	0.2400	0.2667	0.2933
Top Event	0.9300	1.0000	1.0000
	0.0600	0.0667	0.0733
	0.9120	1.0000	1.0000
	0.0600	0.0667	0.0733
	0.6240	0.6889	0.7529
	0.8376	0.8933	0.9416
	0.6360	0.6889	0.7382
	0.6600	0.7333	0.8067
	0.8400	0.9333	1.0000
	0.2400	0.2667	0.2933

4.1.5. Set Reference Points and Values

In the BRB, it needed to set the reasonable reference values for the program to work properly. In this case, this paper set four reference points and reference values for each attribute, noting that the first reference value is an upper bound and the last reference value is a lower bound. The setting of reference values in BRB is shown in Table 3 above. The four numbers from left to right indicate the Very High(G), High(H), Middle(M), and Low(L) possibility of an event. The reference setting of BRB is shown in Table 3.

Table 3. Reference value of data in BRB.

BRB_id	Base Event 1	Base Event 2	Top Event
BRB 1	$[1.0, 0.6, 0.3, \lim_{x_{11} \to 0}(x_{11})]$	$[1.0, 0.8, 0.4, \lim_{x_{12} \to 0}(x_{12}), \lim_{x_{12} \to 0}(x_{12})]$	$[1.0, 0.3, 0.2, 0.0]$
BRB 2	$[1.0, 0.8, 0.4, \lim_{x_{21} \to 0}(x_{21})]$	$[1.0, 0.5, 0.3, \lim_{x_{22} \to 0}(x_{22}), \lim_{x_{22} \to 0}(x_{22})]$	$[1.0, 0.8, 0.6, 0.0]$

Remark 4. *When the median value of triangle fuzzy number interval of event occurrence probability is 0, the reference value of the lower bound of the interval is set as a number approaching 0, because the probability of an event cannot be negative.*

4.2. Training and Optimization of the FFBRB Model

4.2.1. Optimized Parameters and Results

Data show the optimized data of BRB2 ($b1$, $b2$, b), and the optimized parameters in BRB are shown in Table 4.

In Tables 4–6, the optimized rule weights are expressed as RuleWF and the optimized output belief degree is expressed as BeliefF. The results of the optimization of the upper and lower bounds of the interval and the median of the interval are listed.

Table 4. Optimized parameters table in BRB2r.

BRB2r_id	Attribute1	Attribute2	RuleWF	BeliefF
1	L	L	0.1771	(0.0733, 0.4983, 0.2373, 0.1910)
2	L	M	0.0709	(0.2110, 0.5779, 0.0343, 0.1768)
3	L	H	0.0062	(0.2813, 0.3703, 0.1696, 0.1788)
4	L	G	0.8472	(0.9886, 0.0137, 0.0000, 0.0000)
5	M	L	0.0396	(0.0315, 0.7699, 0.1906, 0.0080)
6	M	M	0.5838	(0.8332, 0.0979, 0.0702, 0.0000)
7	M	H	0.9296	(0.2924, 0.4950, 0.1712, 0.0414)
8	M	G	0.5178	(0.0923, 0.2010, 0.2374, 0.4694)
9	H	L	0.8063	(0.9973, 0.0000, 0.0000, 0.0075)
10	H	M	0.8488	(0.4543, 0.0084, 0.2880, 0.2493)
11	H	H	0.4081	(0.1555, 0.1741, 0.4458, 0.2246)
12	H	G	0.2917	(0.0619, 0.3106, 0.0903, 0.5372)
13	G	L	0.0002	(0.5614, 0.4062, 0.0150, 0.0174)
14	G	M	0.1367	(0.0560, 0.0149, 0.1501, 0.7790)
15	G	H	0.2903	(0.0047, 0.0063, 0.3829, 0.6062)
16	G	G	0.5334	(0.0000, 0.0102, 0.0000, 0.9960)

Table 4 is the optimal value of the upper bound of the interval, Table 5 is the optimal value of the ideal value of the interval, and Table 6 is the ideal value of the lower bound of the interval.

Table 5. Optimized parameters table in BRB2m.

BRB2m_id.	Attribute1	Attribute2	RuleWF	BeliefF
1	L	L	0.6018	(0.2020, 0.2635, 0.3770, 0.1575)
2	L	M	0.3155	(0.2320, 0.0685, 0.3231, 0.3764)
3	L	H	0.6173	(0.1893, 0.2777, 0.2197, 0.3133)
4	L	G	0.5771	(0.3145, 0.0551, 0.2492, 0.3811)
5	M	L	0.2627	(0.3164, 0.4002, 0.2218, 0.0616)
6	M	M	0.9665	(0.2234, 0.0333, 0.5372, 0.2061)
7	M	H	0.1127	(0.0023, 0.3528, 0.5186, 0.1263)
8	M	G	0.3443	(0.5425, 0.0730, 0.0759, 0.3085)
9	H	L	0.5466	(0.5419, 0.0308, 0.1200, 0.3073)
10	H	M	0.6745	(0.1283, 0.2672, 0.1916, 0.4129)
11	H	H	0.8846	(0.0487, 0.0797, 0.5155, 0.3561)
12	H	G	0.5213	(0.0568, 0.0764, 0.3596, 0.5072)
13	G	L	0.3741	(0.1902, 0.0219, 0.4706, 0.3173)
14	G	M	0.7260	(0.1378, 0.1024, 0.1934, 0.5663)
15	G	H	0.3316	(0.1201, 0.1004, 0.0978, 0.6817)
16	G	G	0.8969	(0.0382, 0.1119, 0.0657, 0.7842)

To avoid data redundancy, only four bits of data are reserved in Tables 4–6. As the same, the optimized rule weights are expressed as RuleWF and the optimized output belief degree is expressed as BeliefF.

Table 6. Optimized parameters table in BRB2l.

BRB2l_id	Attribute1	Attribute2	RuleWF	BeliefF
1	L	L	0.5453	(0.2353, 0.1349, 0.5162, 0.1137)
2	L	M	0.5036	(0.1352, 0.3365, 0.4205, 0.1078)
3	L	H	0.1688	(0.1922, 0.0713, 0.4822, 0.2543)
4	L	G	0.9502	(0.1944, 0.2135, 0.0504, 0.5417)
5	M	L	0.7318	(0.2970, 0.3715, 0.1161, 0.2154)
6	M	M	0.6618	(0.3590, 0.0935, 0.3166, 0.2310)
7	M	H	0.3964	(0.1935, 0.2580, 0.2173, 0.3313)
8	M	G	0.6569	(0.3582, 0.1651, 0.2443, 0.2324)
9	H	L	0.3200	(0.1115, 0.3012, 0.5681, 0.0192)
10	H	M	0.6779	(0.2005, 0.1247, 0.2703, 0.4044)
11	H	H	0.9339	(0.1083, 0.2752, 0.1753, 0.4412)
12	H	G	0.3865	(0.2076, 0.0799, 0.1489, 0.5635)
13	G	L	0.3149	(0.2779, 0.1064, 0.1853, 0.4304)
14	G	M	0.8496	(0.0266, 0.2361, 0.2723, 0.4650)
15	G	H	0.3898	(0.0570, 0.0798, 0.0547, 0.8085)

4.2.2. Experimental Fitting Images

The fitting images of experimental results and real results of interval lower bound ($a1$, $a2$, a), interval median ($m1$, $m2$, m) and interval upper bound ($b1$, $b2$, b) are listed below. In this paper, the fitting images of the three groups are drawn, respectively, as shown in Figure 11. The results of the three groups were processed by BRB, respectively, and compared with the real value to obtain the error, and finally unified analysis and summary.

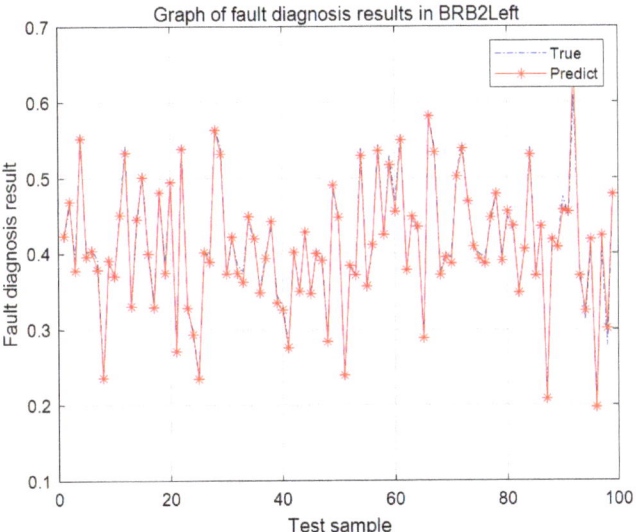

Figure 11. Cont.

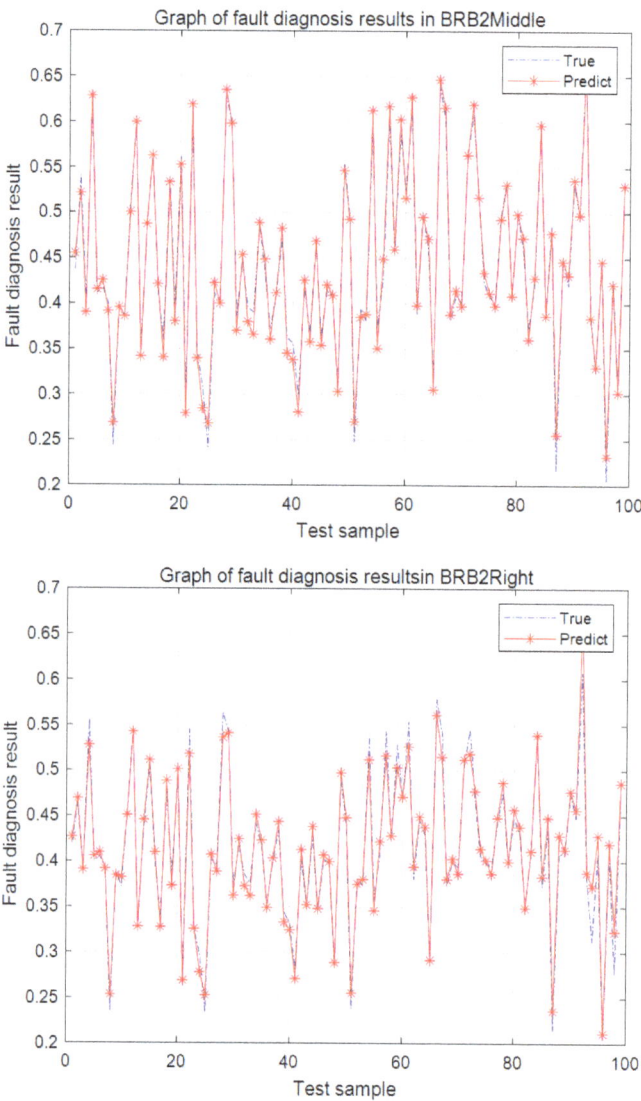

Figure 11. Fitting diagram right of experimental results and real values.

It can be seen that the results of the three groups of experiments fit well with real data. It could obtain the accuracy of each group through experiments, and then obtain the fluctuation range of experimental accuracy of the case. Then, this paper performed 10 experiments to find out the accuracy and, in this experiment, the accuracy of the three groups was 97.98%, 98.99% and 100.00%, the average accuracy of this experiment is 98.99%. It can be concluded that the accuracy of this experiment fluctuates in the range of 97.98% to 100%. In general, the FFBRB model established in this paper has a good processing effect. The experimental diagnosis results are shown in Figure 11.

4.2.3. Other Comparative Experiments

In this paper, ELM and BP neural networks, as the other two comparison methods of this experiment, are also used in flywheel fault diagnosis. This paper also drew the fitting images of the two control experiments, and it can be seen that the ELM and BP neural network methods are feasible, but still not as accurate as the FFBRB scheme. Among them, the difference between ELM and FFBRB schemes is relatively large, and the difference between BP neural network and FFBRB is not very large.

Two other groups of comparison experiments were conducted in this paper to compare with the FFBRB model method used in this paper, and the experimental results are shown in Figure 12 below.

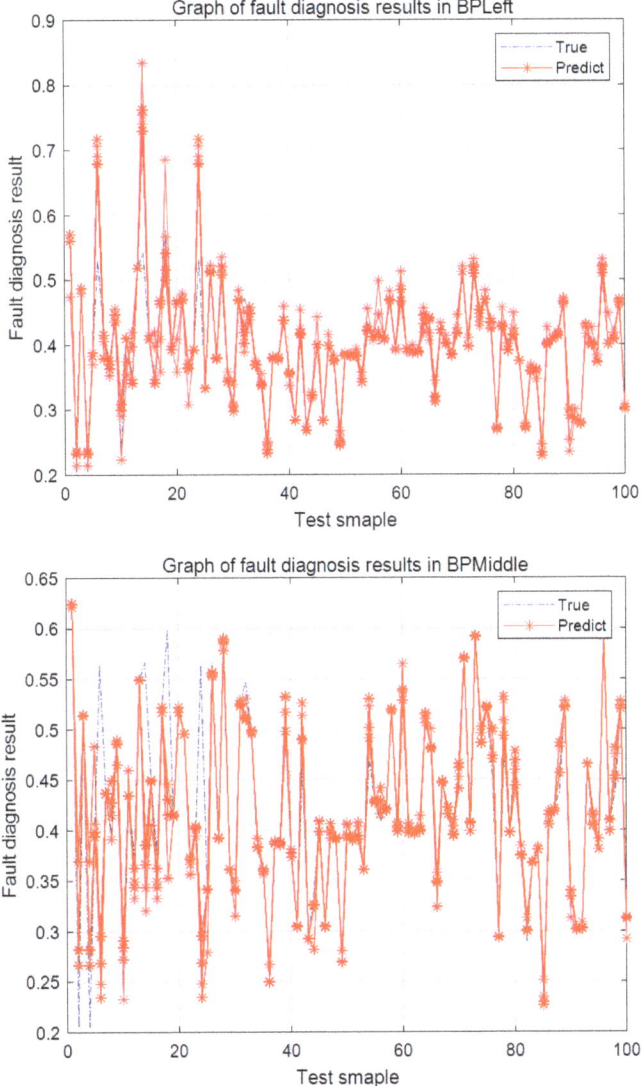

Figure 12. *Cont.*

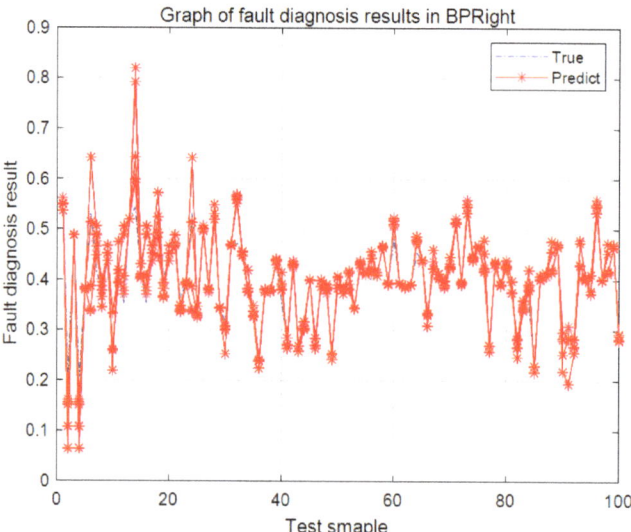

Figure 12. Fitting diagram of experimental results by BP method.

Figure 13 shows the diagnosis results obtained in ELM mode.

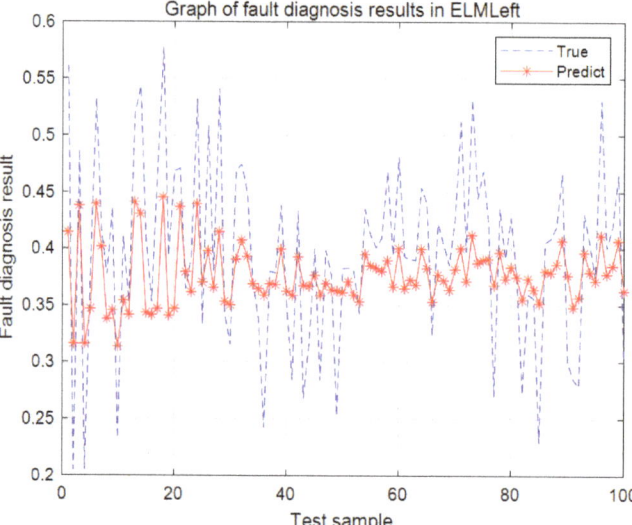

Figure 13. *Cont.*

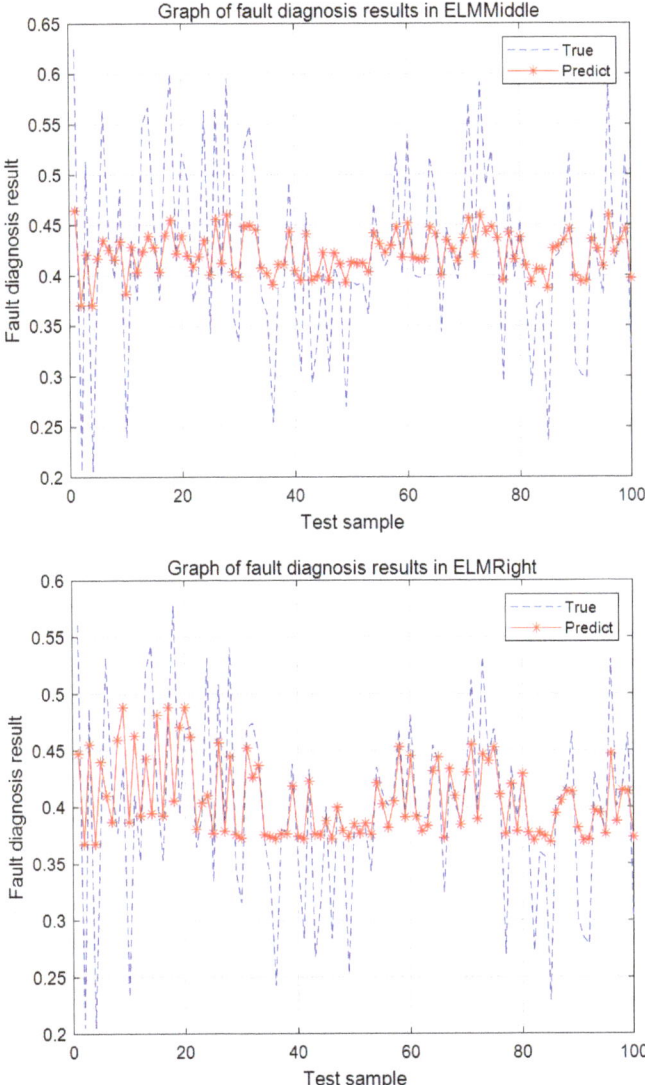

Figure 13. Fitting diagram of experimental results by ELM method.

In this experiment, the accuracy of 10 groups of data is taken, and the average of their probability is taken as the final result. The floating line chart of the accuracy of these 10 groups is shown in Figure 14.

In the three groups of the BP method, the average accuracy of the experimental fault diagnosis value compared with the real value is 85.90%, 91.30% and 85.50%, respectively. In the three groups of the ELM method, the average accuracy of the experimental fault diagnosis value obtained by us compared with the real value is 54.40%, 63.20% and 65.50%, respectively.

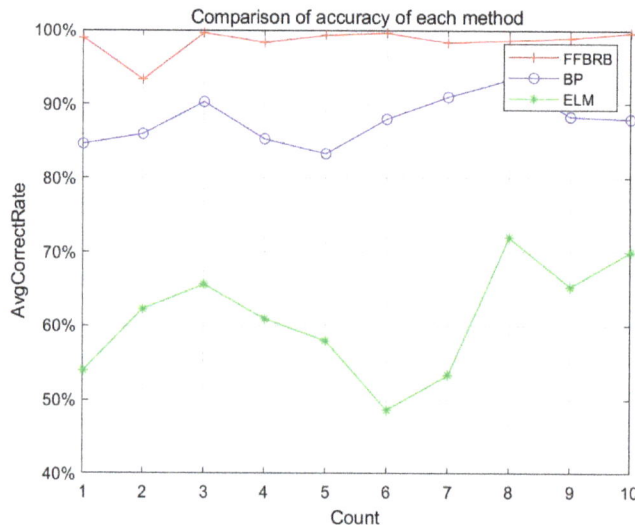

Figure 14. Comparison of experimental accuracy of different methods.

In the three groups of the FFBRB method, the average accuracy of the experimental fault diagnosis value obtained by us compared with the real value is 99.7%, 98.18% and 99.39%, respectively. This paper took the total average accuracy of the three groups of the three methods, and after calculation, the average accuracy of the BP method is 87.57%, the ELM method is 61.03%, the FFBRB method is 99.09%.

To facilitate intuitive observation, this paper sorted these data into a table, as shown in Table 7 below:

Table 7. Comparison of results of different methods.

	BP	ELM	FFBRB
Ave_Group_left	85.90%	54.40%	99.70%
Ave_Group_middle	91.30%	63.20%	98.18%
Ave_Group_right	85.50%	65.50%	99.39%
Average_times_group	87.57%	61.03%	99.09%

4.3. Experimental Conclusion

The experiment verifies the feasibility of the FFBRB model proposed in this paper, and it can be seen from the experimental results that the FFBRB model experiment is superior to the other two methods.

In particular, the BP neural network method is used to obtain the experimental diagnosis value and the real value of the image fitting, high accuracy, but there is still a little gap compared with the FFBRB method, and the BP method cannot explain its process. The experimental results obtained by the ELM method are much different from the real values, the image fitting effect of the experimental results is relatively poor, the accuracy is relatively low, and there is a big gap compared with the FFBRB scheme. The FFBRB fault diagnosis scheme in this paper is relatively optimal among the three, and its experimental results have a good image fitting effect and high accuracy, showing advantages compared with the other two schemes.

5. Conclusions

Based on BRB, a new fault diagnosis model (FFBRB) based on fuzzy fault tree analysis theory is proposed. The FFBRB model expands the expert knowledge base of BRB based on the FFTA mechanism, uses the improved BRB as a fault diagnosis tool, and incorporates an optimization algorithm to further reduce the influence of uncertain factors in the model. The model has the following characteristics:

The FFBRB model has a stronger ability to acquire expert knowledge. The FFBRB model integrates an FFTA mechanism analysis into the BRB expert knowledge base, which makes the model more capable of describing problems.

The FFBRB model has stronger analytical and reasoning ability. By training and optimizing the sample data, the model further improves the accuracy of the data, and thus makes the model more accurate.

The FFBRB model has high accuracy. Compared with traditional data-driven methods the FFBRB processing results have higher accuracy.

The feasibility of the FFBRB model is verified by experiments, and its advantages are compared with the other two methods. Based on the FFBRB model proposed in this paper, the following two aspects can be further studied in the future: (a) the theoretical transformation of the FFTA and interval BRB; (b) other methods could be used to expand the expert knowledge base in the flywheel fault diagnosis; (c) the BRB is an interpretable modeling method, which provided an effective support for the construction of interpretable deep learning models. How to effectively construct a fault diagnosis model based on a deep BRB will be the main work in the next step.

Author Contributions: X.C. and S.L. contributed equally to this work. Conceptualization, X.C. and S.L.; methodology, X.C. and S.L.; software, Y.X.; validation, X.C., S.L. and W.H.; formal analysis, X.C. and S.L; investigation, J.S.; data curation, P.Z.; writing—original draft preparation, X.C.; writing—review and editing, X.C. and W.H.; visualization, X.C.; supervision, W.H. and B.X. All authors have read and agreed to the published version of the manuscript.

Funding: This work was supported in part by the Postdoctoral Science Foundation of China under grant no. 2020M683736, in part by the Natural Science Foundation of Heilongjiang Province of China under Grant No. LH2021F038, in part by the innovation practice project of college students in Heilongjiang Province under grant no. 202010231009, 202110231024, 202110231155, in part by the graduate quality training and improvement project of Harbin Normal University under grant no. 1504120015, in part by the graduate academic innovation project of Harbin Normal University under grant no. HSDSSCX2021-120, HSDSSCX2021-29.

Institutional Review Board Statement: Not applicable.

Informed Consent Statement: Not applicable.

Data Availability Statement: Data sharing not applicable.

Conflicts of Interest: The authors declare no conflict of interest.

References

1. Jiang, L.; Wu, C. Topology optimization of energy storage flywheel. *Struct. Multidiscip. Optim.* **2016**, *55*, 1917–1925. [CrossRef]
2. Chen, C.; Tian, H.; Wu, D.; Pan, Q.; Wang, H.; Liu, X. A Fault Diagnosis Method for Satellite Flywheel Bearings Based on 3D Correlation Dimension Clustering Technology. *IEEE Access* **2018**, *6*, 78483–78492. [CrossRef]
3. Zhang, X.; Luo, W.; Li, X.; Yan, B. A Transfer Fault Diagnosing Method for Protocol Conformance Test. Based on FSMs. In Proceedings of the 2009 Asia-Pacific Conference on Information Processing, Shenzhen, China, 18–19 July 2009; pp. 173–177.
4. Lin, J.W.; Lee, C.L.; Su, C.C.; Chen, J.-E. Fault Diagnosis for Linear Analog Circuits. *J. Electron. Test.* **2001**, *17*, 483–494. [CrossRef]
5. Chen, B.; Wang, C.; Gao, X. Research on the Intelligent Agent of Distributed Fault Diagnose System. In Proceedings of the 2006 1st International Symposium on Systems and Control in Aerospace and Astronautics, Harbin, China, 19–21 January 2006; p. 1255.
6. Qiao, Z.; Elhattab, A.; Shu, X.; He, C. A second-order stochastic resonance method enhanced by fractional-order derivative for mechanical fault detection. *Nonlinear Dyn.* **2021**, *106*, 707–723. [CrossRef]
7. Sun, W.; Shao, S.; Zhao, R.; Yan, R.; Zhang, X.; Chen, X. A sparse auto-encoder-based deep neural network approach for induction motor faults classification. *Measurement* **2016**, *89*, 171–178. [CrossRef]

8. Yao, C.; Wang, R.; Xu, M.; Yang, J. The Combined Diagnosis Approach for the Satellite Attitude Control System Based on Observer Redundancy. In Proceedings of the 2013 International Conference on Quality, Reliability, Risk, Maintenance, and Safety Engineering (QR2MSE), Chengdu, China, 15–18 July 2013; pp. 1785–1789.
9. Cheng, C.; Wang, W.; Ran, G.; Chen, H. Data-Driven Designs of Fault Identification Via Collaborative Deep Learning for Traction Systems in High-Speed Trains. *IEEE Trans. Transp. Electrif.* **2021**, 1. [CrossRef]
10. Yang, J.-B.; Liu, J.; Wang, J.; Sii, H.-S.; Wang, H.-W. Belief rule-base inference methodology using the evidential reasoning Approach-RIMER. *IEEE Trans. Syst. Man Cybern. Part. A Syst. Hum.* **2006**, *36*, 266–285. [CrossRef]
11. Fu, Y.; Yin, Z.; Su, M.; Wu, Y.; Liu, G. Construction and Reasoning Approach of Belief Rule-Base for Classification Base on Decision Tree. *IEEE Access* **2020**, *8*, 138046–138057. [CrossRef]
12. Cheng, C.; Wang, W.J.; Chen, H.; Zhou, Z.; Teng, W.; Zhang, B. Health Status Assessment for LCESs Based on Multi-discounted Belief Rule Base. *IEEE Trans. Instrum. Meas.* **2021**, *70*, 3514213. [CrossRef]
13. Cheng, C.; Qiao, X.; Teng, W.; Gao, M.; Zhang, B.; Yin, X.; Luo, H. Principal component analysis and belief-rule-base aided health monitoring method for running gears of high-speed train. *Sci. China Inf. Sci.* **2020**, *63*, 1–3. [CrossRef]
14. Cheng, C.; Wang, J.; Zhou, Z.; Teng, W.; Sun, Z.; Zhang, B. A BRB-Based Effective Fault Diagnosis Model for High-Speed Trains Running Gear Systems. *IEEE Trans. Intell. Transp. Syst.* **2020**, *23*, 110–121. [CrossRef]
15. Fujino, T.; Hadipriono, F.C. Fuzzy Fault Tree Analysis for Structural Safety. *J. Intell. Fuzzy Syst.* **1996**, *4*, 269–280. [CrossRef]
16. Yiu, T.W.; Cheung, S.O.; Lok, C.L. A Fuzzy Fault Tree Framework of Construction Dispute Negotiation Failure. *IEEE Trans. Eng. Manag.* **2015**, *62*, 171–183. [CrossRef]
17. Wen, L.; Gao, L. A new deep transfer learning based on sparse auto-encoder for fault diagnosis. *IEEE Trans. Syst. Man Cybern. Syst.* **2017**, *49*, 136–144. [CrossRef]
18. Xiao, D.Y.; Huang, Y.X.; Qin, C.; Liu, Z.; Li, Y. Transfer learning with convolutional neural networks for small sample size problem in machinery fault diagnosis. *Proc. Inst. Mech. Eng. Part C J. Mech. Eng. Sci.* **2019**, *233*, 5131–5143. [CrossRef]
19. Cao, Y.; Zhou, Z.J.; Hu, C.H.; Tang, S.W.; Wang, J. A new approximate belief rule base expert system for complex system modelling. *Decis. Support. Syst.* **2021**, *150*, 113558. [CrossRef]
20. Cao, Y.; Zhou, Z.; Hu, C.; He, W.; Tang, S. On the Interpretability of Belief Rule-Based Expert Systems. *IEEE Trans. Fuzzy Syst.* **2020**, *29*, 3489–3503. [CrossRef]
21. Tabesh, M.; Roozbahani, A.; Hadigol, F.; Ghaemi, E. Risk Assessment of Water Treatment Plants Using Fuzzy Fault Tree Analysis and Monte Carlo Simulation. *Iran. J. Sci. Technol. Trans. Civ. Eng.* **2021**, 1–16. [CrossRef]
22. Zhang, M.; Xu, T.; Sun, H.; Meng, X. Mine Hoist Fault Diagnosis Based on Fuzzy Fault Tree and Bayesian network. *Ind. Mine Autom.* **2020**, *46*, 1–5, 45.
23. Yang, J.B.; Xu, D.L. A study on generalizing Bayesian inference to evidential reasoning. In *Belief Functions: Theory and Applications*; Springer: Cham, Switzerland, 2014; pp. 180–189.
24. Yang, J.-B.; Liu, J.; Wang, J.; Sii, H. The Evidential Reasoning approach for Inference in rule-based systems. In Proceedings of the 2003 IEEE International Conference on Systems, Man and Cybernetics. Conference Theme—System Security and Assurance (Cat. No.03CH37483), Washington, DC, USA, 8 October 2003; pp. 2461–2468.
25. Fenz, S. An ontology-based approach for constructing Bayesian networks. *Data Knowl. Eng.* **2011**, *73*, 73–88. [CrossRef]
26. Li, M.; Feng, X.; Chen, J. Research of Threat Identification Based on Bayesian Networks. In Proceedings of the 2009 5th International Conference on Wireless Communications, Networking and Mobile Computing, Beijing, China, 24–26 September 2009; pp. 1–3.
27. Jarraya, A.; Leray, P.; Masmoudi, A. Discrete Exponential Bayesian Networks: An Extension of Bayesian Networks to Discrete Natural Exponential Families. In Proceedings of the 2011 IEEE 23rd International Conference on Tools with Artificial Intelligence, Boca Raton, FL, USA, 7–9 November 2011; pp. 205–208.
28. Yang, J.B.; Liu, J.; Wang, J. An Optimal Learning Method for Constructing Belief Rule Bases. In Proceedings of the 2004 IEEE International Conference on Systems, Man and Cybernetics (IEEE Cat. No.04CH37583), The Hague, The Netherlands, 10–13 October 2004; pp. 994–999.
29. Zhou, Z.-J.; Hu, G.-Y.; Zhang, B.-C.; Hu, C.-H.; Qiao, P.-L. A Model for Hidden Behavior Prediction of Complex Systems Based on Belief Rule Base and Power Set. *IEEE Trans. Syst. Man Cybern. Syst.* **2017**, *48*, 1649–1655. [CrossRef]

Article

Intelligent Fault Diagnosis Method for Blade Damage of Quad-Rotor UAV Based on Stacked Pruning Sparse Denoising Autoencoder and Convolutional Neural Network

Pu Yang [1,2,*], Chenwan Wen [1,2], Huilin Geng [2] and Peng Liu [2]

1. Key Laboratory of Advanced Aircraft Navigation, Control and Health Management, Ministry of Industry and Information Technology, Nanjing University of Aeronautics and Astronautics, Nanjing 211106, China; wenchenwan@nuaa.edu.cn
2. College of Automation Engineering, Nanjing University of Aeronautics and Astronautics, Nanjing 211106, China; hlgeng@nuaa.edu.cn (H.G.); llpeng@nuaa.edu.cn (P.L.)
* Correspondence: ppyang@nuaa.edu.cn

Abstract: This paper introduces a new intelligent fault diagnosis method based on stack pruning sparse denoising autoencoder and convolutional neural network (sPSDAE-CNN). This method processes the original input data by using a stack denoising autoencoder. Different from the traditional autoencoder, stack pruning sparse denoising autoencoder includes a fully connected autoencoding network, the features extracted from the front layer of the network are used for the operation of the subsequent layer, which means that some new connections will appear between the front and rear layers of the network, reduce the loss of information, and obtain more effective features. Firstly, a one-dimensional sliding window is introduced for data enhancement. In addition, transforming one-dimensional time-domain data into the two-dimensional gray image can further improve the deep learning (DL) ability of models. At the same time, pruning operation is introduced to improve the training efficiency and accuracy of the network. The convolutional neural network model with sPSDAE has a faster training speed, strong adaptability to noise interference signals, and can also suppress the over-fitting problem of the convolutional neural network to a certain extent. Actual experiments show that for the fault of unmanned aerial vehicle (UAV) blade damage, the sPSDAE-CNN model we use has better stability and reliable prediction accuracy than traditional convolutional neural networks. At the same time, For noise signals, better results can be obtained. The experimental results show that the sPSDAE-CNN model still has a good diagnostic accuracy rate in a high-noise environment. In the case of a signal-to-noise ratio of −4, it still has an accuracy rate of 90%.

Keywords: intelligent fault diagnosis; stacked pruning sparse denoising autoencoder; convolutional neural network; anti-noise

1. Introduction

UAVs are very suitable for performing tasks in spacious indoor and outdoor environments, such as personnel search and rescue, material transportation, military patrol and surveillance, pesticide spraying, crop seeding, etc. Due to the increasing complexity of the tasks performed by drones, the sensors and actuators on the drone are becoming more and more complex, and the reliability requirements of the drone are getting higher and higher during the mission. Once the drone has a serious fault in flight, it will cause more serious property losses, and in more serious cases, it may cause casualties [1]. During the flight of the drone, any minor fault can easily cause the drone itself to malfunction, thereby affecting the sensors, actuators, and other related equipment on the drone. Therefore, the safety and reliability of UAVs is now an issue worthy of study and discussion. At the same time, we also need to specifically consider the different types of faults of different types of UAV [2].

For various faults on the drone, the drone control system can respond to the faults on the drone only when the system identifies and diagnoses each fault, respectively, so as to minimize the loss of personnel and property in the case of UAV faults. One of the main issues is the identification of faults in drones. The identification of faults is mainly divided into knowledge-based fault methods, model-based fault diagnosis methods, and data-based fault diagnosis methods.

The main knowledge-based fault diagnosis methods are symbolic expert systems [3], symbolic directed graph (SDG) methods, and fault tree methods. In [4], the symbolic directed graph is introduced, the symbolic directed graph is mainly a graphical model based on causality. In [5], the fault diagnosis method based on the fault tree is mainly introduced, the fault tree uses a graphical method for fault diagnosis. A fault tree is formed by connecting the fault in the system and the cause of the system fault. When the system fails, the cause of the system fault is deduced from the current fault state of the system from the bottom to the top. As a knowledge-based fault diagnosis method, the diagnosis model is simple, and the diagnosis results are easier to apply in practical engineering. However, because knowledge-based fault diagnosis requires learning the types of faults to be diagnosed, when a fault that is not in the knowledge base occurs in the system, the system will not be able to provide the correct diagnosis result.

The model-based fault diagnosis method [6] is based on the accurate mathematical model of the system. In the analytical model of the system, the residual signal between the input and output of the system is obtained by observation and measurement. By analyzing the residual signal in the system, the difference between the actual output and the expected output of the system can be obtained. Therefore, the system can be diagnosed based on these.

The data-driven fault diagnosis method is to classify and identify all the non-faulty and faulty data of the system, so the system's fault diagnosis can be realized without obtaining the precise mathematical model of the system. Data-based fault diagnosis methods mainly include machine learning methods [7], signal processing methods [8], information fusion methods [9], rough set methods [10], multivariate statistical analysis methods [11], etc. Because the data-based fault diagnosis method does not rely on the accurate model of the system for diagnosis, it is better to use the data-based method for fault diagnosis for complex high-level systems that are difficult to accurately model. However, because the data-based fault diagnosis method does not depend on the internal structure of the system, the interpretability of the results of system fault diagnosis is not very good [12].

At present, many intelligent fault diagnosis methods have been proposed in various research fields. In literature [13,14], the bearing is taken as the research object to study the relationship between the data collected by the bearing in different types of damage; in article [15,16], the fault diagnosis of drill is realized by analyzing the thermal image and vibration data of drill; in [17,18], the researchers took the battery pack as the research object and applied the intelligent fault diagnosis algorithm proposed by themselves to the actual battery system to diagnose the battery pack; in the research field of gearbox and high-speed train, a large number of fault diagnosis methods have also been proposed; in [19,20], it was studied how to judge the fault type through the collected signal when the gearbox fails; several new intelligent fault diagnosis methods are mainly proposed in [21–23], and good results have been achieved in the fault diagnosis of high-speed trains. Although many fault diagnosis methods have been proposed, there are still few intelligent fault diagnosis methods for UAVs. Therefore, we choose the quad-rotor UAVs as the research object in this paper.

During the operation of the quad-rotor UAV, the actuator or structure of the drone malfunctioned due to the operation problem of the pilot or due to some non-human reasons. In the literature [24], the researchers collected the vibration signal of the aircraft frame through the analysis of these data to diagnose whether the motor is malfunctioning. In [25], the researchers artificially damaged the rotor of the drone, and then collected the noise of the drone during the flight, and used the deep learning method to analyze and process the noise to realize the fault diagnosis of the system. The collection of sound signals has strict requirements on the environment, so this method cannot be applied in practice. In the literature [26], the author introduced the convolutional neural network with a wide convolution kernel into the fault diagnosis method, and diagnosed the bearing data through the convolutional neural network. A wide convolution kernel can improve the anti-interference ability of convolutional neural network to some extent. In reference to the problem of inaccurate diagnosis results for data with large noise signals, the literature [27] proposed to denoise the data based on the stack denoising autoencoder, and achieved good results, but due to the introduction of a new network structure, the convergence speed of the training network has been adversely affected to a large extent. Most of the existing fault diagnosis algorithms need to preprocess the data to eliminate the noise interference in the data, thereby improving the accuracy of classification, but there are few methods to directly classify the original noisy data and obtain a good classification accuracy.

In response to the above-mentioned problems, we adopted a method called Stacked Pruning Sparse Denoising Autoencoder and Convolutional Neural Network (sPSDAE-CNN) to identify and classify the actuator damage fault of the UAV. The main contribution of this paper is as follows:

1. We use a new and improved convolutional neural network method, which can be directly applied to the original UAV data collected in practice. Compared with the traditional method, it does not require separate data preprocessing. The comparison is shown in Figure 1;
2. The method uses a stack denoising autoencoder as the first layer of the convolutional neural network, which is very robust against data with much noise in the data, and still has a relatively high fault diagnosis accuracy rate under high noise conditions;
3. Directly convert the sensor data collected by the drone into a gray sampling map. Expanding the dimensionality of the sample can further improve the feature extraction ability of the DL model;
4. This method is aimed at the problem that enough data cannot be collected during neural network training. We use a one-dimensional sliding window for overlapping sampling to enhance the data, increase the data scale, and improve the generalization of the neural network ability;
5. We use the feature maps learned by visualizing sPSDAE-CNN to explore the actual feature learning and classification mechanism of the sPSDAE-CNN model. At the same time, the pruning operation is introduced to speed up the training of SDAE.

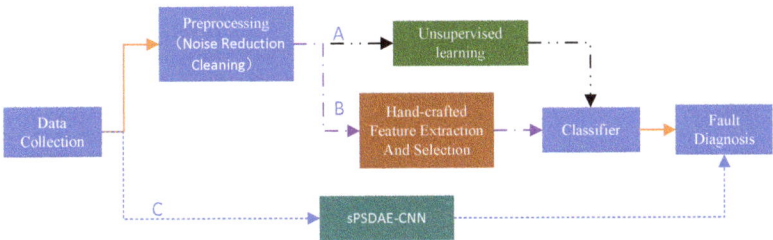

Figure 1. Three kinds of intelligent fault diagnosis framework. (**A**) The feature extraction of unsupervised learning [28]. (**B**) The traditional method. (**C**) The method used in this article.

At present, there is much research on sensor fault and actuator fault of four-rotor UAVs. In article [29,30], it is mainly studied to diagnose the actuator fault of four-rotor UAV by using the traditional model class method, including hybrid observer and adaptive neural network observer. In [31], Kalman filter is mainly used to process the sensor data of UAV and then to diagnose the possible sensor faults. In [32], researchers proposed a disturbance observer to observe the faults in the system and then realized diagnosis and fault-tolerant control through sliding mode control method.

However, there is little research on the fault of UAV blade damage. In the process of a UAV mission, when the UAV blade is damaged to a certain extent, when the damage does not exceed the threshold, the UAV may still be able to perform the mission in the environment of small interference. However, at this time, the stability of the UAV has been greatly damaged, and there may be some risks during the mission. Therefore, we need to evaluate the blade damage of UAV through the proposed method, and timely evaluate the health state of UAV, so as to prevent UAV crashes. At the same time, we also introduce a sparse pruning stack noise reduction autoencoder to improve the adaptability of the model to high noise data. In addition, pruning operation is added to improve the algorithm complexity of the model. At present, most fault diagnosis methods for four-rotor UAVs are verified by numerical simulation. This paper collects experimental data on the actual aircraft and verifies the algorithm, which has good practicability.

There is not a simple linear relationship between the damage of the drone blades and the sensor data of the drone. Therefore, the sensor data of the drone blades under different damage conditions are analyzed by using the deep learning method, and a deep learning model about the relationship between the sensor data of the drone and the damage degree of the blades is obtained, and the model is optimized.

The remaining organizational structure of this article is as follows: Section 2 briefly introduces the convolutional neural network and the stack denoising autoencoder. Section 3 introduces the intelligent fault diagnosis method based on sPSDAE-CNN. In Section 4, we use experiments to verify the sPSDAE-CNN method, and compare and analyze it with some commonly used methods. At the end of Section 5, we draw conclusions and propose future work by summarizing the work.

2. Introduction to the Convolutional Neural Network and Stack Denoising Autoencoder

2.1. A Brief Introduction to Convolutional Neural Networks

In this part, we will briefly introduce the convolutional neural network and the stack denoising autoencoder. For more details about the neural network, please refer to the literature [33]. Convolution neural network is a multilevel deep neural network [34]. Its basic structure consists of the input layer, convolution layer, activation layer, pooling layer, full connection layer, and output layer. Generally, there are several convolution layers and pooling layers, and the general structure is a convolution layer connected with a pooling layer. Each neuron in the input is locally connected to the input, and the weighted summation with the local input through the corresponding connection weight and the bias is added to obtain the input of the neuron. This process is equivalent to the convolution process, so it is called a convolutional neural network.

2.1.1. Convolutional Layer

The convolutional layer uses a convolution kernel to perform convolution operations on our input data or local regions of features, and extract relevant features from the data. Figure 2 shows the structure diagram of the convolutional layer and the pooling layer. The top layer is the pooling layer, the middle is the convolutional layer, and the bottom is the input layer [34]. In Figure 2, convolution neurons are organized into feature planes, and each neuron in the convolution layer is locally connected to the feature surface in its input layer. The output of each neuron in the convolution layer can be obtained by passing the local weighting and transfer to the activation function.

An important feature of convolutional neural networks is weight sharing. The weights of convolutional neural networks in the plane of the same input feature and the same output feature are shared. Weight sharing also reduces the complexity of the network model to a certain extent. It also avoids the over-fitting problem caused by too many parameters. In actual operations, most of the related operations can be replaced by convolution operations, which can avoid the problem of reversing the convolution kernel during backpropagation. The formula for convolution operation is shown in (1):

$$y_i^{k+1}(j) = \kappa_i^k \times \chi^k(j) + b_i^k, \tag{1}$$

where κ_i^k and b_i^k respectively represent the weight and bias of the kth filter kernel of the ith layer of the neural network, and use $\chi^k(j)$ to represent the jth local region of the kth layer. Where $\times$ is used to calculate the inner product of the kernel and the local area, and $y_i^{k+1}(j)$ represents the input of the j neuron in the frame i of the $k + l$ layer.

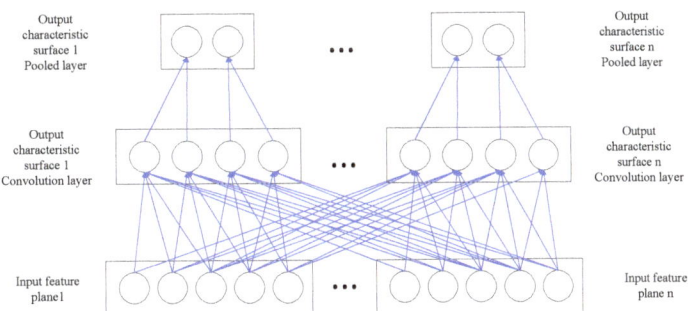

Figure 2. Schematic diagram of the convolutional layer and the pooling layer structure.

2.1.2. Activation Layer

After the convolutional layer, we need to use the activation function to introduce nonlinear modeling capabilities to our neural network, eliminate redundant data in the data, and enhance the learning ability of the neural network, so that the features in the data can be further processed for segmentation. The commonly used activation functions mainly include sigmoid function, tanh function, ReLu function, ELU function, etc. For details, please refer to the literature [35]. In our convolutional neural network, we choose to use the ReLu function as the activation function. Its main feature is compared with linear functions. ReLu has better expression ability compared with nonlinear functions, as ReLu does not have the problem of gradient disappearance and can maintain the convergence rate of the model in a stable state. The ReLu function is expressed as follows (2):

$$\alpha_i^{k+1}(j) = \text{ReLu}(y_i^{k+1}(j)) = \max\{0, y_i^{k+1}(j)\}, \tag{2}$$

where $y_i^{k+1}(j)$ represents the output of the first convolutional layer, and $\alpha_i^{k+1}(j)$ represents the result of $y_i^{k+1}(j)$ activated by ReLu.

2.1.3. Pooling Layer

The pooling layer is also one of the most common and basic mechanisms of convolutional neural networks. It is actually a form of downsampling, and there are many forms of nonlinear pooling functions in convolutional neural networks. Max pooling function is the most common one. The principle of this mechanism is that when a feature of data is discovered, its exact location is far less important than its relative location with other features. Pooling reduces the size of the data space by constantly reducing the number of network parameters and the amount of computation. Overfitting can also be suppressed to some extent. The max-pooling operation can be expressed as shown in Figure 3:

The expression is (3):

$$a^k_{(nh,nw,c)} = \max(a^{k-1}_{(nh \times stride:nh \times stride+f, nw \times stride:nw \times stride+f, c)}), \tag{3}$$

where nh represents the height in the current pixel, nw represents the width of the current pixel, and c represents the channel, f represents the size of the pooling core, and $stride$ represents the step size of the pooling core movement.

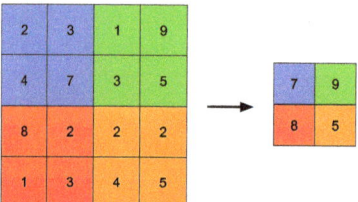

Max Pool with 2 by 2 filter and stride 2

Figure 3. Maximum pooling operation.

2.1.4. Batch Normalization

Batch standardization was proposed in [36] to accelerate the training speed of deep neural networks by reducing the transfer of internal covariates. The batch normalization layer is usually added after the convolutional layer or the fully connected layer, and before the activation layer. Given p-dimensional data into the BN layer $X = (x^{(1)}, \ldots, x^{(p)})$ the operation of the BN layer can be expressed as the following expression (4):

$$\hat{x}^{(i)} = \frac{x^{(i)} - E(x^{(i)})}{\sqrt{Var[x^{(i)}]}}$$
$$y^{(i)} = \gamma^{(i)} \hat{x}^{(i)} + \beta^{(i)}, \quad (4)$$

where $y^{(i)}$ represents the p-dimensional output of the BN layer, and $\gamma^{(i)}$ and $\beta^{(i)}$ are the scaling and bias that the BN layer needs to learn, which need to be learned in the neural network training.

2.2. Stacked Denoising Autoencoder

The encoder is a commonly used learning model in deep learning. The structure of this model is shown in Figure 4. The stack noise reduction autoencoding network is based on the encoder. The encoder must learn to obtain noise-free input from the noisy data. Unlike the supervised learning model CNN and Recurrent Neural Networks (RNN) [34], it combines unsupervised data feature extraction with supervised overall fine-tuning, and it can mainly realize the noise reduction and dimensionality reduction of the features of high-noise information. The structure is shown in Figure 5. Stack noise reduction autoencoder and encoder are mainly composed of encoder and decoder, which can be used to extract hidden features of samples and reconstruct input.

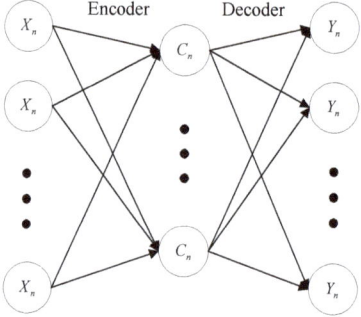

Figure 4. The structure of the encoder.

Assuming that $C(x|\hat{x})$ represents the error between the original data x and the noisy data $\hat{x}$ the DAE parameters are optimized and adjusted by using back propagation and gradient descent methods. After training DAE, the hidden layer can be regarded as the input of the next DAE, and this multiple DAE can form the model of the stack denoising autoencoder [37].

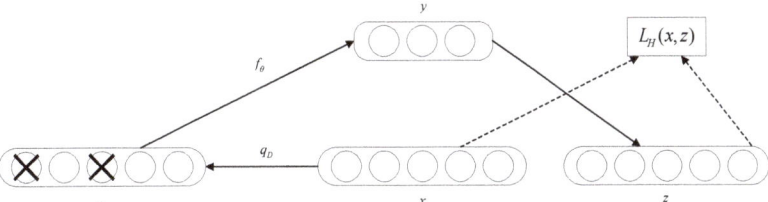

Figure 5. The structure of the denoising autoencoder.

3. Proposed Convolutional Neural Network with Stacked Pruning Sparse Denoising Autoencoder

In this paper, an intelligent quadrotor UAV fault diagnosis method based on stacked pruning sparse noise reduction autoencoder and convolutional neural network is proposed. We mainly use sPSDAE as the first layer of the neural network to reduce noise and dimensionality of the original data. The introduction of stack pruning sparse noise reduction autoencoder can improve the model generalization ability of the neural network and suppress the over-fitting problem. Secondly, convolutional neural network (CNN) is used to extract and classify system features. The algorithm model is shown in Figure 6:

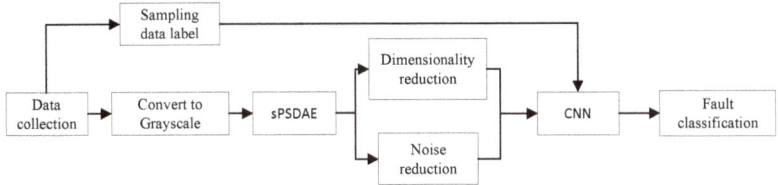

Figure 6. sPSDAE-CNN algorithm model.

Firstly, collect the flight data of the drone. In order to simulate the damage of the blades of the drone in the actual flight, we collect the drone data by artificially damaging the blades of the quadrotor rotor drone in a laboratory environment. The individual blades of the UAV are set to have different degrees and types of damage. The main types and degrees of damage are shown in Table 1 below:

Table 1. Main types and degrees of damage.

Types of Damage to the Blades	Damage Degree of the Blade
No damage	0%
Broken blade	5%
Broken blade	10%
Broken blade	15%
Broken blade	20%
Blade crack	Slightly deformation
Blade crack	General deformation
Blade crack	Severely deformation

Eight different types and degrees of damage to the blades are shown in Figure 7:

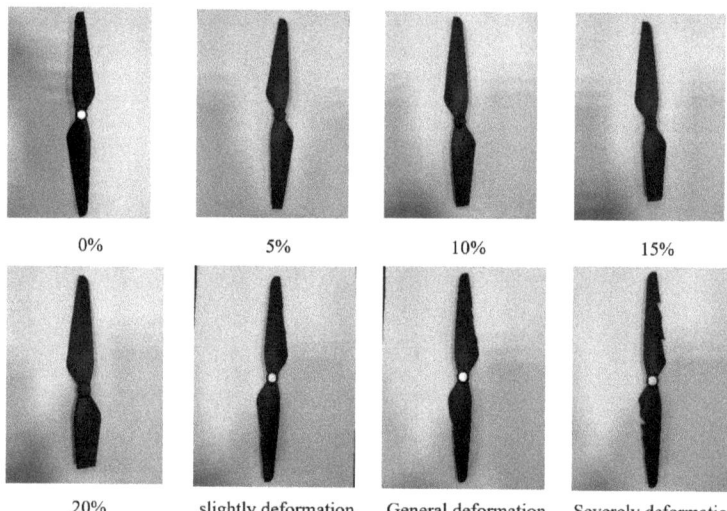

Figure 7. Eight different types of blade damage.

This paper chooses to use a quad-rotor drone with pixhawk4 flight control as the main control board for data collection. We let quad-rotor drones conduct flight experiments in different health states, collect data, and convert the collected data into a two-dimensional grayscale image. The paper selects the output of the four actuators in the flight log of the drone, the quaternion representing the attitude of the drone, the angular velocity on the three coordinate axes of the drone, and the position information, velocity information and acceleration information of the flight on the three coordinate axes of XYZ. Taking 20 sampling periods as a data state, a 20 × 20 two-dimensional matrix is formed, which is converted into a 20 × 20 grayscale image. As shown in Figure 8.

Figure 8. Converting one-dimensional time-domain signals to two-dimensional gray-scale images.

3.1. Proposed sPSDAE-CNN Model Structure

We convert the drone flight data after batch normalization (BN) into a grayscale image. Using stacked pruning sparse denoising autoencoders to reduce the dimensionality and denoising of the original data, it can also initially extract data features. The data processed by the sparse noise reduction autoencoder will be directly used as the input of the convolutional neural network. On the whole, the structure of the sPSDAE-CNN proposed in this paper is roughly the same as the structure of the traditional convolutional neural network. The main difference is that the stack noise reduction autoencoder is introduced, but the introduction of the noise reduction encoder further increases the complexity of the network and increases the computational cost, so the sparse pruning operation is added to reduce the complexity of the network. The noise reduction autoencoder improves the adaptability of the network to high-noise data, and the pruning operation greatly improves the calculation efficiency of the encoder. The specific structure of sPSDAE-CNN is shown in Figure 9.

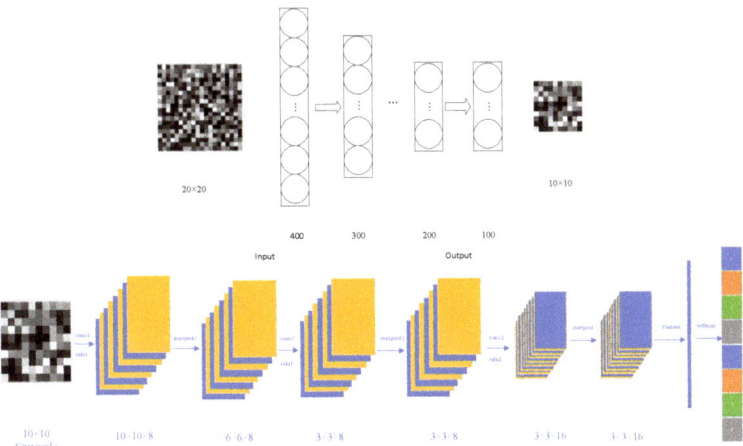

Figure 9. The specific structure of sPSDAE-CNN.

Finally, in the classification stage of the model, the softmax function is used to perform logit transformation on the classification results, and eight different four-rotor UAV health state probability distributions are obtained (5).

$$q(z_j) = \frac{e^{z_j}}{\sum_{k}^{8} e^{z_k}},\qquad(5)$$

where z represents the logical value of the jth neuron.

3.2. Construction of Sparse Noise Reduction Autoencoding Network

In order to explore the deep-level features in the time-domain sequence signal, we convert the one-dimensional time-domain sequence signal into a two-dimensional gray-scale image by using a matrix transformation method. Figure 9 shows the structure of a stacked noise reduction autoencoder with four hidden layers. Since each layer of a traditional stacked noise reduction encoder has an impact on its subsequent network levels, we use the pruning method to cut off the layers that have no effect on the training of the next layer of the network, while ensuring the maximum information flow in the network. Therefore, the latter layer can obtain the maximum effective information of the previous layer, which improves the training speed and feature extraction performance. The schematic diagram of constructing the stacked pruning sparse denoising autoencoder(sUPSDAE) fully connected network model based on the DAE model is shown in Figure 10:

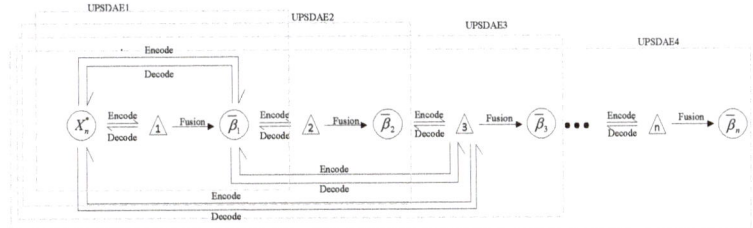

Figure 10. Schematic diagram of sUPSDAE fully connected network model.

sUPSDE adopts the feature fusion method for information sharing, which reduces the loss of information and broadens the transmission level of the network. As the number of training layers increases, the number of network calculations will increase sharply, and it is also prone to the problem of overfitting. We reduce the amount of calculation by introducing sparse pruning operations while suppressing overfitting.

In Figure 10, we can get that the model of the ith layer, which is related to the first i unit nodes when it is trained. In order to introduce sparse operations into sUPSDEA, this paper randomly selects some features of the input layer in the training loop, and uses Formula (6) [38] to randomly discard it, and then periodically introduce sparse operations in subsequent node training until all units have been trained.

$$\begin{aligned} v &= Berboulli(1 - p_1) \\ \overline{\beta_i}^* &= v \times \overline{\beta_i} \end{aligned} \qquad (6)$$

where p_1 represents the probability of the current training unit being discarded, and $\overline{\beta_i}$ represents the input matrix before discarding. $\overline{\beta_i}^*$ is the input matrix after random discarding in one cycle.

After the sUPSDEA training is over, backpropagation is performed by using Back Propagation Neural Network (BPNN) [39], and the parameters and weights of the network are fine-tuned. In this process, the discarded units are added through Equation (7) to further reduce the possible overfitting of the model.

$$\begin{aligned} \tau &= Berboulli(1 - p_2) \\ \overline{X_i}^* &= \tau \times \overline{X_i} \end{aligned} \qquad (7)$$

where p_2 is the probability of discarding irrelevant nodes in the fine-tuning process, X_i is the output of the network in the fine-tuning process, and $\overline{X_i}^*$ is the input data randomly discarded in one cycle of the fine-tuning process.

3.3. The Influence of Various Parts of the Model on the Results

3.3.1. The Effect of Sparse Pruning and Noise Reduction Autoencoder on the Results

The stack sparse noise reduction autoencoder transforms the original two-dimensional 20×20 grayscale images into 10×10 grayscale images by dimensionality reduction, which dramatically reduces the computational cost of the subsequent convolutional neural network. At the same time, the noise signal contained in the data can be filtered out, which also realizes the prediction of the original signal of the signal destroyed by the noise. By training the model parameters of the model, the model can finally achieve an accurate prediction of the original signal and eliminate the interference of noise to the original signal to a large extent, which can effectively improve the final diagnosis effect of the model.

3.3.2. The Effect of Convolutional Neural Networks on Results

The convolutional neural network uses the output of the dimensionality reduction of the stack sparse noise reduction autoencoder as the input of the convolutional neural network, and uses the convolutional neural network to extract the characteristics of the data collected by the drone. By combining the high-dimensional input data, the feature is mapped to the low-dimensional UAV health status, which can easily convert the original data into the UAV health status. At the same time, it has a very good non-linear fitting ability, which is very beneficial to the fault diagnosis of the quad-rotor UAV, which improves the adaptive ability of the model to a certain extent.

3.4. Data Augmentation

In order to recognize images using in-depth learning, a large amount of image data needs to be prepared for model training, especially when using neural network model algorithms. For example, most common data collections in in-depth learning contain a large amount of image data, including 60,000 training data and 10,000 test set data in the Mixed National Institute of Standards and Technology (MNIST) dataset. There are 60,000 color images in the Canadian Institute For Advanced Research-10 (CIFAR-10) dataset, of which 50,000 are training data and 10,000 are test data. Therefore, in order to train our own neural network model, we need to prepare a large amount of experimental data for the training model. However, the experimental data cannot meet the actual training requirements of the neural network, and data enhancement methods need to be introduced to increase the amount of sample data. In the field of computer vision, data enhancement is usually achieved by introducing operations such as flip, rotation, clipping, distortion, scaling, etc. However, such methods cannot be used in time domain sequence signals. We enhanced the data in fault diagnosis by introducing a fixed-length sliding window to slice sequential time-domain signals in turn, as shown in Figure 11.

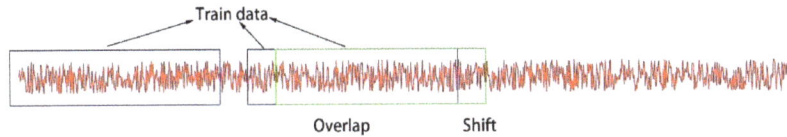

Figure 11. Sliding window for data enhancement.

Using this method, we will get 79,980 training samples from 80,000 original data collected by UAV. This method can effectively solve the problem of insufficient training samples in actual training, but this method has been ignored in many articles [40–43], because they do not use overlapping sampling methods for data enhancement. As a result, there are only hundreds or thousands of training samples during model training. At the bottom of the article, we will verify the necessity of data enhancement through actual experiments.

4. Validation of the sPSDAE-CNN Model

4.1. Data Description

The training of a neural network model requires a lot of data to be collected from a laboratory P200 quad-rotatory UAV. The main control panel of the UAV is pixhawk4, which is also equipped with jeson tx2, binocular camera, and other sensors, As shown in Figure 12.

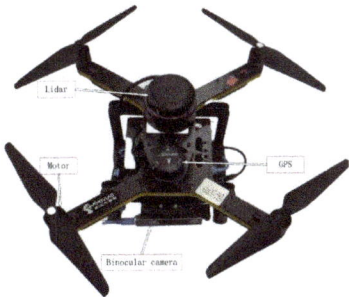

Figure 12. P200 drone.

Over 90,000 data were collected in flight, of which 80,000 were valid. The training data is pre-processed and divided into four datasets, of which there are eight types of pre-defined faults, and eight types are considered to be the eight states of the UAV. The actual experimental data are shown in the Table 2.

Table 2. Description of UAV datasets.

Types of Damage to the Blades		No Damage	Broken Blade				Blade Crack		
Data Set		0	5%	10%	15%	20%	slightly deformation	General deformation	Severely deformation
A	Train	10,000	10,000	10,000	10,000	10,000	10,000	10,000	10,000
	Test	200	200	200	200	200	200	200	200
B	Train	14,000	14,000	14,000	14,000	14,000	14,000	14,000	14,000
	Test	280	280	280	280	280	280	280	280
C	Train	18,000	18,000	18,000	18,000	18,000	18,000	18,000	18,000
	Test	360	360	360	360	360	360	360	360
D	Train	20,000	20,000	20,000	20,000	20,000	20,000	20,000	20,000
	Test	400	400	400	400	400	400	400	400

4.2. Experimental Settings

This paper compares our methods with traditional convolution neural networks, SVM [44] and traditional unsupervised learning stack autoencoders. Consider the impact of different size datasets on the performance of the neural network, then compare the changes of the performance of the neural network before and after sparse pruning operation. Finally, experiments are carried out in different noise levels to compare and analyze the anti-noise ability of the sPSDAE-CNN model.

4.2.1. Parameters of the Proposed Network

The sPSDAE-CNN network model proposed by this paper consists of an sPSDAE sparse pruning noise reduction autoencoder and a convolution neural network. The sPSDAE consists of one input layer, one output layer, and four hidden layers. The specific structure is shown in Figure 9, The specific structural parameters of convolutional neural networks are shown in Table 3.

Table 3. Structural parameters of convolutional neural networks.

No	Layer Type	KernelSize Stride	Output Size (Width × Depth)	Padding
1	Convolution1	4 × 4/1	10 × 10 × 8	Yes
2	Pooling1	4 × 4/1	3 × 3 × 8	No
3	Convolution2	2 × 2/1	6 × 6 × 8	No
4	Pooling2	3 × 3/1	3 × 3 × 8	Yes
5	Convolution3	1 × 1/1	3 × 3 × 16	No
6	Pooling3	2 × 2/1	3 × 3 × 16	Yes
7	Fully-connected	144	144 × 1	/
8	Softmax	8	8	/

The introduced pruning operation also improves the training speed of the network. The output of sPSDAE is used as input of CNN. The main structure of a convolution network is three convolution layers and pooling layers. Then there is a hidden layer of full connection layer. Finally, the output layer is reached by a softmax layer. The convolution cores of the system select small convolution cores to convolute, the pooling layer chooses maximum pooling, and the activation function of the neural network chooses RELU function. In order to improve the performance of the network, a batch normalization operation is added behind each convolution layer and the full connection layer. The batch normalization operation can accelerate the convergence speed of the training of the neural network and suppress the over-fitting phenomenon in the network. The convolution and pooling parameters of convolution neural networks are detailed in Table 3.

4.2.2. Hyperparameter Optimization of the Proposed Network

We use PyTorch, a deep learning framework developed by an American company called Facebook, to conduct our actual experiment. To minimize our loss function, this paper uses the random gradient descent method to optimize our convolution neural network model. In the actual experiment, we choose the Adam optimizer as the final hyperparameter optimization method. The Adam optimizer combines the advantages of AdaGrad and RMSProp algorithms, and calculates the update step by step, comprehensively considering the first-order moment estimation and the second-order moment estimation of the loss function. Adam has the advantages of simple implementation, high computational efficiency, and low requirement of memory. In addition, the parameter update of this method is not affected by the scaling transformation of gradient, and it is also very suitable for the application of large-scale data and parameters in practice. Finally, it has good performance even in the case of large gradient noise. We set the learning rate of Adam's optimizer to 0.001. At the same time, the cross-entropy loss function is trained as the objective function. Reference in detail [45].

4.2.3. The Effect of the Number of Training Data on the Results

As a type of convolution network, there are a large number of parameters in the sPSDAE-CNN model that need to be determined during the training process of the model. In order to improve the recognition accuracy of the network and suppress over-fitting in the system, a large amount of experimental data is needed to train the network. To study the training results of the neural network under different training samples, the number of training data of the neural network is set to 100, 200, 300, 900, 1500, 3000, 6000, 12,000, 15,000, and 20,000 training samples to study the performance of sPSDAE-CNN. In deep learning, there are balanced and unbalanced data collections. In Table 2, our data is fully balanced, so accuracy can still be used to evaluate the algorithm.

Because the data set is completely balanced, the data samples of UAVs under each fault condition are the same. In the actual experiment, the first three datasets do not use the sliding window method to enhance and expand the data. To reduce the influence of the random initial values of the neural network on the training results of the network, 30 repeated experiments were performed on each sample to calculate the average value. The paper uses AMD Ryzen™ 5 4600H processor, NVIDIA GTX1650 graphics card, and 16GB of memory. The test data collection is tested using DataSet D in Table 3, and the test results are shown in Figure 13.

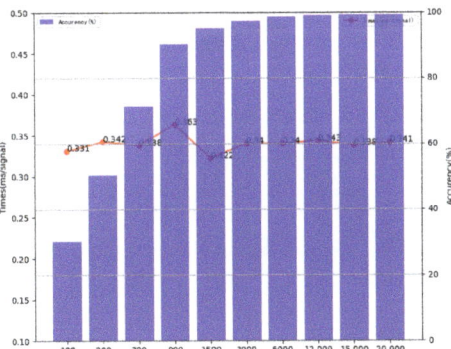

Figure 13. Diagnostic results for different data volumes.

In Figure 13, it is clear that the accuracy of the test dataset increases significantly as the training data goes from a smaller number of samples to a more significant number of samples. When the training data increases from 100 to 300, the accuracy of the test set data increases by about 20%. With the increase of training data, the accuracy of the neural network gradually approaches and converges to 100%. When the training data is increased from 100 to 300, the accuracy of the test set data is improved by about 20%. As the training data increases, the accuracy of the neural network approaches and converges to 100%, and the standard deviation converges to 0. Secondly, we can observe that the average time of a signal diagnosed by the training model of the neural network is 4 ms, which meets the requirements of test data. By comparing the training time of different test sets, we can find that the increase of the number of training data has little effect on the test time. In Figure 14, the points of different colors represent that the blades of the UAV are in different fault states. In the beginning, due to the small number of training samples, it is not easy to segment the characteristics between different types of data. With the increase of the number of training samples of the model, the data in the same fault situation begin polymerization, and the characteristics of different types of fault data become easier to segment. At the same time, this shows that by using the data enhancement method to enhance the original data collected by UAV, we can greatly increase the data scale and data diversity of neural network training samples, which can further improve the generalization ability of the model. Therefore, in subsequent experiments, this paper selected as many as 20,000 samples as possible for training.

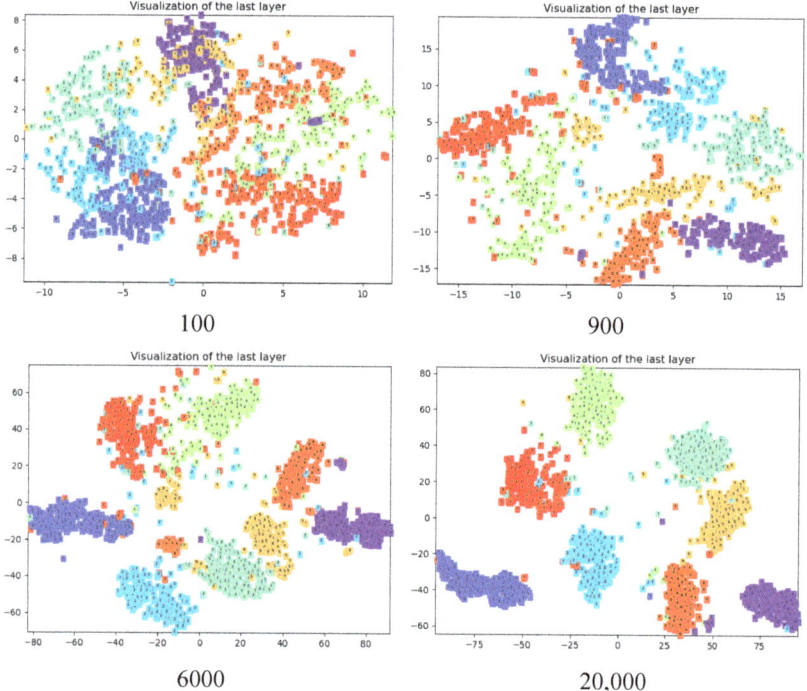

Figure 14. Visualizing test samples from the last hidden fully connected layer with t-SNE under different training data numbers.

In the subsequent model training, we choose 20,000 training samples. The parameters in the neural network model are determined through the training set, and then we use t-SNE visualization to make the t-SNE diagram of the neural network of each layer, as shown in Figure 15. It can be seen from the figure that the separability of different features in the unprocessed original data is very poor. After successively passing through each layer of the neural network, different features in the data begin to separate. In the last layer of the neural network, we can clearly see that different types of features in the data have been completely separated. Finally, different types of faults of UAV are diagnosed through the softmax layer.

In order to evaluate the accuracy of the model for different types of fault diagnosis, we introduce the Confusion Matrix. The Confusion Matrix can evaluate the performance of the classification model by counting the number of correct and wrong classifications. The Confusion Matrix of the model is shown in Figure 16. It can be seen from the figure that the accuracy of unmanned fault diagnosis of different types of four rotors remains basically the same, At the same time, the final model can be obtained through the Confusion Matrix, and the accuracy of fault diagnosis for four rotor UAV can reach about 98%, which can meet the needs of our actual projects and experiments.

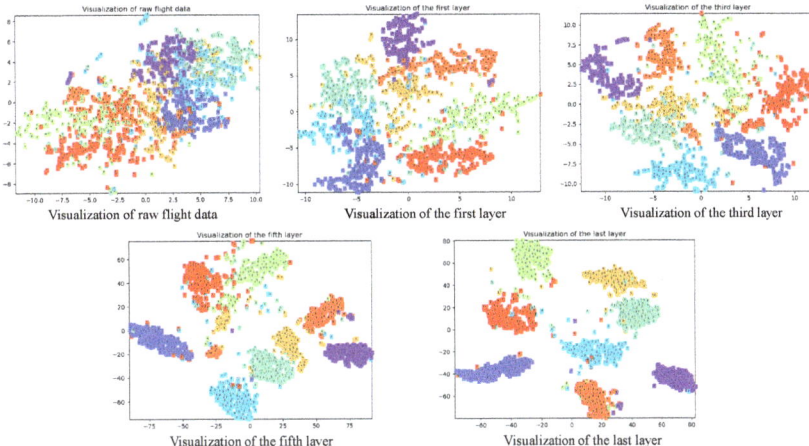

Figure 15. t-SNE Visualization of each layer of neural network.

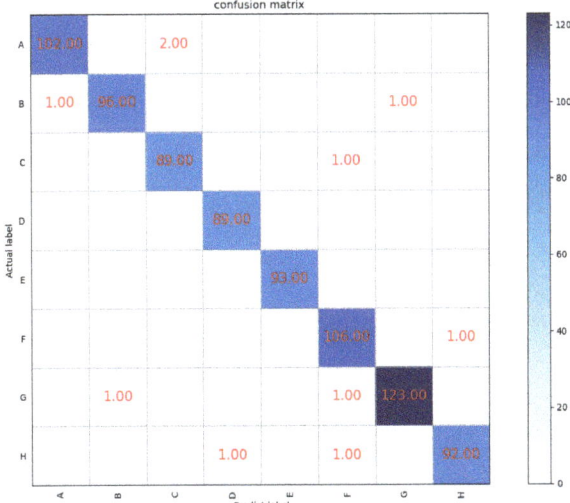

Figure 16. Confusion Matrix of the proposed model.

4.2.4. Training Speed of sPSDAE-CNN

Because this paper adds a stack denoising autoencoder in the front part of the neural network, it will not only improve the performance of the neural network, but also increase the time and cost of model training. Pruning operation is proposed for the stack denoising autoencoder, which not only introduces the noise reduction performance of stack denoising autoencoder, but also reduces the time cost of neural network training as much as possible. In the training of the model, we can find that under the same amount of data, the training speed of the neural network with pruning operation is basically the same as that without stack noise reduction encoder, but its training speed is much better than that without pruning operation. The specific network training speed is shown in Figure 17.

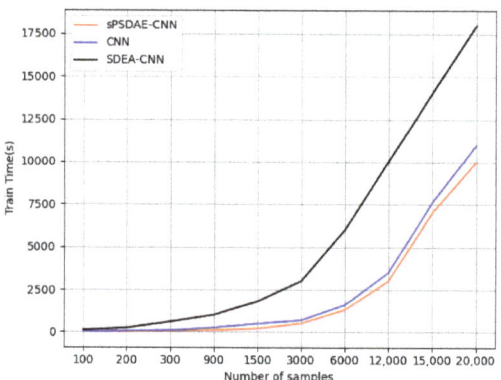

Figure 17. Comparison of training time required by different neural network models under different data scales.

4.2.5. Performance under Different Noise Interferences

Taking the collected UAV data as the original data, the drone may be disturbed by various signals during the execution of relevant tasks, so as to introduce noise signals to the data of sensors of the drone. It is impossible to obtain all the noisy data through experiments, so Gaussian white noise is artificially added to the original collected data to simulate the noise interference signals that may appear in the actual drone, and the signals with different signal-to-noise ratios are obtained. SNR is defined as follows (8):

$$SNR_{db} = 10\log_{10}(\frac{P_{signal}}{P_{noise}}), \tag{8}$$

where P_{noise} and P_{signal} represent the energy of signal and noise, respectively. It can be seen in Figure 18 that the UAV data collected in the laboratory environment is ideal and contains relatively little noise. In order to simulate the flight data under different interference in the actual flight environment, Gauss white noise of different degrees is added to the data, because Gauss white noise is the most common noise signal in nature. Therefore, we obtain aircraft data with different sizes of Gaussian white noise, that is, data with different signal-to-noise ratios. Finally, it can be seen from Figure 18 that the data after adding Gaussian white noise is closer to the UAV data in the actual flight environment. We evaluate the performance of the proposed model in different noise environments by studying the performance of the algorithm model with a signal-to-noise ratio of −4 dB to 10 dB.

In order to verify the efficiency of our proposed algorithm, we use the same test data to test the performance of CNN, SVM ,and SDAE, as shown in Figure 19:

As can be seen from Figure 19 that, firstly, because the sparse pruning noise reduction autoencoding convolutional neural network proposed by us has good noise reduction characteristics, it can be clearly seen that when the noise in the signal is considerable, the fault diagnosis effect of the model is obviously better than several other intelligent fault diagnosis methods. Secondly, due to the introduction of sparse pruning operation in the stack noise reduction autoencoder, this operation can improve the computational efficiency of the network to a certain extent, and make our proposed model still have very good performance in the case of low noise.

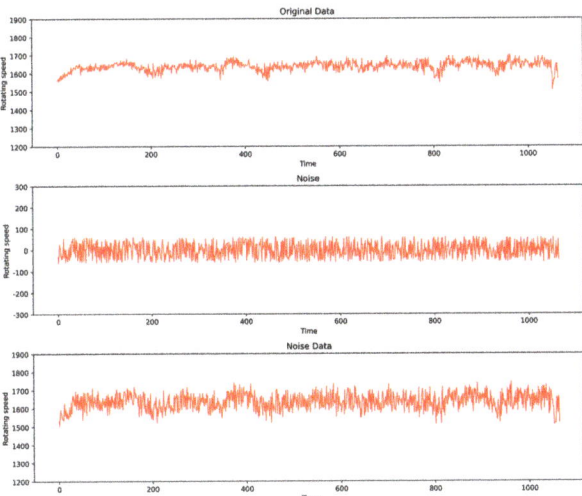

Figure 18. Original UAV data, noise data to be added, and final synthesized data containing Gaussian white noise.

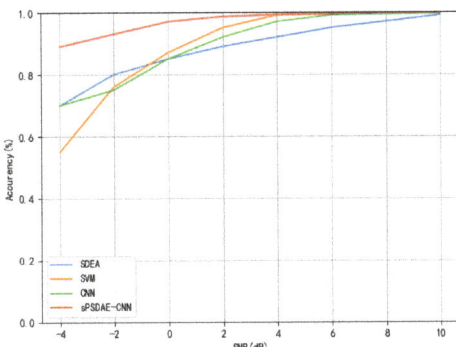

Figure 19. Comparison of accuracy of different fault diagnosis algorithms under different noise levels.

The experimental conditions are mainly flight experiments outdoors, and fault diagnosis is carried out by using the UAV data collected and saved by pixhawk4 flight control board. We artificially add different degrees of Gaussian white noise signals to the collected data to simulate the actual noise signals, and obtain the experimental data with different degrees of noise. SVM is used for classification. The 20 × 20 gray image obtained from UAV data processing is transformed into a feature vector with a length of 400. Multi classification support vector machine is used, in which the radial basis function (RBF) kernel function is selected as the kernel function. Gamma is set to the best value of 0.001 through many experiments. Finally, the experimental results in this paper are obtained. Secondly, in the use of convolutional neural network, we directly use the convolutional neural network proposed in the article, add a convolution layer to the previous layer of convolutional neural network to extract the data features, reduce the dimension, convert the original 20 × 20 graphics into 10 × 10 gray images, and carry out subsequent operations and classification. When DEA is used for recognition and classification, DAE is used for dimensionality reduction and optimization of the original data. BCE error is used in the training process. Adam optimizer is used, and then convolutional neural network is used for data feature extraction and classification.

5. Conclusions

In this paper, we adopt a new intelligent fault diagnosis method based on sPSDAE-CNN. Through a matrix transformation of the data collected from the UAV flight experiment, the one-dimensional time-series signal is transformed into two-dimensional gray image data, which expands the dimension of the sample and enhances the processing ability of the DL model. Secondly, by introducing a sparse pruning stack noise reduction autoencoder, the accuracy of a fault diagnosis algorithm in a high noise environment can be improved, and the input dimension of CNN data can also be reduced. In addition, pruning operation is used to reduce the complexity of the encoder, which can make the encoder converge quickly when minimizing the loss function. The combination of sPSDAE and the convolutional neural network can greatly improve the robustness and generalization ability of the fault diagnosis model. In order to verify the effectiveness of the model, this paper chooses CNN, SVM, and SDAE to compare. The experimental results show that under the condition of normal experimental data, sPSDAE-CNN has good results compared with other algorithms, but when the noise signal in the signal gradually begins to increase, the performance of other algorithms decreases significantly. Among them, when the signal-to-noise ratio reaches -4 dB, sPSDAE-CNN still has an accuracy of about 90%, the accuracy of the other three algorithms decreased to less than 80%, and SVM is less than 60%. Therefore, the fault diagnosis sPSDAE-CNN algorithm used in this paper can be used as a fault diagnosis method of four-rotor UAV in an actual high noise environment.

The method proposed in the article first converts a one-dimensional time-domain signal into a two-dimensional grayscale image, which expands the dimensionality of the data and can improve the ability of subsequent algorithms to extract features from the data. Secondly, the method of resampling was used to enhance the flight data of the quad-rotor UAV, which greatly improved the problem of the insufficient data set. Finally, the sparse pruning noise reduction autoencoder is introduced to perform noise reduction, dimensionality reduction, and feature extraction on the data. After processing, the noise in the original data can be filtered to a large extent, and the pruning operation can also improve the model—the calculation efficiency and noise reduction performance. All the data used in the article are balanced data sets. In the actual environment, it is impossible for all data to be unbalanced data sets. In the follow-up research, the application scope of unbalanced data sets will be further expanded.

In addition, in this paper, balanced data sets are used, but during the actual UAV mission, the data we collect can not be completely balanced data sets. Therefore, in future research, we will improve and expand the application scope of the algorithm based on the performance of sPSDAE-CNN on unbalanced data sets.

Secondly, the data used in this paper are all offline data collected at the end of the UAV flight. At present, it is not possible to collect the data of four-rotor UAV in real-time in the actual flight process to realize fault diagnosis. In future research, we can try to diagnose the fault of UAV in real-time and online with the algorithm used in this paper; this problem needs to be further studied and solved.

Author Contributions: Conceptualization, C.W.; methodology, P.Y.; software, C.W.; validation, C.W.; formal analysis, C.W.; investigation, H.G.; resources, P.Y.; data curation, H.G.; writing—original draft preparation, C.W. and P.Y.; writing—review and editing, P.Y.; visualization, P.L.; supervision, P.L.; project administration, P.Y.; funding acquisition, P.Y. All authors have read and agreed to the published version of the manuscript.

Funding: This work is supported by Key Laboratories for National Defense Science and Technology (6142605200402), the Aeronautical Science Foundation of China (20200007018001), the National Natural Science Foundation of China (61922042), the Aero Engine Corporation of China Industry University Research Cooperation Project (HFZL2020CXY011), and the Research Fund of State Key Laboratory of Mechanics and Control of Mechanical Structures (Nanjing University of Aeronautics and Astronautics) MCMS-I-0121G03. Any opinions, findings, and conclusions, or recommendations expressed in this material are those of the author(s) and do not necessarily reflect the views of the sponsoring agency.

Institutional Review Board Statement: Not applicable.

Informed Consent Statement: Not applicable.

Data Availability Statement: Not applicable.

Conflicts of Interest: The authors declare no conflict of interest.

References

1. Bateman, F.; Noura, H.; Ouladsine, M. Fault diagnosis and fault-tolerant control strategy for the aerosonde UAV. *IEEE Trans. Aerosp. Electron. Syst.* **2011**, *47*, 2119–2137. [CrossRef]
2. Yu, X.; Jiang, J. A survey of fault-tolerant controllers based on safety-related issues. *Annu. Rev. Control* **2015**, *39*, 46–57. [CrossRef]
3. Haupt, F.; Berding, G.; Namazian, A.; Wilke, F.; Böker, A.; Merseburger, A.; Geworski, L.; Kuczyk, M.A.; Bengel, F.M.; Peters, I. Expert system for bone scan interpretation improves progression assessment in bone metastatic prostate cancer. *Adv. Ther.* **2015**, *34*, 986–994. [CrossRef] [PubMed]
4. Zhu, Y.; Geng, L. Research on SDG fault diagnosis of ocean shipping boiler system based on fuzzy granular computing under data fusion. *Pol. Marit. Res.* **2018**, *25*, 92–97. [CrossRef]
5. Hu, G.; Phan, H.; Ouache, R.; Gandhi, H.; Hewage, K.; Sadiq, R. Fuzzy fault tree analysis of hydraulic fracturing flowback water storage failure. *J. Nat. Gas Sci. Eng.* **2019**, *72*, 103039. [CrossRef]
6. Liu, D.; Gu, X.; Li, H. A complete analytic model for fault diagnosis of power systems. *Proc. Chin. Soc. Electr. Eng.* **2011**, *31*, 85–92.
7. Chen, X.; Qi, X.; Wang, Z.; Cui, C.; Wu, B.; Yang, Y. Fault diagnosis of rolling bearing using marine predators algorithm-based support vector machine and topology learning and out-of-sample embedding. *Measurement* **2021**, *176*, 109116. [CrossRef]
8. Chen, F.; Wyer, R.S., Jr. The effects of affect, processing goals and temporal distance on information processing: Qualifications on temporal construal theory. *J. Consum. Psychol.* **2015**, *25*, 326–332. [CrossRef]
9. Xiao, Y.; Li, C.; Song, L.; Yang, J.; Su, J. A multidimensional information fusion-based matching decision method for manufacturing service resource. *IEEE Access* **2021**, *9*, 39839–39851. [CrossRef]
10. Chady, T.; Sikora, R.; Misztal, L.; Grochowalska, B.; Grzywacz, B.; Szydłowski, M.; Waszczuk, P.; Szwagiel, M. The application of rough sets theory to design of weld defect classifiers. *J. Nondestruct. Eval.* **2017**, *36*, 40. [CrossRef]
11. Esteki, M.; Farajmand, B.; Kolahderazi, Y.; Simal-Gandara, J. Chromatographic fingerprinting with multivariate data analysis for detection and quantification of apricot kernel in almond powder. *Food Anal. Methods* **2017**, *10*, 3312–3320. [CrossRef]
12. Jiang, Y.; Zhiyao, Z.; Haoxiang, L.; Quan, Q. Fault detection and identification for quadrotor based on airframe vibration signals: A data-driven method. In Proceedings of the 2015 34th Chinese Control Conference (CCC), Hangzhou, China, 28–30 July 2015; pp. 6356–6361.
13. Tang, S.; Yuan, S.; Zhu, Y. Convolutional neural network in intelligent fault diagnosis toward rotatory machinery. *IEEE Access* **2020**, *8*, 86510–86519. [CrossRef]
14. Tao, H.; Wang, P.; Chen, Y.; Stojanovic, V.; Yang, H. An unsupervised fault diagnosis method for rolling bearing using STFT and generative neural networks. *J. Frankl. Inst.* **2020**, *357*, 7286–7307. [CrossRef]
15. Glowacz, A. Fault diagnosis of electric impact drills using thermal imaging. *Measurement* **2021**, *171*, 108815. [CrossRef]
16. Polat, K. The fault diagnosis based on deep long short-term memory model from the vibration signals in the computer numerical control machines. *J. Inst. Electron. Comput.* **2020**, *2*, 72–92. [CrossRef]
17. Xiong, R.; Sun, W.; Yu, Q.; Sun, F. Research progress, challenges and prospects of fault diagnosis on battery system of electric vehicles. *Appl. Energy* **2020**, *2*, 72–92. [CrossRef]
18. Hu, X.; Zhang, K.; Liu, K.; Lin, X.; Dey, S.; Onori, S. Advanced fault diagnosis for lithium-ion battery systems: A review of fault mechanisms, fault features, and diagnosis procedures. *IEEE Ind. Electron. Mag.* **2020**, *14*, 65–91. [CrossRef]
19. Azamfar, M.; Singh, J.; Bravo-Imaz, I.; Lee, J. Multisensor data fusion for gearbox fault diagnosis using 2-D convolutional neural network and motor current signature analysis. *Mech. Syst. Signal Process.* **2020**, *144*, 106861. [CrossRef]
20. He, Z.; Shao, H.; Wang, P.; Lin, J.J.; Cheng, J.; Yang, Y. Deep transfer multi-wavelet auto-encoder for intelligent fault diagnosis of gearbox with few target training samples. *Knowl.-Based Syst.* **2021**, *191*, 105313. [CrossRef]
21. Chen, H.; Jiang, B.; Ding, S.X.; Huang, B. Data-driven fault diagnosis for traction systems in high-speed trains: A survey, challenges, and perspectives. *IEEE Trans. Intell. Transp. Syst.* **2020**, *64*, 1–3. [CrossRef]
22. Chen, H.; Jiang, B. A review of fault detection and diagnosis for the traction system in high-speed trains. *Trans. Intell. Transp. Syst.* **2019**, *21*, 450–465. [CrossRef]
23. Huang, D.; Li, S.; Qin, N.; Zhang, Y. Fault diagnosis of high-speed train bogie based on the improved-CEEMDAN and 1-D CNN algorithms. *IEEE Trans. Instrum. Meas.* **2021**, *70*, 1–11.
24. Iannace, G.; Ciaburro, G.; Trematerra, A. Fault diagnosis for UAV blades using artificial neural network. *Robotics* **2019**, *8*, 59. [CrossRef]
25. Liu, W.; Chen, Z.; Zheng, M. An Audio-Based Fault Diagnosis Method for Quadrotors Using Convolutional Neural Network and Transfer Learning. In Proceedings of the 2020 American Control Conference (ACC), Denver, CO, USA, 1–3 July 2020; pp. 1367–1372.
26. Zhang, W.; Peng, G.; Li, C.; Chen, Y.; Zhang, Z. A new deep learning model for fault diagnosis with good anti-noise and domain adaptation ability on raw vibration signals. *Sensors* **2015**, *17*, 425. [CrossRef] [PubMed]

27. Che, C.; Wang, H.; Ni, X.; Fu, Q. Intelligent fault diagnosis method of rolling bearing based on stacked denoising autoencoder and convolutional neural network. *Ind. Lubr. Tribol.* **2020**, *72*, 947–953. [CrossRef]
28. Cheng, Y.; Lin, M.; Wu, J.; Zhu, H.; Shao, X. Intelligent fault diagnosis of rotating machinery based on continuous wavelet transform-local binary convolutional neural network. *Knowl.-Based Syst.* **2021**, *216*, 106796. [CrossRef]
29. Okada, K.F.Á.; de Morais, A.S.; Oliveira-Lopes, L.C.; Ribeiro, L. Neuroadaptive Observer-Based Fault-Diagnosis and Fault-Tolerant Control for Quadrotor UAV. In Proceedings of the 2021 14th IEEE International Conference on Industry Applications, São Paulo, Brazil, 15–18 August 2021; pp. 285–292.
30. Guo, J.; Qi, J.; Wu, C. Robust fault diagnosis and fault-tolerant control for nonlinear quadrotor unmanned aerial vehicle system with unknown actuator faults. *Int. J. Adv. Robot. Syst.* **2021**, *18*, 17298814211002734. [CrossRef]
31. Patan, M.G.; Caliskan, F. Sensor fault–tolerant control of a quadrotor unmanned aerial vehicle. *Proc. Inst. Mech. Eng. Part G J. Aerosp. Eng.* **2021**. [CrossRef]
32. Wang, B.; Huang, P.; Zhang, W. A Robust Fault-Tolerant Control for Quadrotor Helicopters against Sensor Faults and External Disturbances. *Complexity* **2021**, *2021*, 6672812. [CrossRef]
33. Kalchbrenner, N.; Grefenstette, E.; Blunsom, P. A convolutional neural network for modelling sentences. *arXiv* **2014**, arXiv:1404.2188.
34. LeCun, Y.; Bengio, Y.; Hinton, G. Deep learning. *Nature* **2015**, *521*, 436–444. [CrossRef] [PubMed]
35. Sibi, P.; Jones, S.A.; Siddarth, P. Analysis of different activation functions using back propagation neural networks. *J. Theor. Appl. Inf. Technol.* **2013**, *47*, 1264–1268.
36. Ioffe, S.; Szegedy, C. Batch normalization: Accelerating deep network training by reducing internal covariate shift. *Int. Conf. Mach. Learn.* **2015**, *37*, 448–456.
37. Lu, C.; Wang, Z.Y.; Qin, W.L.; Ma, J. Fault diagnosis of rotary machinery components using a stacked denoising autoencoder-based health state identification. *Signal Process.* **2017**, *130*, 377–388. [CrossRef]
38. Heys, J.J.; Holyoak, N.; Calleja, A.M.; Belohlavek, M.; Chaliki, H.P. Revisiting the simplified Bernoulli equation. *Open Biomed. Eng. J.* **2010**, *4*, 123. [CrossRef]
39. Werbos, P.J. Backpropagation through time: What it does and how to do it. *Proc. IEEE* **1990**, *78*, 1550–1560. [CrossRef]
40. Guo, X.; Chen, L.; Shen, C. Hierarchical adaptive deep convolution neural network and its application to bearing fault diagnosis. *Measurement* **2016**, *93*, 490–502. [CrossRef]
41. Kiranyaz, S.; Ince, T.; Abdeljaber, O.; Avci, O.; Gabbouj, M. 1-d convolutional neural networks for signal processing applications. In Proceedings of the ICASSP 2019—2019 IEEE International Conference on Acoustics, Speech and Signal Processing (ICASSP), Brighton, UK, 12–17 May 2019; pp. 8360–8364.
42. Avci, O.; Abdeljaber, O.; Kiranyaz, S.; Inman, D. Structural damage detection in real time: Implementation of 1D convolutional neural networks for SHM applications. *Struct. Health Monit. Damage Detect.* **2017**, *7*, 49–54.
43. Janssens, O.; Slavkovikj, V.; Vervisch, B.; Stockman, K.; Loccufier, M.; Verstockt, S.; Van de Walle, R.; Van Hoecke, S. Convolutional neural network based fault detection for rotating machinery. *J. Sound Vib.* **2016**, *377*, 331–345. [CrossRef]
44. Tan, Y.; Wang, J. A support vector machine with a hybrid kernel and minimal Vapnik-Chervonenkis dimension. *IEEE Trans. Knowl. Data Eng.* **2004**, *16*, 385–395.
45. Kingma, D.P.; Ba, J. Adam: A method for stochastic optimization. *arXiv* **2014**, arXiv:1412.6980.

Article

Disturbance Detection of a Power Transmission System Based on the Enhanced Canonical Variate Analysis Method

Shubin Wang [1], Yukun Tian [1,2], Xiaogang Deng [1,*], Qianlei Cao [2], Lei Wang [2] and Pengxiang Sun [2]

1. College of Control Science and Engineering, China University of Petroleum, Qingdao 266580, China; shubinw@126.com (S.W.); tianyk42085@hundsun.com (Y.T.)
2. Qingdao Topscomm Communication Co., Ltd., Qingdao 266109, China; caoqianlei@topscomm.com (Q.C.); wanglei5@topscomm.com (L.W.); sunpengxiang@topscomm.com (P.S.)
* Correspondence: dengxiaogang@upc.edu.cn

Citation: Wang, S.; Tian, Y.; Deng, X.; Cao, Q.; Wang, L.; Sun, P. Disturbance Detection of a Power Transmission System Based on the Enhanced Canonical Variate Analysis Method. *Machines* **2021**, *9*, 272. https://doi.org/10.3390/machines9110272

Academic Editors: Hongtian Chen, Kai Zhong, Guangtao Ran and Chao Cheng

Received: 11 October 2021
Accepted: 4 November 2021
Published: 6 November 2021

Publisher's Note: MDPI stays neutral with regard to jurisdictional claims in published maps and institutional affiliations.

Copyright: © 2021 by the authors. Licensee MDPI, Basel, Switzerland. This article is an open access article distributed under the terms and conditions of the Creative Commons Attribution (CC BY) license (https://creativecommons.org/licenses/by/4.0/).

Abstract: Aiming at the characteristics of dynamic correlation, periodic oscillation, and weak disturbance symptom of power transmission system data, this paper proposes an enhanced canonical variate analysis (CVA) method, called SLCVAkNN, for monitoring the disturbances of power transmission systems. In the proposed method, CVA is first used to extract the dynamic features by analyzing the data correlation and establish a statistical model with two monitoring statistics T^2 and Q. Then, in order to handling the periodic oscillation of power data, the two statistics are reconstructed in phase space, and the k-nearest neighbor (kNN) technique is applied to design the statistics nearest neighbor distance DT^2 and DQ as the enhanced monitoring indices. Further considering the detection difficulty of weak disturbances with the insignificant symptoms, statistical local analysis (SLA) is integrated to construct the primary and improved residual vectors of the CVA dynamic features, which are capable to prompt the disturbance detection sensitivity. The verification results on the real industrial data show that the SLCVAkNN method can detect the occurrence of power system disturbance more effectively than the traditional data-driven monitoring methods.

Keywords: canonical variate analysis; disturbance detection; power transmission system; k-nearest neighbor analysis; statistical local analysis

1. Introduction

With the increasing demand on the power energy in the modern industry, power transmission systems are becoming more and more large-scale and complicated [1,2]. Due to the system complexity, anomalies and disturbances are often unavoidable in real power systems. If these unexpected events are not handled timely, they may cause huge accident risks and even the widespread power outages, which are companied by the huge economic loss and severe life inconvenience. Therefore, it is of great value to detect the abnormal events quickly and maintain the safe running of power systems [3]. In recent years, the wide area measurement system (WAMS) based on synchronous phaser technology has been successfully applied in the power industry. The phasor measurement units in WAMS provide the basic data support for the real-time dynamic monitoring of the power system [4]. Accordingly, safety monitoring and disturbance detection of power systems based on the measurement data analysis has been a hot topic in academic and engineering fields [5–7].

Aiming at the power system disturbance detection task, researchers have conducted a lot of studies, which can be roughly divided into two categories: time/frequency domain analysis and multivariate statistical analysis. The time/frequency domain analysis investigates the power system changes from the perspective of the signal processing, which involves the time domain, frequency domain, or time-frequency domain. In consideration of the good time-frequency localization property, Huang et al. [8] discussed the application of the Morelet wavelets method in power system disturbance detection. The Hilbert Huang

Transform is another time-frequency signal analysis tool. Manglik et al. [9] applied it to the disturbance detection for the electric power system. Ghaderi et al. [10] proposed the time-frequency analysis method assisted by current waveform energy and normalized joint time-frequency moment and demonstrated its performance in the high-impedance ground fault detection. Salehi et al. [11] designed a morphological edge detection filter to obtain the transient features of fault signals. Liu et al. [12] used the wavelet packet Tsallis singularity entropy algorithm for disturbance detection. In general, the time/frequency domain analysis methods mainly analyze the single signal and fail to fully consider the correlation between different parameters. In response to this shortcoming, some scholars started their work by applying multivariate statistical analysis. Multivariate statistical analysis (MSA) methods can realize the simultaneous detection of multiple parameter changes and have outstanding advantages in the complex industrial systems. However, most of the present MSA studies focus on the system modeling and disturbance detection in the chemical process, steel industry, and high-train system [13–17], but MSA's application to power system monitoring is very rare. Barocio et al. [18] first introduced the principal component analysis (PCA) method into the field of power system monitoring and discussed the detection and visualization of power system disturbances based on PCA. Guo et al. [19] built a transmission line fault detection method by combining PCA and support vector machine. Considering the masking influence caused by the oscillation trend and strong noise of power system data, Cai et al. [20] further proposed a PCAkNN method, which is superior to the basic PCA method in the numerical model testing and New England power system model data. These research articles point out that the multivariate statistical analysis has great application potential in the field of power system monitoring.

Although PCA and PCAkNN methods have achieved significant success in the power system monitoring field, they have some shortcomings deserving further studies. On the one hand, these methods do not take into account the dynamic characteristics of power system data, which easily leads to a high missing detection rate. Different from the other industrial process data with the steady operation mode, the power system data, such as the voltage and current, are with obvious dynamic trends. On the other hand, the present methods do not consider how to enhance the detection of weak disturbances. In real applications, some disturbances may be with small amplitudes, slow changes, unclear disturbance characteristics, and are easy to be covered by noises [21,22]. How to enhance the detection capability on these weak disturbances is one challenging task.

Aiming at the aforementioned problems, this paper proposes a SLCVAkNN-based disturbance detection method for power transmission system monitoring by combining canonical variate analysis (CVA), kNN, and statistical local analysis (SLA). Compared with the traditional PCA-based power system monitoring methods, CVA has a stronger dynamic feature extraction ability [23–25], which provides a new and powerful tool for power system data analysis. Referring to the present PCAkNN method, the CVAkNN statistical model is developed to deal with the dynamic periodic oscillation signals. Furthermore, in order to enhance the detection of weak faults, SLA is integrated for SLCVAkNN modeling, which mines the local statistical information for better weak disturbance monitoring.

The rest of the paper's content is arranged as follows. The principle of the proposed SLCVAkNN methodology is given in the Section 2, while the corresponding disturbance detection procedure is detailed in Section 3. One case study on the actual industrial data is used to verify the effectiveness of the proposed method.

2. The Proposed Methodology

This section clarifies the proposed SLCVAkNN-based power system disturbance detection method. The whole methodology involves three parts: dynamic system modeling using canonical variate analysis, monitoring index construction based on kNN, and weak disturbance detection by statistical local analysis.

2.1. CVA Monitoring Model

A power system is a classical dynamic process [26,27], where the measurement data demonstrate the clear trend along the sampling time. The measured three-phase electric field and current waveform change with time, and the current data point has a certain correlation with the historical samples. Therefore, it is more reasonable to apply the dynamic data analysis tool to extract the process features.

Canonical variate analysis (CVA) is an effective dynamic data analysis tool, which has been applied to the model identification and control in the multivariate dynamic system [28,29]. This paper introduces it to deal with power system data. For a certain power transmission line, the data points under normal system operation have a fixed correlation along the time dimension. When a disturbance occurs, this correlation may be destroyed. By monitoring the correlation among the time series data, CVA can find the system disturbance effectively. When CVA is applied to data modeling, the training data are firstly divided into the historical data set and the future data set, and the CVA optimization problem is designed to find the maximum correlation between these two data sets for describing the data dynamic features. The algorithm details are clarified as follows.

For the power system measurement data vector $x_h \in R^m$ at the h-th sampling instant, its corresponding historical data vector p_h and future data vector f_h are constructed as

$$p_h = [x_h^T, x_{h-1}^T, \cdots, x_{h-l+1}^T]^T \in R^M \tag{1}$$

$$f_h = [x_{h+1}^T, x_{h+2}^T, \cdots, x_{h+l}^T]^T \in R^M \tag{2}$$

where $M = m \times l$, and l represents the time lag order.

Given the projection vectors a and b, they are used to transform the historical and future vectors into their respective projections $d = a^T p_h$ and $v = b^T f_h$. CVA is to optimize the vector pair a and b so that the correlation between d and v is maximized, which are also called canonical variates. This can be described by the mathematical expression as

$$\begin{cases} \max_{a,b} \rho(d,v) = a^T \Sigma_{pf} b \\ \text{s.t. } var(d) = a^T \Sigma_{pp} a = 1 \\ var(v) = b^T \Sigma_{ff} b = 1 \end{cases} \tag{3}$$

where Σ_{pf} represents the cross-covariance matrix of the historical and future data vectors, and Σ_{pp}, Σ_{ff} denote the covariance matrix of the historical and future data vectors, respectively.

Suppose that the training data set includes n samples as $X = [x_1^T, x_2^T, \cdots, x_n^T]^T \in R^{n \times m}$, then the historical and future data matrix can be expressed by

$$P = [p_l^T, p_{l+1}^T, \cdots, p_{n-l}^T]^T \in R^{N \times M} \tag{4}$$

$$F = [f_l^T, f_{l+1}^T, \cdots, f_{n-l}^T]^T \in R^{N \times M} \tag{5}$$

where $N = n - 2l + 1$ is the sample number of the historical and future data matrix. Then the covariance matrices defined in Equation (3) can be calculated by

$$\Sigma_{pf} = \frac{1}{N-1} P^T F \tag{6}$$

$$\Sigma_{pp} = \frac{1}{N-1} P^T P \tag{7}$$

$$\Sigma_{ff} = \frac{1}{N-1} F^T F \tag{8}$$

Solving the optimization problem described by Equation (3) leads to a singular value decomposition on the matrix $\Xi = \Sigma_{pp}^{-1/2} \Sigma_{pf} \Sigma_{ff}^{-1/2}$, which is expressed by

$$\Xi = U \Lambda V^T \tag{9}$$

The solution of Equation (9) is further used to build the a series of the projection vectors a_i and $b_i (1 \leq i \leq M)$, which are computed by

$$a_i = \Sigma_{pp}^{-1/2} U(:,i), \quad b_i = \Sigma_{ff}^{-1/2} V(:,i), \tag{10}$$

where $(:,i)$ represents the i-th column of the matrix. The vectors a_i and b_i are ordered by the corresponding correlation degree, which is given in the diagonal elements of matrix Λ, also meaning the correlation coefficients. The first s pairs of projection vectors $\{a_i, b_i, 1 \leq i \leq s\}$ describe the stronger correlation and indicate the close relationship between the historical data and the future data. Therefore, a projection matrix $A_s = [a_1 a_2 \cdots a_s]$ is defined to extract the canonical variate vector d_h as

$$d_h = A_s^T p_h. \tag{11}$$

which describes the main dynamic features of process data. Here, s is determined so that the corresponding canonical variate vectors describe a cumulative percentage of 90% of correlation coefficients.

As A_s only involves the first s projection directions, it cannot cover all the data information. The rest information in the CVA model can be described by the CVA residual vector e_h as

$$e_h = (I - A_s A_s^T) p_h \tag{12}$$

Based on the canonical variate vector and CVA residual vector, two monitoring statistics T^2 and Q are often used to judge the process state. The T^2 statistic describes the changes of principal dynamic states, while the Q statistic monitors the changes of the residual information. For the h-th sample, the statistics are written by

$$T_h^2 = d_h^T d_h \tag{13}$$

$$Q_h = e_h^T e_h \tag{14}$$

In the normal operation, these two statistics should satisfy $T_h^2 \leq T_{h,lim}^2$ and $Q_h \leq Q_{h,lim}$, where $T_{h,lim}^2$ and $Q_{h,lim}$ are the corresponding confidence limits. In some literature, the confidence limits of these two statistics can be obtained by assuming the prior distribution [30]. However, these distribution assumptions are often difficult to satisfy. Therefore, this paper applies the data-driven kernel density estimation to determine the confidence limit [31,32].

2.2. CVAkNN Model Based on kNN Monitoring Index

As the measurement data of power transmission systems have the periodic fluctuation characteristic, the traditional CVA monitoring statistics T^2 and Q behave unsteadily with the periodic changes. In this case, disturbance detection by directly monitoring the amplitudes of monitoring statistics cannot discover the disturbance signals effectively and may lead to a high disturbance missing rate.

In order to overcome this defect, this paper introduces the k-nearest neighbor analysis (kNN) to enhance the basic monitoring statistics. kNN is one effective multimodal data analysis tool and does not depend on the amplitude changes before and after the disturbance. In the literature [33,34], kNN was introduced and adapted for real-time detection of system disturbances. By combining the CVA model and the kNN-based monitoring statistics, the improved method, which is called CVAkNN, has a stronger capability of dealing with the periodic oscillation data property. The main idea of CVAkNN is to first reconstruct the monitoring statistic in the phase space and then build the monitoring index based on the distance between the reconstructed statistic vector and its k-nearest neighbor.

Phase space reconstruction is a good method to deal with time series analysis. This method regards one-dimensional time series as the result of nonlinear dynamic system motion and constructs the phase vectors by re-arranging the time series. This theory has been successfully applied in the fields of chaotic time series prediction and equipment failure data analysis [35,36]. Here it is introduced to deal with the CVA monitoring statistics for the further kNN analysis.

For the training data set with n samples $x_1, x_2, \ldots, x_n$, the corresponding statistics vectors are obtained by the CVA modeling as

$$T^2 = [T_l^2 \ T_{l+1}^2 \ \cdots \ T_n^2] \tag{15}$$

$$Q = [Q_l \ Q_{l+1} \ \cdots \ Q_n] \tag{16}$$

Further, the phase reconstruction statistics matrix can be formulated as follows:

$$MT^2 = \begin{bmatrix} T_l^2 & \cdots & T_{l+L-2}^2 & T_{l+L-1}^2 \\ T_{l+1}^2 & \cdots & T_{l+L-1}^2 & T_{l+L}^2 \\ \vdots & \vdots & \vdots & \vdots \\ T_{n-L+1}^2 & \cdots & T_{n-1}^2 & T_n^2 \end{bmatrix} \tag{17}$$

$$MQ = \begin{bmatrix} Q_l & \cdots & Q_{l+L-2} & Q_{l+L-1} \\ Q_{l+1} & \cdots & Q_{l+L-1} & Q_{l+L} \\ \vdots & \vdots & \vdots & \vdots \\ Q_{n-L+1} & \cdots & Q_{n-1} & Q_n \end{bmatrix} \tag{18}$$

where L is the embedding dimension defining the length of the reconstructed phase vector. Based on the results of the phase space reconstruction, the dynamic behavior of the statistics can be better described, which is conducive to the detection of power system disturbances.

In the online monitoring stage, a new testing sample x_t is collected at the t-th sampling instant. Then the monitoring statistics can be computed by applying Equations (13) and (14), and the reconstructed phase vectors are described as

$$NT_t^2 = \begin{bmatrix} T_{t-L+1}^2 & \cdots & T_{t-1}^2 & T_t^2 \end{bmatrix} \tag{19}$$

$$NQ_t = [Q_{t-L+1} \ \cdots \ Q_{t-1} \ Q_t] \tag{20}$$

To determine whether the test data x_t is normal, it is necessary to compare the similarity between NT_t^2, NQ_t^2, and the reconstructed statistics matrix in Equations (17) and (18). If the reconstructed statistics NT_t^2, NQ_t^2 are strongly similar to one column of the training statistics vectors in Equations (17) and (18), then the test data x_t describe the normal working condition. Otherwise, it means that some faults occur in the power transmission system. Therefore, the key is how to perform this similarity comparison. This paper introduces the k-nearest neighbor (kNN) analysis to construct a kNN-based distance measurement indicator: statistical nearest neighbor distance (SNND).

The idea of SNND is to find the first k-th nearest neighbors of the test vector in the given matrix data and compute the distance between the test vector and the k-th nearest neighbors as a disturbance detection criterion. The SNND index for NT_t^2 is defined as

$$DT_t^2 = \left\| NT_t^2 - MT^2(j_k,:) \right\|, \tag{21}$$

where $MT^2(j_k,:)$ represents the j_k-th row in the MT^2 matrix, which corresponds to the k-th nearest neighbor of NT_t^2, and $||.||$ represents the L2 norm calculation. By analogy, the SNND indicator of NQ_t can be established as

$$DQ_t = ||NQ_t - MQ(j_k,:)||. \tag{22}$$

Under normal operating conditions, the above two indicators should fluctuate within a relatively small range. That means $DT_t^2 \leq DT_{lim}^2$ and $DQ_t \leq DQ_{lim}$ for the normal running status. Once the threshold is exceeded, it means that there is a system disturbance. The threshold can be obtained by the kernel density estimation method.

2.3. SLCVAkNN Model Assisted by Statistical Local Analysis

In the power transmission system, some weak disturbances are often difficult to detect, such as the high-impendence single-phase ground fault. When this kind of disturbance occurs, the changes reflected by the measure voltage and current variables are very small. Further, considering the influence of modeling error and process noise, this kind of disturbance may be concealed and viewed as the normal process changes. Therefore, enhacning the weak disturbance detection is of great value to ensure the safety of power transmission systems. In this paper, we integrate the statistical local analysis (SLA) with CVAkNN and propose an improved SLCVAkNN monitoring model for better weak disturbance monitoring performance.

SLA was originally proposed by Basseville [37] for inspecting the process parameter changes. In recent years, some researchers have introduced it into the chemical process fault detection and demonstrated its effectiveness [38–40]. In this paper, we will perform the statistical local analysis on the CVA model. To look back into the CVA monitoring statistics in Equations (13) and (14), it is found that the monitoring statistics used to indicate the process status are composed of the canonical variate vector d_h and the CVA residual vector e_h. Therefore, if we attempt to improve the weak disturbance monitoring of CVA statistics, the vectors d_h and e_h must be improved with stronger disturbance sensitivity.

According to the statistical local analysis theory, given the system observation z_j and the system parameter ϑ, a primary residual vector $\varphi(z_j, \vartheta)$ can be defined for disturbance detection if it meets the following conditions: [37,38]

- $E\{\varphi(z_j, \vartheta)\} = 0$, if $\vartheta = \vartheta_0$;
- $E\{\varphi(z_j, \vartheta)\} \neq 0$, if ϑ is in the neighborhood of ϑ_0, but $\vartheta \neq \vartheta_0$;
- $\varphi(z_j, \vartheta)$ is differentiable with ϑ;
- $\varphi(z_j, \vartheta)$ exists for ϑ in the neighborhood of ϑ_0.

Here ϑ_0 represents the parameters under the normal condition.

By investigating the i-th element in the vector d_h, which is denoted as $d_{h,i}$, it is easily derived by Equation (11) that $d_{h,i} = a_i^T p_h$. Naturally, the variance of $d_{h,i}$ can be computed as

$$E\{d_{h,i}^2\} = a_i^T E\{p_h p_h^T\} a_i \tag{23}$$

For the statistical samples, $E\{p_h p_h^T\}$ is factually equal to the covariance matrix Σ_{pp}. Further combining the first constraint on the vector a in Equation (3), it is known that $a_i^T E\{p_h p_h^T\} a_i = 1$. Therefore, we build the SLA primary residual of the canonical variate as

$$\varphi_{d_{h,i}} = d_{h,i}^2 - 1. \tag{24}$$

which meets the condition $E\{\varphi_{d_{h,i}}\} = 0$ in the normal condition.

Similarly, we analyze the variance of $e_{h,i}$ to obtain

$$E\{e_{h,i}^2\} = A_r(i,:)E\{p_h p_h^T\}A_r(i,:)^T \tag{25}$$

As A_r can be obtained in the model training procedure, the above expression must be equal to a fixed value, which is denoted as $\sigma_i = A_r(i,:)E\{p_h p_h^T\}A_r(i,:)^T$. Therefore, the SLA primary residual of the CVA residual can be built as

$$\varphi_{e_{h,i}} = e_{h,i}^2 - \sigma_i. \tag{26}$$

which meets the condition $E\{\varphi_{e_{h,i}}\} = 0$ for the normal data.

For a more sensitive disturbance detection, the SLA improved residual is applied in a moving window with the width of w, which is expressed by

$$\psi_{d_{h,i}} = \begin{cases} \frac{1}{\sqrt{h}} \sum_{j=1}^{h} \varphi_{d_{-j,i}}, & h < w \\ \frac{1}{\sqrt{w}} \sum_{j=h-w+1}^{h} \varphi_{d_{j,i}}, & h \geq w \end{cases} \tag{27}$$

$$\psi_{e_{h,i}} = \begin{cases} \frac{1}{\sqrt{h}} \sum_{j=1}^{h} \varphi_{e_{j,i}}, & h < w \\ \frac{1}{\sqrt{w}} \sum_{j=h-w+1}^{h} \varphi_{e_{j,i}}, & h \geq w \end{cases} \tag{28}$$

Up to now, we can obtain the SLA improved residual vectors $\psi_{d,h} = [\psi_{d_{h,1}} \ \psi_{d_{h,2}} \ \cdots \ \psi_{d_{h,s}}]^T$ and $\psi_{e,h} = [\psi_{e_{h,1}} \ \psi_{e_{h,2}} \ \cdots \ \psi_{e_{h,M}}]^T$. These residual vectors are used to replace the original CVA features d_h and e_h so that the monitoring model is modified to the SLCVAkNN model.

With the SLA improved residual vectors, the monitoring statistics are constructed as follows:

$$T_h^2 = \psi_{d,h}^T \psi_{d,h} \tag{29}$$

$$Q_h = \psi_{e,h}^T \psi_{e,h} \tag{30}$$

3. Disturbance Detection Procedure Based on SLCVAkNN

Power transmission system disturbance detection based on SLCVAkNN method is divided into two stages: offline modeling stage and online detection stage. The corresponding flowchart is shown in Figure 1.

Stage 1: offline modeling stage

1. Acquire the normal condition data to constitute the training data set $X = [x_1^T, x_2^T, \cdots, x_n^T]^T \in R^{n \times m}$ and perform data normalization processing. Here, the mentioned normal condition data mean the data from a section of transmission line between two adjacent nodes. For different lines, the corresponding modelings are needed separately.
2. Construct historical data sets P and future data sets F according to Equations (4) and (5), calculate the covariance matrices by Equations (6)–(8), and solve the CVA optimization by the SVD as Equation (9).
3. Extract the canonical variate vector d_h and the CVA residual vector e_h, as shown in Equations (11) and (12).
4. Perform Equations (24) and (26) to obtain the SLA primary residual vectors and further calculate the SLA improved residual vectors by Equations (27) and (28).
5. Compute the monitoring statistics T_h^2 and Q_h for all the training samples according to Equations (29) and (30).
6. Construct the statistics matrix in the phase space according to Equations (17) and (18).
7. Calculate the SNND monitoring indices DT^2 and DQ for all the training samples and determine the 95% confidence limits DT_{lim}^2 and DQ_{lim} by kernel density estimation.

Stage 2: online detection stage

1. Obtain online new data x_t and normalize it with the training data.
2. Construct the corresponding historical vector p_t and project the p_t to the CVA model and obtain the canonical variate vector d_t and e_t according to Equations (11) and (12).
3. Apply Equations (24), (26)–(28) to compute the SLA primary residual vector and the improved residual vector orderly.
4. Compute the monitoring statistics T_t^2 and Q_t for the online new sample x_t according to Equations (29) and (30).
5. Construct the phase space statistics vector NT_t^2 and NQ_t^2, and calculate the SNND index DT_t^2 and DQ_t^2 by Equations (21) and (22).
6. Compare the SNND indices with the corresponding confidence limits DT_{lim}^2 and DQ_{lim}. If any one exceeds the confidence limit, a disturbance sample is indicated.

Here, it is pointed out that the local neighborhood standardization (LNS) [41] may be used to enhance the traditional z-score standardization. Compared with the traditional z-score method, LNS has better capability to deal with the non-steady data with the periodic oscillations.

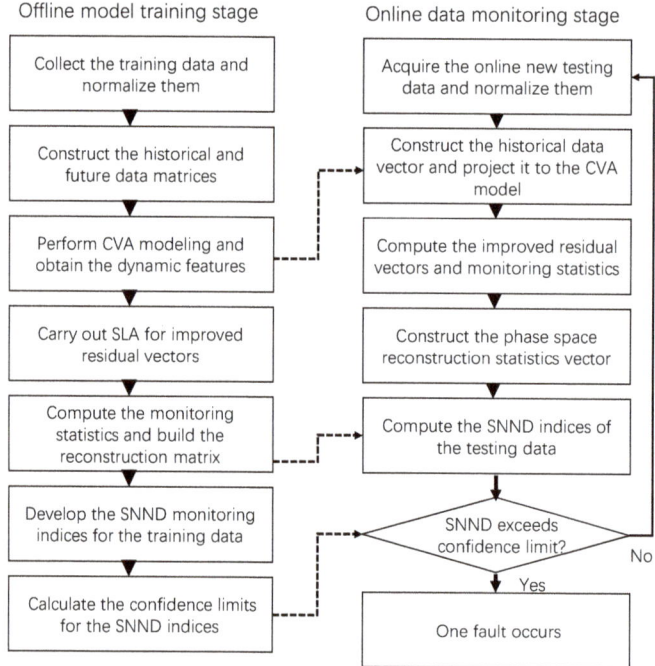

Figure 1. Flow chart of disturbance detection by SLCVAkNN.

4. Case Analysis

In order to verify the advantages of the SLCVAkNN method in the power transmission system disturbance detection, this section gives the case study on the real industrial data collected from the actual power transmission system. For method comparison, four methods, including the proposed SLCVAkNN method and three other methods of PCA, PCAkNN, and CVAkNN, are all applied to build the monitoring models for disturbance detection. The PCA method has two monitoring statistics T^2 and Q, while the other three methods are with the kNN-based statistics DT^2 and DQ. When these methods are used, they indicate the system status by the monitoring charts, where the monitoring indices of normal and faulty samples are given by black and blue solid lines, respectively, while the detection threshold, that is the 95% confidence limit of the monitoring index, is plotted by the red dashed line. One evaluation index, called the disturbance detection rate (DDR), is used to evaluate the different monitoring methods. DDR is the percentage of the abnormal samples exceeding the detection threshold over all the abnormal samples.

The used real industrial data are collected from the seven transmission lines in a power supply station in August 2018. These lines are radially connected. Their data are collected because all of them involve the ground fault. The data acquisition units, designed by Qingdao Topscomm Communication CO. LTD, are used to collect the electric field intensity and current. Here, the real line voltage is up to 110 KV so that the existing equipment can not directly measure it. Therefore, the electric field intensity is applied to reflect the voltage trend. For each transmission line, one corresponding data set is recorded that involves the normal state and the abnormal state. The data set has the length of about 1300 samples, where the disturbance starting time (DST) is different in different transmission lines. The detailed information about the acquired data sets are listed in Table 1, where DST data record the sample number corresponding to the disturbance starting time. A demonstration of the collected data for the DATA-A is given in Figure 2, where six measured variables, including the electric field intensities of phase A, B, and C, and the currents of phase A, B, and C, are involved. Due to the existence of the harmonic load, the current sine wave distortion can be seen in these curves.

Table 1. The collected industrial data sets.

No.	Description	DST
DATA-1	Data set collected from line 904 exit	446
DATA-2	Data set collected from line 906, pole 116-3	456
DATA-3	Data set collected from line 906, pole 90-2	445
DATA-4	Data set collected from line 906, pole 151-5	458
DATA-5	Data set collected from line 906, pole 97-1	452
DATA-6	Data set collected from line 906 exit	493
DATA-7	Data set collected from line 907 exit	420

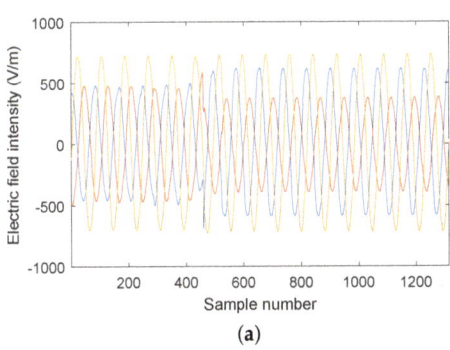
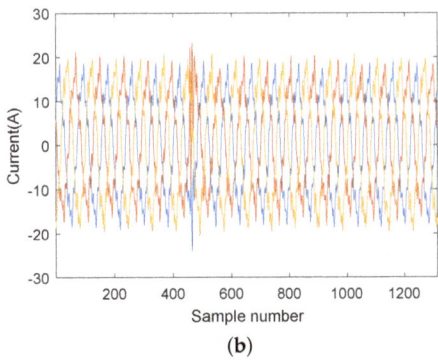

Figure 2. Data waveform collected from 904 line exit. (a) Three-phase electric field intensity; (b) Three-phase current.

Taking the data set DATA-2 as one example, it is collected from the pole 116-3 of the line 906. This data set includes 1312 samples. To investigate it with the help of on-site engineers, it is known that the disturbance occurs from the 456th sample. Although engineers can find this disturbance by careful analysis, this manual way is very time-consuming and inefficient, so it is difficult to implement in large-scale transmission system monitoring. Therefore, building an automatic multivariate data analysis tool is very necessary. In this section, we apply four MSA methods, which are PCA, PCAkNN, CVAkNN, and SLCVAkNN, to perform the automatic fault detection. When the statistical models are developed, the model parameters are set as follows: $k = 3, L = 10, l = 2, w = 20$. For the data set DATA-2, the first 320 sampling point are considered to be in a normal operating state, they can be utilized as the training data set for model development, while monitoring charts of PCA, PCAkNN, CVAkNN, and SLCVAkNN are demonstrated in the Figures 3–6, respectively. By the PCA monitoring results shown in Figure 3, it can be seen that the disturbance cannot be detected very effectively. The DDR of PCA T^2 is 4.43%, while the Q is a little better with the DDR of 29.52%. When PCAkNN is used, the DT^2 has a similarly poor detection rate, but the DQ statistic achieves clear improvement with the DDR of 57.76%. These results demonstrate that the PCAkNN method proposed by Cai et al. [20] can deal with the power system data with oscillation characteristic effectively. However, from these figures, the monitoring statistics do not exceed the confidence limits significantly. This may lead to the uncertain judgement on the occurrence of disturbance. When the CVAkNN is applied in Figure 5, the DQ statistic performs a little better with the DDR of 49.71%. However, its DT^2 indicator clearly improves the DDR to 92.51%, which means a significant detection rate improvement of about 70% in contrast with the PCAkNN's DQ index. The best monitoring results on this data set is provided by SLCVAkNN, which are shown in Figure 6. By this figure, it is observed that the disturbance is detected very clearly with the DDRs of 97.25% and 96.80% for DT^2 and DQ, respectively. This case gives a comprehensive comparison on the four methods of PCA, PCAkNN, CVAkNN, and SLCVAkNN. The applications show that PCAkNN does better than PCA due to the use of kNN, while SLCVAkNN further prompts the disturbance detection performance with the integration of CVA and SLA.

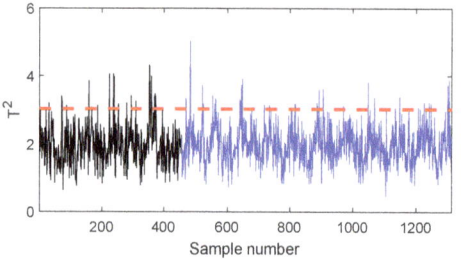 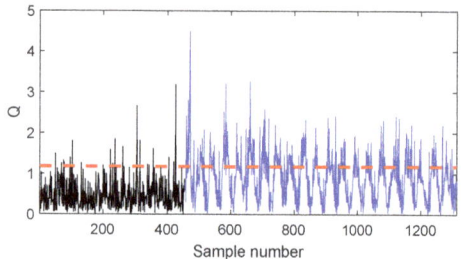

Figure 3. PCA monitoring results on the DATA-2 case.

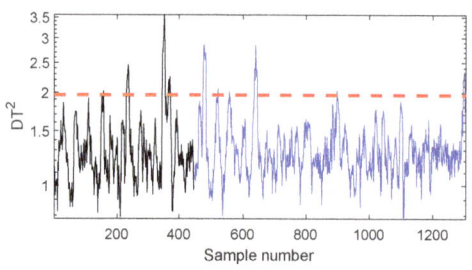 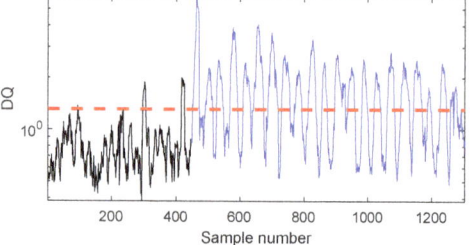

Figure 4. PCAkNN monitoring results on the DATA-2 case.

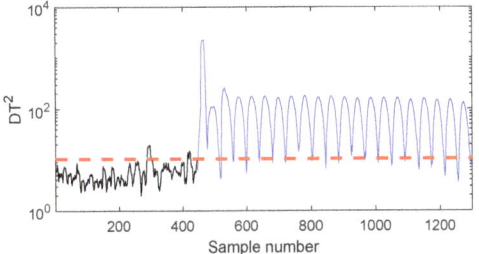
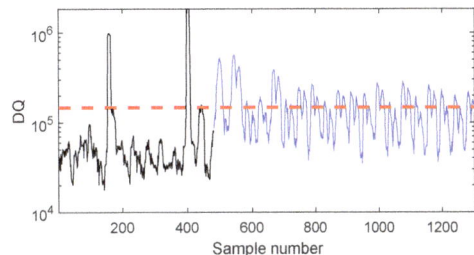

Figure 5. CVAkNN monitoring results on the DATA-2 case.

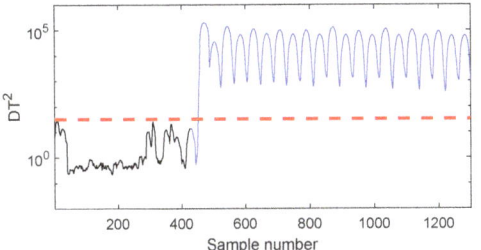
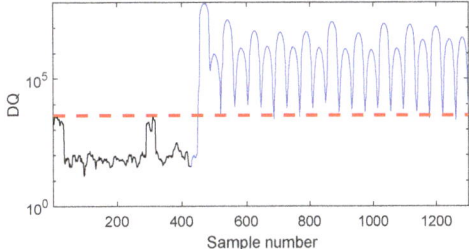

Figure 6. SLCVAkNN monitoring results on the DATA-2 case.

Another example on the data set DATA-6 is illustrated, which corresponds to the line 906 exit. The modeling procedure is similar to the above case. Here we only give the monitoring charts of CVAkNN and SLCVAkNN, as shown in the Figures 7 and 8. With the consideration of system dynamics, the CVAkNN DT^2 monitoring chart gives a higher DDR of 88.51%. Compared with the CVAkNN method, which has only one effective monitoring statistic, SLCVAkNN has two well-behaved monitoring statistics. The DT^2 and DQ have the DDRs of 97.37% and 97.25%, respectively. The testing results on DATA-6 further verify the advantage of the proposed method over the CVAkNN method.

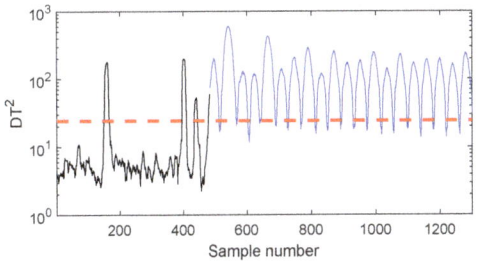
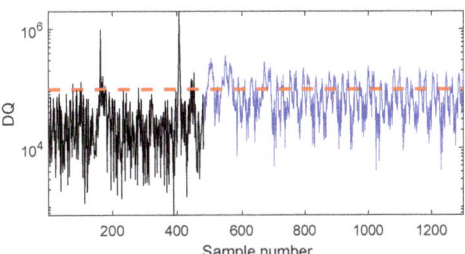

Figure 7. CVAkNN monitoring results on the DATA-6 case.

The summary of disturbance detection rates for all seven data sets are shown in Table 2. From this table, it is shown that the faults in DATA-2 and DATA-4 are difficult to detect by PCA, whose DDRs are all lower than 30%. By the use of PCAkNN, these two faults are detected with higher DDRs, which are 57.76% and 26.78%, respectively. By contrast, CVAkNN does better on the two faults. In particular, its DT^2 statistic gives the DDR higher than 90%. When SLCVAkNN is used, its two monitoring statistics have the higher DDRs than 95%. For the sets of DATA-1, DATA-3, DATA-6, and DATA-7, PCA can detect these faults with about 70–80% DDR on one statistic. That means PCA can alarm these faults, but the alarm degree is not very sufficient. The PCAkNN and CVAkNN improve the DDR to about 90%. Further combining the SLA technique, SLCVAkNN achieves higher DDR than CVAkNN on these four sets. As to DATA-5, all these four methods give a similarly good performance with the DDRs higher than 95%. Considering all seven of these data sets, we observe that the average detection rates of CVAkNN outperforms the PCA and PCAkNN method, while the ones of SLCVAkNN statistics can reach 97.46% and 96.29%, which are the highest among these four methods.

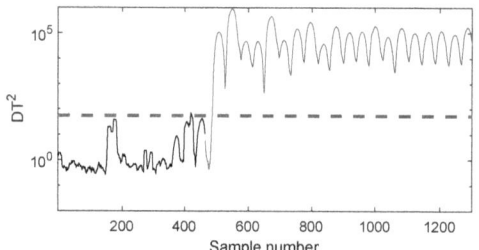 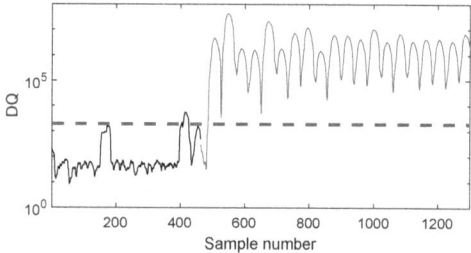

Figure 8. SLCVAkNN monitoring results on the DATA-6 case.

Table 2. The disturbance detection rate of PCA, PCAkNN, CVAkNN, and SLCVAkNN for the tested data sets.

NO.	PCA		PCAkNN		CVAkNN		SLCVAkNN	
	T^2	Q	DT^2	DQ	DT^2	DQ	DT^2	DQ
DATA-1	7.50%	70.47%	26.18%	90.89%	96.76%	64.39%	96.83%	89.48%
DATA-2	4.43%	29.52%	3.85%	57.76%	92.51%	49.71%	97.25%	96.80%
DATA-3	5.99%	83.06%	19.93%	97.81%	100.00%	100.00%	97.85%	97.97%
DATA-4	5.15%	10.18%	8.54%	26.78%	96.60%	51.93%	97.36%	96.79%
DATA-5	48.78%	96.05%	82.81%	99.54%	100.00%	100.00%	98.06%	98.29%
DATA-6	4.39%	77.44%	5.49%	92.20%	88.51%	41.20%	97.37%	97.25%
DATA-7	7.95%	84.99%	11.09%	95.97%	93.49%	55.33%	97.47%	97.47%
Average	12.03%	64.53%	22.56%	80.13%	95.41%	66.08%	97.46%	96.29%

To sum up, the applications on real industrial data verify the effectiveness of the proposed SLCVAkNN in the power transmission system monitoring. All the tested faults are about the ground faults. Although this paper does not provide the results on the other disturbances such as 1,3-phase short circuits, overvoltages, the presented algorithm is also suitable for these cases because they similarly lead to the changes of voltage and current. However, one related issue should be noted. In this article, this method detects all the occurred disturbances, including normal disturbances such as load power variations. To judge whether the disturbance is a fault or a normal disturbance is a further job. In fact, as to this issue, one solution is to enrich the modeling data with different normal changes. As the kNN used in this method can deal with the multimodal data case, the trained model can distinguish the faults and normal disturbances effectively when the normal changing data are considered in the model training procedure.

5. Conclusions

This paper proposes a power transmission system disturbance detection method based on SLCVAkNN. The real industrial data collected from the field transformer station are applied to verify the proposed method. By investigating the application results, we can draw the following conclusions.

- **CVA-based monitoring method can provide better dynamic information mining.** The dynamic data analysis tool CVA is introduced to deal with the power transmission system data. By observing the application results, we find that CVAkNN has a higher detection rate than PCAkNN.
- **The statistical local analysis can further enhance the disturbance monitoring.** Considering that many high-impendence ground faults in the real power systems are with insignificant symptoms, the weak disturbance detection methods are very important in improving the disturbance detection sensitivity. By focusing on the statistical local information of CVA features, the proposed SLCVAkNN method outperforms the CVAkNN method.

Author Contributions: Conceptualization, S.W. and X.D.; methodology, S.W. and Y.T.; software, Y.T. and X.D.; validation, Y.T. and L.W.; formal analysis, S.W. and Y.T.; investigation, Y.T. and P.S.; resources, L.W. and Q.C.; data curation, Y.T., L.W., and Q.C.; writing—original draft preparation, Y.T. and P.S.; writing—review and editing, S.W. and X.D.; visualization, L.W. and Q.C.; supervision, X.D.; project administration, X.D.; funding acquisition, X.D. All authors have read and agreed to the published version of the manuscript.

Funding: This work was funded by the Shandong Provincial Natural Science Foundation (Grant No. ZR2020MF093), the Major Scientific and Technological Projects of CNPC (Grant No. ZD2019-183-003), and the Fundamental Research Funds for the Central Universities (that is, the Opening Fund of National Engineering Laboratory of Offshore Geophysical and Exploration Equipment, Grant No. 20CX02310A).

Institutional Review Board Statement: Not applicable.

Informed Consent Statement: Not applicable.

Data Availability Statement: Not applicable.

Conflicts of Interest: The authors declare no conflict of interest.

Nomenclature

CVA	canonical variate analysis
MSA	multivariate statistical analysis
SLA	statistical local analysis
WAMS	wide area measurement system
A_s	projection matrix
d_h	canonical variate vector
DQ_t	SNND monitoring index
f_h	future data vector at the h-th sample instant
MT^2	phase reconstruction statistics matrix
NT_t^2	reconstructed statistics vector
p_h	historical data vector at the h-th sample instant
Q_h	monitoring statistic at the h-th sample instant
x_h	data vector at the h-th sample instant
$\phi_{d_{h,i}}$	SLA primary residual of canonical variate
$\psi_{d_{h,i}}$	SLA improved residual of canonical variate
Σ	Covariance matrix
kNN	k-nearest neighbor
PCA	principal component analysis
SNND	statistical nearest neighbor distance
a	projection vector
b	projection vector
DT_t^2	SNND monitoring index
e_h	residual vector at the h-th sample instant
F	future data matrix

MQ	phase reconstruction statistics matrix
NQ_t	reconstructed statistics vector
P	historical data matrix
T_h^2	monitoring statistic at the h-th sample instant
X	data matrix
$\phi_{e_{h,i}}$	SLA primary residual of CVA residual
$\psi_{e_{h,i}}$	SLA improved residual of CVA residual

References

1. Samuelsson, O.; Hemmingsson, M.; Nielsen, A.H.; Pedersen, K.O.H.; Rasmussen, J. Monitoring of power system events at transmission and distribution level. *IEEE Trans. Power Syst.* **2006**, *21*, 1007–1008. [CrossRef]
2. Patel, B.; Bera, P. Fast fault detection during power swing on a hybrid transmission line using WPT. *IET Gener. Transm. Distrib.* **2019**, *13*, 1811–1820. [CrossRef]
3. Musa, M.H.H.; He, Z.; Fu, L.; Deng, Y. Linear regression index-based method for fault detection and classification in power transmission line. *IEEJ Trans. Electr. Electron. Eng.* **2018**, *13*, 979–987. [CrossRef]
4. Chang, H.-H.; Linh, N.V.; Lee, W.-J. A novel nonintrusive fault identification for power transmission networks using power-spectrum-based hyperbolic s-transform-part i: Fault classification. *IEEE Trans. Ind. Appl.* **2018**, *54*, 5700–5710. [CrossRef]
5. Costa, F.B. Fault-induced transient detection based on real-time analysis of the wavelet coefficient energy. *IEEE Trans. Power Deliv.* **2013**, *29*, 140–153. [CrossRef]
6. Math, H.J.B.; Das, R.; Djokic, S.; Ciufo, P.; Meyer, J.; Ronnberg, S.K.; Zavoda, F. Power quality concerns in implementing smart distribution-grid applications. *IEEE Trans. Smart Grid* **2016**, *8*, 391–399.
7. Cheng, C.; Wang, W.; Chen, H.; Zhang, B.; Shao, J.; Teng, W. Enhanced fault diagnosis usign broad learning for traction systems in high-speed trains. *IEEE Trans. Power Electron.* **2021**, *36*, 7461–7469. [CrossRef]
8. Huang, S.; Hsieh, C.; Huang, C. Application of morlet wavelets to supervise power system disturbances. *IEEE Trans. Power Deliv.* **1999**, *14*, 235–243. [CrossRef]
9. Manglik, A.; Li, W.; Ahmad, S.U. Fault Detection in power system using the Hilbert-Huang transform. In Proceedings of the 2016 IEEE Canadian Conference On Electrical And Computer Engineering (CCECE), Vancouver, BC, Canada, 14–18 May 2016.
10. Ghaderi, A.; Mohammadpour, H.A.; Ginn, H.L.; Shin, Y.-J. High-impedance fault detection in the distribution network using the time-frequency-based algorithm. *IEEE Trans. Power Deliv.* **2015**, *30*, 1260–1268. [CrossRef]
11. Salehi, M.; Namdari, F. Fault classification and faulted phase selection for transmission line using morphological edge detection filter. *IET Gener. Transm. Distrib.* **2018**, *12*, 1595–1605. [CrossRef]
12. Liu, Z.; Hu, Q.; Cui, Y.; Zhang, Q. A new detection approach of transient disturbances combining wavelet packet and tsallis entropy. *Neurocomputing* **2014**, *142*, 393–407. [CrossRef]
13. Zhang, X.; Kano, M.; Song, Z. Optimal weighting distance-based similarity for locally weighted PLS modeling. *Ind. Eng. Chem. Res.* **2020**, *59*, 11552–11558. [CrossRef]
14. Deng, X.; Du, K. Efficient batch process monitoring based on random nonlinear feature analysis. *Canadian J. Chem. Eng.* **2021**, in press, 1–12. [CrossRef]
15. Zhang, X.; Kano, M.; Matsuzaki, S. A comparative study of deep and shallow predictive techniques for hot metal temperature prediction in blast furnace ironmaking. *Comput. Chem. Eng.* **2019**, *130*, 106575. [CrossRef]
16. Chen, H.; Jiang, B.; Ding, S.X.; Huang, B. Data-driven fault diagnosis for traction systems in high-speed trains: A survey, challenges, and perspectives. *IEEE Trans. Intell. Transp. Syst.* **2020**, in press, 1–17. [CrossRef]
17. Chen, H.; Jiang, B. A review of fault detection and diagnosis for the traction system in high-speed trains. *IEEE Trans. Intell. Transp. Syst.* **2020**, *21*, 450–465. [CrossRef]
18. Barocio, E.; Pal, B.C.; Fabozzi, D.; Thornhill, N.F. Detection and visualization of power system disturbances using principal component analysis. In Proceedings of the 2013 IREP Symposium Bulk Power System Dynamics and Control-IX Optimization, Security and Control of the Emerging Power Grid, Rethymnon, Greece, 25–30 August 2013.
19. Guo, Y.; Li, K.; Liu, X. Fault diagnosis for power system transmission line based on PCA and SVMs. In Proceedings of the 2nd International Conference on Intelligent Computing for Sustainable Energy and Environment (ICSEE), Shanghai, China, 12–13 September 2012.
20. Cai, L.; Thornhill, N.F.; Kuenzel, S.; Pal, B.C. Wide-area monitoring of power systems using principal component analysis and k-nearest neighbor analysis. *IEEE Trans. Power Syst.* **2018**, *33*, 4913–4923. [CrossRef]
21. Chen, H.; Wu, J.; Jiang, B.; Chen, W. A modified neighborhood preserving embedding-based incipient fault detection with applications to small-scale cyber-physical systems. *ISA Trans.* **2020**, *104*, 175–183. [CrossRef] [PubMed]
22. Chen, H.; Jiang, B.; Zhang, T.; Lu, N. Data-driven and deep learning-based detection and diagnosis of incipient faults with application to electrical traction systems. *Neurocomputing* **2020**, *396*, 429–437. [CrossRef]
23. Jiang, B.; Braatz, R.D. Fault detection of process correlation structure using canonical variate analysis-based correlation features. *J. Process. Control* **2017**, *58*, 131–138. [CrossRef]
24. Li, X.; Yang, Y.; Bennett, I.; Mba, D. Condition monitoring of rotating machines under time-varying conditions based on adaptive canonical variate analysis. *Mech. Syst. Signal Process.* **2019**, *131*, 348–363. [CrossRef]

25. Chen, Z.; Yang, C.; Peng, T.; Dan, H.; Li, C.; Gui, W. A cumulative canonical correlation analysis-based sensor precision degradation detection method. *IEEE Trans. Ind. Electron.* **2019**, *66*, 6321–6330. [CrossRef]
26. Han, S.; Xu, Z.; Sun, B.; He, L. Dynamic characteristic analysis of power system interarea oscillations using HHT. *Int. J. Electr. Power Energy Syst.* **2010**, *32*, 1085–1090. [CrossRef]
27. Hu, Z. Method considering the dynamic coupling characteristic in power system for stability assessment. *IET Gener. Transm. Distrib.* **2017**, *11*, 2534–2539. [CrossRef]
28. Zhang, S.; Zhao, C.; Huang, B. Simultaneous static and dynamic analysis for fine-scale identification of process operation statuses. *IEEE Trans. Ind. Inform.* **2019**, *15*, 5320–5329. [CrossRef]
29. Li, X.; Duan, F.; Bennett, I.; Mba, D. Canonical variate analysis, probability approach and support vector regression for fault identification and failure time prediction. *J. Intell. Fuzzy Syst.* **2018**, *34*, 3771–3783. [CrossRef]
30. Chiang, L.H.; Russell, E.L.; Braatz, R.D. *Fault Detection and Diagnosis in Industrial Systems*; Springer: London, UK, 2001.
31. Odiowei, P.E.P.; Cao, Y. Nonlinear dynamic process monitoring using canonical variate analysis and kernel density estimations. *IEEE Trans. Ind. Inform.* **2010**, *6*, 36–45. [CrossRef]
32. Zhang, Z.; Deng, X. Anomaly detection using improved deep SVDD model with data structure preservation. *Pattern Recognit. Lett.* **2021**, *148*, 1–6. [CrossRef]
33. Zhang, X.; Li, Y. Multiway principal polynomial analysis for semiconductor manufacturing process fault detection. *Chemom. Intell. Lab. Syst.* **2018**, *181*, 29–35. [CrossRef]
34. Cai, L.; Thornhill, N.F.; Kuenzel, S.; Pal, B.C. Real-time detection of power system disturbances based on k-nearest neighbor analysis. *IEEE Access* **2017**, *5*, 5631–5639. [CrossRef]
35. Zhang, A.; Xu, Z. Chaotic time series prediction using phase space reconstruction based conceptor network. *Cogn. Neurodynamics* **2020**, *14*, 849–857. [CrossRef]
36. Yu, P.; Yan, X. Stock price prediction based on deep neural networks. *Neural Comput. Appl.* **2020**, *32*, 1609–1628. [CrossRef]
37. Basseville, M. On-board component fault detection and isolation using the statistical local approach. *Automatica* **1998**, *34*, 1391–1415. [CrossRef]
38. Kruger, U.; Kumar, S.; Littler, T. Improved principal component monitoring using the local approach. *Automatica* **2007**, *43*, 1532–1542. [CrossRef]
39. Ge, Z.; Yang, C.; Song, Z. Improved kernel PCA-based monitoring approach for nonlinear processes. *Chem. Eng. Sci.* **2009**, *64*, 2245–2255. [CrossRef]
40. Deng, X.; Cai, P.; Deng, J.; Cao, Y.; Song, Z. Primary-auxiliary statistical local kernel principal component analysis and its application to incipient fault detection of nonlinear industrial processes. *IEEE Access* **2019**, *7*, 122192–122204. [CrossRef]
41. Ma, H.; Hu, Y.; Shi, H. A novel local neighborhood standardization strategy and its application in fault detection of multimode processes. *Chemom. Intell. Lab. Syst.* **2012**, *118*, 287–300. [CrossRef]

Article

Auxiliary Model-Based Multi-Innovation Fractional Stochastic Gradient Algorithm for Hammerstein Output-Error Systems

Chen Xu [1] and Yawen Mao [2,*]

[1] Key Laboratory of Advanced Process Control for Light Industry (Ministry of Education), Jiangnan University, Wuxi 214122, China; chenxu@jiangnan.edu.cn
[2] School of Science, Jiangnan University, Wuxi 214122, China
* Correspondence: myw0530@163.com

Abstract: This paper focuses on the nonlinear system identification problem, which is a basic premise of control and fault diagnosis. For Hammerstein output-error nonlinear systems, we propose an auxiliary model-based multi-innovation fractional stochastic gradient method. The scalar innovation is extended to the innovation vector for increasing the data use based on the multi-innovation identification theory. By establishing appropriate auxiliary models, the unknown variables are estimated and the improvement in the performance of parameter estimation is achieved owing to the fractional-order calculus theory. Compared with the conventional multi-innovation stochastic gradient algorithm, the proposed method is validated to obtain better estimation accuracy by the simulation results.

Keywords: hammerstein output-error systems; auxiliary model; multi-innovation identification theory; fractional-order calculus theory

1. Introduction

The accuracy of a system model affects the performance and safety of industrial control systems [1–5], and system identification is a theory and method for constructing mathematical model of systems and has been widely implemented in practice [6–9]. The behavior of most modern industrial control systems and synthetic systems are nonlinear by nature. Presently, an important research field in modern signal processing is the research of parameter identification for nonlinear systems, in which the block-structure systems, such as the Hammerstein model, are among the most current nonlinear systems due to their efficiency and accuracy to model complex nonlinear systems [10–12]. The representative feature of a Hammerstein model is that its architecture consists of two blocks: a static nonlinear model followed by a linear dynamic model. The simplicity in structure makes it provide a good compromise between the accuracy of nonlinear systems and the tractability of linear systems, and thus promoting its use in different nonlinear applications such as automatic control [13–15], fault detection and diagnosis [16–18], and so on.

Recently, several new system identification methods and theories have been developed for nonlinear models in the literature, including the least squares methods [19], the gradient-based methods [20], the iterative methods [21],the subspace identification methods [22], the hierarchical identification theory [23], the auxiliary model and the multi-innovation (MI) identification theories [24]. One well-known algorithm is the stochastic gradient (SG) algorithm, which has lower computational cost and complexity than the recursive least squares algorithm, whereas slow-convergence phenomena are often observed. Therefore, different modifications of the SG algorithm were developed to enhance its performance [25–30]. In particular, by extending scalar innovation into innovation vectors, the MI identification theory was proposed to improve the convergence speed and estimation accuracy in [31], and the fractional-order calculus method was introduced to show that it can achieve more satisfactory performance in [32,33].

To the best of our knowledge, different fractional-order gradient methods have been produced [34–36]. For example, in [37], a fractional-order SG algorithm was designed to identify the Hammerstein nonlinear ARMAX systems by an improved fractional-order gradient method. Based on the MI theory and the fractional-order calculus, an MI fractional least mean squares identification algorithm was presented for the Hammerstein controlled autoregressive systems, where the update mechanism was composed of the first-order gradient and the fractional gradient [38]. However, the above-discussed papers only consider the Hammerstein equation-error systems, and the cross-products between the parameters in the linear block and nonlinear block can lead to many redundant parameters. When the dimensions of parameter vectors are large, it will cause high computational complexity and deteriorate the identification accuracy.

In this work, we study the identification problem of the Hammerstein output-error moving average (OEMA) systems, which have been less studied due to the difficulty in identification [39,40]. To avoid estimating the redundant parameters, the Hammerstein model is parameterized using the key-term separation principle [41]. Furthermore, based on the identification model, the fractional-order SG algorithm is extended to the identification of Hammerstein OEMA systems and an auxiliary model-based multi-innovation fractional stochastic gradient (AM-MIFSG) algorithm is presented by the auxiliary model identification idea. The proposed algorithm can generate higher estimation accuracy than the common multi-innovation stochastic gradient (MISG) algorithm, with fewer parameters required to be estimated.

The paper is structured as follows. Section 2 gives a description for Hammerstein OEMA systems. Section 3 introduces the multi-innovation identification theory and drives an auxiliary model-based multi-innovation stochastic gradient (AM-MISG) identification algorithm for a comparison purpose. Section 4 presents the AM-MIFSG identification algorithm for the Hammerstein OEMA systems. Section 5 gives the convergence analysis of the proposed AM-MIFSG algorithm. Section 6 verifies the results in this paper using a simulation example. Finally, concluding remarks are given in Section 7.

2. The System Description

Consider the Hammerstein OEMA systems shown in Figure 1,

$$y_k = \frac{B(z)}{A(z)}\bar{u}_k + D(z)v_k, \tag{1}$$

$$\bar{u}_k = c_1 f_1(u_k) + c_2 f_2(u_k) + \cdots + c_m f_m(u_k), \tag{2}$$

where $\{u_k\}$ and $\{y_k\}$ are the input and output sequences of the system, $\{\bar{u}_k\}$ is the output sequence of the nonlinear block, and it can be represented as a linear combination of a known basis $f(u_k) := [f_1(u_k), f_2(u_k), \cdots, f_m(u_k)]$ with unknown coefficients c_i ($i = 1, 2, \cdots, m$), $\{v_k\}$ is a stochastic white noise sequence with zero mean and variance σ^2, $A(z)$, $B(z)$ and $D(z)$ are the polynomials in the unit backward shift operator z^{-1} [$z^{-1}y_k = y_{k-1}$], and defined as

$$A(z) := 1 + a_1 z^{-1} + a_2 z^{-2} + \cdots + a_{n_a} z^{-n_a},$$
$$B(z) := 1 + b_1 z^{-1} + b_2 z^{-2} + \cdots + b_{n_b} z^{-n_b},$$
$$D(z) := 1 + d_1 z^{-1} + d_2 z^{-2} + \cdots + d_{n_d} z^{-n_d}.$$

Assume that the orders of these polynomials n_a, n_b and n_d are known and $u_k = 0$, $y_k = 0$ and $v_k = 0$ for $k \leqslant 0$.

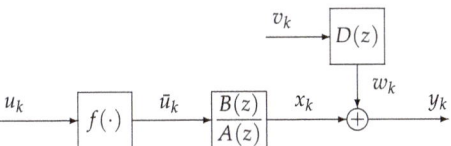

Figure 1. The Hammerstein OEMA systems.

Define the intermediate variables x_k and w_k as follows:

$$\begin{aligned}
x_k &:= \frac{B(z)}{A(z)}\bar{u}_k \\
&= [1-A(z)]x_k + B(z)\bar{u}_k \\
&= \bar{u}_k - \sum_{i=1}^{n_a} a_i x_{k-i} + \sum_{i=1}^{n_b} b_i \bar{u}_{k-i},
\end{aligned} \quad (3)$$

$$\begin{aligned}
w_k &:= D(z)v_k \\
&= \sum_{i=1}^{n_d} d_i v_{k-i} + v_k.
\end{aligned} \quad (4)$$

Take the first variable $\bar{u}_k$ on the right-hand side of (3) as a separated key-term. Based on the principle of key-term separation [42,43], substituting $\bar{u}_k$ in (2) into (3) gives

$$x_k = \sum_{i=1}^{m} c_i f_i(u_k) - \sum_{i=1}^{n_a} a_i x_{k-i} + \sum_{i=1}^{n_b} b_i \bar{u}_{k-i}. \quad (5)$$

Define the following related parameter vectors:

$$\boldsymbol{\theta} := \begin{bmatrix} \boldsymbol{\theta}_s \\ \boldsymbol{d} \end{bmatrix} \in \mathbb{R}^n, \quad n := n_a + n_b + n_d + m,$$

$$\boldsymbol{\theta}_s := [\boldsymbol{a}^\mathrm{T}, \boldsymbol{b}^\mathrm{T}, \boldsymbol{c}^\mathrm{T}]^\mathrm{T} \in \mathbb{R}^{n_a+n_b+m},$$

$$\boldsymbol{a} := [a_1, a_2, \cdots, a_{n_a}]^\mathrm{T} \in \mathbb{R}^{n_a}, \quad \boldsymbol{b} := [b_1, b_2, \cdots, b_{n_b}]^\mathrm{T} \in \mathbb{R}^{n_b},$$

$$\boldsymbol{c} := [c_1, c_2, \cdots, c_m]^\mathrm{T} \in \mathbb{R}^m, \quad \boldsymbol{d} := [d_1, d_2, \cdots, d_{n_d}]^\mathrm{T} \in \mathbb{R}^{n_d},$$

and the information vectors:

$$\boldsymbol{\varphi}_k := \begin{bmatrix} \boldsymbol{\varphi}_{s,k} \\ \boldsymbol{\varphi}_{n,k} \end{bmatrix} \in \mathbb{R}^n,$$

$$\boldsymbol{\varphi}_{s,k} := [-x_{k-1}, -x_{k-2}, \cdots, -x_{k-n_a}, \bar{u}_{k-1}, \bar{u}_{k-2}, \cdots, \bar{u}_{k-n_b}, f(u_k)]^\mathrm{T} \in \mathbb{R}^{n_a+n_b+m},$$

$$\boldsymbol{\varphi}_{n,k} := [v_{k-1}, v_{k-2}, \cdots, v_{k-n_d}]^\mathrm{T} \in \mathbb{R}^{n_d}.$$

From (1)–(5), we have

$$\begin{aligned}
y_k &= x_k + w_k \\
&= \boldsymbol{\varphi}_{s,k}^\mathrm{T} \boldsymbol{\theta}_s + \boldsymbol{\varphi}_{n,k}^\mathrm{T} \boldsymbol{d} + v_k \\
&= \boldsymbol{\varphi}_k^\mathrm{T} \boldsymbol{\theta} + v_k.
\end{aligned} \quad (6)$$

Equation (6) is the identification model of the Hammerstein OEMA system. Please note that the parameter vector $\boldsymbol{\theta}$ contains all the parameters of the system in (1)–(2), and the parameters in the linear and nonlinear blocks are separated. This means there is no need to identify redundant parameters. This paper aims to present an AM-MIFSG algorithm for Hammerstein OEMA systems to improve the parameter estimation accuracy.

3. The AM-MISG Algorithm

In this section, we introduce the auxiliary model and multi-innovation identification theories briefly, and derive the AM-MISG algorithm for the Hammerstein OEMA system.

Let $\hat{\boldsymbol{\theta}}_k$ denote the estimate of $\boldsymbol{\theta}$. Based on the search principle of negative gradient, defining and minimizing the cost function

$$J(\boldsymbol{\theta}) := \frac{1}{2}\sum_{j=1}^{k}[y_j - \boldsymbol{\varphi}_j^\mathsf{T}\boldsymbol{\theta}]^2,$$

the following SG algorithm can be obtained for estimating the parameter vector $\boldsymbol{\theta}$:

$$\hat{\boldsymbol{\theta}}_k = \hat{\boldsymbol{\theta}}_{k-1} - \mu_1 \frac{\partial J(\boldsymbol{\theta})}{\partial \boldsymbol{\theta}} = \hat{\boldsymbol{\theta}}_{k-1} + \frac{\boldsymbol{\varphi}_k}{s_k} e_k, \tag{7}$$

$$e_k = y_k - \boldsymbol{\varphi}_k^\mathsf{T}\hat{\boldsymbol{\theta}}_{k-1}, \tag{8}$$

$$s_k = s_{k-1} + \|\boldsymbol{\varphi}_k\|^2. \tag{9}$$

where μ_1 is the step size for the SG algorithm, which is taken as $\mu_1 = \frac{1}{s_k}$, and $s_0 = 1$.

However, it is worth noting that the variables x_{k-i}, $\bar{u}_{k-i}$ and v_{k-i} in $\boldsymbol{\varphi}_k$ are unknown, and thus the algorithms in (7)–(9) cannot be implemented directly. The solution is to use the idea of the auxiliary model to build the following auxiliary models based on the parameter estimate $\hat{\boldsymbol{\theta}}_k$:

$$\hat{x}_k = \hat{\boldsymbol{\varphi}}_{s,k}^\mathsf{T}\hat{\boldsymbol{\theta}}_{s,k},$$

$$\hat{\bar{u}}_k = \hat{c}_{1,k}f_1(u_k) + \hat{c}_{2,k}f_2(u_k) + \cdots + \hat{c}_{m,k}f_m(u_k),$$

$$\hat{v}_k = y_k - \hat{\boldsymbol{\varphi}}_k^\mathsf{T}\hat{\boldsymbol{\theta}}_k,$$

and use the outputs $\hat{x}_{k-i}$, $\hat{\bar{u}}_{k-i}$ and $\hat{v}_{k-i}$ of the auxiliary models instead of the unknown variables x_{k-i}, $\bar{u}_{k-i}$ and v_{k-i} to construct the estimates of the information vectors:

$$\hat{\boldsymbol{\varphi}}_k = \begin{bmatrix} \hat{\boldsymbol{\varphi}}_{s,k} \\ \hat{\boldsymbol{\varphi}}_{n,k} \end{bmatrix},$$

$$\hat{\boldsymbol{\varphi}}_{s,k} = [-\hat{x}_{k-1}, -\hat{x}_{k-2}, \cdots, -\hat{x}_{k-n_a}, \hat{\bar{u}}_{k-1}, \hat{\bar{u}}_{k-2}, \cdots, \hat{\bar{u}}_{k-n_b}, f(u_k)]^\mathsf{T},$$

$$\hat{\boldsymbol{\varphi}}_{n,k} = [\hat{v}_{k-1}, \hat{v}_{k-2}, \cdots, \hat{v}_{k-n_d}]^\mathsf{T}.$$

The SG algorithm update the parameter estimate using the current data information, thus its computational complexity is low, but estimation accuracy needs to be improved. Based on the multi-innovation identification theory [44,45], a slide window of length p (i.e., innovation length) is built to improve the estimation performance of the SG algorithm, which contains the data information from the current time k to $k - p + 1$, i.e.,

$$E_{p,k} = [y_k - \hat{\boldsymbol{\varphi}}_k^\mathsf{T}\hat{\boldsymbol{\theta}}_{k-1}, y_{k-1} - \hat{\boldsymbol{\varphi}}_{k-1}^\mathsf{T}\hat{\boldsymbol{\theta}}_{k-2}, \cdots, y_{k-p+1} - \hat{\boldsymbol{\varphi}}_{k-p+1}^\mathsf{T}\hat{\boldsymbol{\theta}}_{k-p}]^\mathsf{T}. \tag{10}$$

Define the stacked output vector $Y_{p,k}$ and information matrix $\hat{\boldsymbol{\Phi}}_{p,k}$ as

$$Y_{p,k} := [y_k, y_{k-1}, \cdots, y_{k-p+1}]^\mathsf{T} \in \mathbb{R}^p,$$

$$\hat{\boldsymbol{\Phi}}_{p,k} := [\hat{\boldsymbol{\varphi}}_k, \hat{\boldsymbol{\varphi}}_{k-1}, \cdots, \hat{\boldsymbol{\varphi}}_{k-p+1}] \in \mathbb{R}^{n \times p}.$$

In principle, the estimate $\hat{\boldsymbol{\theta}}_{t-1}$ is closer to the optimal value $\boldsymbol{\theta}$ than $\hat{\boldsymbol{\theta}}_{t-i}$ for $i = 2, \cdots, p$, then Equation (10) can be approximated by

$$E_{p,k} = Y_{p,k} - \hat{\boldsymbol{\Phi}}_{p,k}^\mathsf{T}\hat{\boldsymbol{\theta}}_{k-1}.$$

In summary, we can obtain the AM-MISG algorithm as follows:

$$\hat{\theta}_k = \hat{\theta}_{k-1} + \frac{\Phi_{p,k}}{s_k} E_{p,k}, \tag{11}$$

$$E_{p,k} = Y_{p,k} - \Phi_{p,k}^\mathrm{T} \hat{\theta}_{k-1}, \tag{12}$$

$$s_k = s_{k-1} + \|\hat{\varphi}_k\|^2, \quad s_0 = 1, \tag{13}$$

$$Y_{p,k} = [y_k, y_{k-1}, \cdots, y_{k-p+1}]^\mathrm{T}, \tag{14}$$

$$\Phi_{p,k} = [\hat{\varphi}_k, \hat{\varphi}_{k-1}, \cdots, \hat{\varphi}_{k-p+1}], \tag{15}$$

$$\hat{u}_k = f(u_k)\hat{c}_k, \tag{16}$$

$$\hat{x}_k = \hat{\varphi}_{s,k}^\mathrm{T} \hat{\theta}_{s,k}, \tag{17}$$

$$\hat{v}_k = y_k - \hat{\varphi}_k^\mathrm{T} \hat{\theta}_k, \tag{18}$$

$$f(u_k) = [f_1(u_k), f_2(u_k), \cdots, f_m(u_k)], \tag{19}$$

$$\hat{\varphi}_k = \begin{bmatrix} \hat{\varphi}_{s,k} \\ \hat{\varphi}_{n,k} \end{bmatrix}, \tag{20}$$

$$\hat{\varphi}_{s,k} = [-\hat{x}_{k-1}, -\hat{x}_{k-2}, \cdots, -\hat{x}_{k-n_a}, \hat{u}_{k-1}, \hat{u}_{k-2}, \cdots, \hat{u}_{k-n_b}, f(u_k)]^\mathrm{T}, \tag{21}$$

$$\hat{\varphi}_{n,k} = [\hat{v}_{k-1}, \hat{v}_{k-2}, \cdots, \hat{v}_{k-n_d}]^\mathrm{T}, \tag{22}$$

$$\hat{\theta}_k = \begin{bmatrix} \hat{\theta}_{s,k} \\ \hat{d}_k \end{bmatrix}, \tag{23}$$

$$\hat{\theta}_{s,k} = [\hat{a}_k^\mathrm{T}, \hat{b}_k^\mathrm{T}, \hat{c}_k^\mathrm{T}]^\mathrm{T}. \tag{24}$$

Please note that the AM-MISG algorithm will reduce to the auxiliary model-based stochastic gradient (AM-SG) algorithm when $p=1$.

4. The AM-MIFSG Algorithm

This section deduces an AM-MIFSG algorithm to improve the parameter estimation performance of above AM-MISG identification algorithm.

In (7), the first-order gradient is used to update the parameter vector. In contrast to the integer order, for the quadratic objective function, the derivative of a fractional-order near a point is uncertain, so its essential property is nonlocal. This excellent property enables the fractional-order gradient method to jump out of local optimum and reach global minimum point more quicker. Here, we propose to add the fractional-order gradient in addition to the first-order gradient, and the final update relation is written as:

$$\hat{\theta}_k = \hat{\theta}_{k-1} - \mu_1 \frac{\partial J(\theta)}{\partial \theta} - \mu_\alpha \frac{\partial^\alpha J(\theta)}{\partial \theta}, \tag{25}$$

where μ_α is the step size for the factional order derivative ∂^α. According to the Caputo and Riemann–Liouville definition [46,47], the fractional derivation of a power function $f(t) = t^n$ ($n > -1$) is defined as:

$$D_t^\alpha t^n = \frac{\Gamma(n+1)}{\Gamma(n+1-\alpha)} t^{n-\alpha}, \tag{26}$$

where D_t^α is the fractional derivative operator of order α and Γ is the gamma function which defined as $\Gamma(n) = (n-1)!$.

According to (26), the fractional-order gradient in Equation (25) can be written as follows:

$$\frac{\partial^\alpha J(\theta)}{\partial \theta} = -\varphi_k \left(\frac{\partial^\alpha \theta}{\partial \theta} \right) = -\varphi_k \left(\frac{\Gamma(2)}{\Gamma(2-\alpha)} \theta^{1-\alpha} \right), \tag{27}$$

where $\Gamma(2) = 1$. Then Equation (25) can be approximated as follows:

$$\hat{\theta}_k = \hat{\theta}_{k-1} + \frac{\varphi_k}{s_k}e_k + \frac{\psi_k}{s_{\alpha,k}}e_k, \quad 0 < \alpha < 1, \tag{28}$$

$$s_{\alpha,k} = s_{\alpha,k-1} + \|\psi_k\|^2, \quad s_{\alpha,0} = 1, \tag{29}$$

$$\psi_k = \frac{\mathrm{diag}(\varphi_k)(|\theta|_{k-1}^{1-\alpha})}{\Gamma(2-\alpha)}. \tag{30}$$

Please note that the absolute value of θ is used to avoid complex values, this is a common way of dealing with fractional-order gradient [38]. The introduction of fractional-order parameter α provides additional degrees of freedom and increases the flexibility of the parameter estimation.

Similar to the AM-MISG algorithm in Section 3, expanding the information vector ψ_k to the information matrix

$$\Psi_{p,k} = [\psi_k, \psi_{k-1}, \cdots, \psi_{k-p+1}],$$

and applying the auxiliary model identification idea, we can obtain the following AM-MIFSG algorithm:

$$\hat{\theta}_k = \hat{\theta}_{k-1} + \left(\frac{\hat{\Phi}_{p,k}}{s_k} + \frac{\hat{\Psi}_{p,k}}{s_{\alpha,k}}\right)E_{p,k}, \tag{31}$$

$$E_{p,k} = Y_{p,k} - \hat{\Phi}_{p,k}^\mathrm{T}\hat{\theta}_{k-1}, \tag{32}$$

$$s_k = s_{k-1} + \|\hat{\varphi}_k\|^2, \quad s_0 = 1, \tag{33}$$

$$s_{\alpha,k} = s_{\alpha,k-1} + \|\hat{\psi}_k\|^2, \quad s_{\alpha,0} = 1, \tag{34}$$

$$Y_{p,k} = [y_k, y_{k-1}, \cdots, y_{k-p+1}]^\mathrm{T}, \tag{35}$$

$$\hat{\Phi}_{p,k} = [\hat{\varphi}_k, \hat{\varphi}_{k-1}, \cdots, \hat{\varphi}_{k-p+1}], \tag{36}$$

$$\hat{\Psi}_{p,k} = [\hat{\psi}_k, \hat{\psi}_{k-1}, \cdots, \hat{\psi}_{k-p+1}], \tag{37}$$

$$\hat{\psi}_j = \frac{\mathrm{diag}(\hat{\varphi}_j)(|\theta|_{k-1}^{1-\alpha})}{\Gamma(2-\alpha)}, \quad j = k, k-1, \cdots, k-p+1, \tag{38}$$

$$\hat{u}_k = f(u_k)\hat{c}_k, \tag{39}$$

$$\hat{x}_k = \hat{\varphi}_{s,k}^\mathrm{T}\hat{\theta}_{s,k}, \tag{40}$$

$$\hat{v}_k = y_k - \hat{\varphi}_k^\mathrm{T}\hat{\theta}_k, \tag{41}$$

$$f(u_k) = [f_1(u_k), f_2(u_k), \cdots, f_m(u_k)], \tag{42}$$

$$\hat{\varphi}_k = \begin{bmatrix} \hat{\varphi}_{s,k} \\ \hat{\varphi}_{n,k} \end{bmatrix}, \tag{43}$$

$$\hat{\varphi}_{s,k} = [-\hat{x}_{k-1}, -\hat{x}_{k-2}, \cdots, -\hat{x}_{k-n_a}, \hat{u}_{k-1}, \hat{u}_{k-2}, \cdots, \hat{u}_{k-n_b}, f(u_k)]^\mathrm{T}, \tag{44}$$

$$\hat{\varphi}_{n,k} = [\hat{v}_{k-1}, \hat{v}_{k-2}, \cdots, \hat{v}_{k-n_d}]^\mathrm{T}, \tag{45}$$

$$\hat{\theta}_k = \begin{bmatrix} \hat{\theta}_{s,k} \\ \hat{d}_k \end{bmatrix}, \tag{46}$$

$$\hat{\theta}_{s,k} = [\hat{a}_k^\mathrm{T}, \hat{b}_k^\mathrm{T}, \hat{c}_k^\mathrm{T}]^\mathrm{T}. \tag{47}$$

Here, the above AM-MIFSG algorithm reduces to the auxiliary model-based fractional stochastic gradient (AM-FSG) algorithm when $p = 1$.

Remark 1. *In general, as the innovation length p increases, the collected data are being used more fully, and therefore the estimation accuracy is gradually improved. However, the computational amount increases at the same time. How to choose optimal innovation p is an open problem to be solved. In practice, we often choose $p < n$.*

Remark 2. *The differential order α is chose in the range of (0,1). The orders may show different characteristics for different systems, and can be adjusted during the procedure as needed.*

The implementation of the AM-MIFSG algorithm is listed as follows.

1. Choose p, α and initialize: let $k = 1$, $\hat{\theta}_0 = \begin{bmatrix} \hat{\theta}_{s,0} \\ \hat{a}_0 \end{bmatrix} = 1_{n/p_0}$, $s_0 = 1$, $s_{\alpha,0} = 1$, and set $\hat{x}_i = 1/p_0$, $\hat{u}_i = 1/p_0$ and $\hat{v}_i = 1/p_0$ for $i \leqslant 0$, $p_0 = 10^6$, and give the base function $f_i(\cdot)$.
2. Collect the input-output data u_k and y_k, form the basis function vector $f(u_k)$ by (42), and the information vectors $\hat{\varphi}_k$ by (43), $\hat{\varphi}_{s,k}$ by (44) and $\hat{\varphi}_{n,k}$ by (45).
3. Compute $\hat{\psi}_j$ by (38). Form the stacked output vector $Y_{p,k}$ by (35), the information matrices $\hat{\Phi}_{p,k}$ and $\hat{\Psi}_{p,k}$ by (36) and (37).
4. Compute the innovation vector $E_{p,k}$ by (32), s_k by (33) and $s_{\alpha,k}$ by (34).
5. Update the parameter estimate $\hat{\theta}_k$ by (31), compute the estimates $\hat{u}_k$ by (39), $\hat{x}_k$ by (40), $\hat{v}_k$ by (41).
6. Increase k by 1, go to step 2.

The algorithm obtained above combined with the method in [48–53] can cope with linear and nonlinear systems with different disturbances. Furthermore, prediction models or soft sensor models can be obtained with the assistance of other parameter estimation algorithms [54–59] and can be applied to process control and other fields [60–65].

5. Convergence Analysis

Theorem 1. *For the system in (1)–(2) and the AM-MIFSG algorithm in (31)–(47), assume that the noise sequence $\{v_k\}$ satisfies*

(A1) $\mathrm{E}[v_k|\mathcal{F}_t] = 0$, a.s., $\mathrm{E}[v_k^2|\mathcal{F}_t] \leqslant \sigma^2 < \infty$, a.s.,

and there exist an integer N_k and a positive constant ϱ independent of k such that the following persistent excitation condition holds,

$$(A2) \sum_{i=0}^{N_k} \frac{\hat{\Phi}_{\alpha,p,k+i}^\top \hat{\Phi}_{\alpha,p,k+i}}{s_{k+i}} \geqslant \varrho I, \text{ a.s.,} \qquad (48)$$

where $\hat{\Phi}_{\alpha,p,k} = [\hat{\varphi}_k \odot \theta_\alpha, \hat{\varphi}_{k-1} \odot \theta_\alpha, \cdots, \hat{\varphi}_{k-p+1} \odot \theta_\alpha]$, $\theta_\alpha := 1_n + \hat{\theta}_{k-1}^{1-\alpha}$, $\odot$ denotes an element-by-element multiplication of vectors. Then the parameter estimation error given by the AM-MIFSG algorithm satisfies $\lim_{k\to\infty} \mathrm{E}[\|\hat{\theta}_k - \theta\|^2] \to 0$.

Proof. Define the parameter estimation error $\tilde{\theta}_k = \hat{\theta}_k - \theta \in \mathbb{R}^n$. To simplify the proof, assuming $s_{\alpha,k} = s_k/\Gamma(2-\alpha)$. Inserting (32) into (31) and rearranging, we have

$$\begin{aligned}
\tilde{\theta}_k &= \tilde{\theta}_{k-1} + \frac{\hat{\Phi}_{p,k}}{s_k}\left[Y_{p,k} - \hat{\Phi}_{p,k}^\top \hat{\theta}_{k-1}\right] \odot \theta_\alpha \\
&= \tilde{\theta}_{k-1} + \frac{\hat{\Phi}_{p,k}}{s_k}\left[\Phi_{p,k}^\top \theta_{k-1} + V_{p,k} - \hat{\Phi}_{p,k}^\top \hat{\theta}_{k-1}\right] \odot \theta_\alpha \\
&=: \tilde{\theta}_{k-1} + \frac{\hat{\Phi}_{p,k}}{s_k}\left[\mu_{p,k} - \varsigma_{p,k} + V_{p,k}\right] \odot \theta_\alpha, \qquad (49)
\end{aligned}$$

where

$$\begin{aligned}
\mu_{q,t} &:= [\Phi_{p,k} - \hat{\Phi}_{p,k}]^\top \theta \in \mathbb{R}^p, \quad \varsigma_{q,t} := \hat{\Phi}_{p,k}^\top \tilde{\theta}_{k-1} \in \mathbb{R}^p, \\
V_{p,k} &:= [v_k, v_{k-1}, \cdots, v_{k-p+1}] \in \mathbb{R}^p.
\end{aligned}$$

Pre-multiplying (49) by $\bar{\theta}_k^\mathsf{T}$ gives

$$\begin{aligned}\bar{\theta}_k^\mathsf{T}\bar{\theta}_k &= \bar{\theta}_{k-1}^\mathsf{T}\bar{\theta}_{k-1} + \frac{2}{s_k}\bar{\theta}_{k-1}^\mathsf{T}\hat{\Phi}_{\alpha,p,k}[\mu_{p,k} - \varsigma_{p,k} + V_{p,k}] \\ &+ \frac{1}{r_k^2}[\mu_{p,k} - \varsigma_{p,k} + V_{p,k}]^\mathsf{T}\hat{\Phi}_{\alpha,p,k}^\mathsf{T}\hat{\Phi}_{\alpha,p,k}[\mu_{p,k} - \varsigma_{p,k} + V_{p,k}].\end{aligned}$$

□

The rest can be proved in a similar to the way in [66].

6. Examples

Consider the following Hammerstein OEMA system:

$$\begin{aligned}y_k &= \frac{B(z)}{A(z)}\bar{u}_k + D(z)v_k,\\ A(z) &= 1 + a_1 z^{-1} + a_2 z^{-2} = 1 + 0.45 z^{-1} + 0.56 z^{-2},\\ B(z) &= 1 + b_1 z^{-1} + b_2 z^{-2} = 1 + 0.25 z^{-1} - 0.35 z^{-2},\\ D(z) &= 1 + d_1 z^{-1} = 1 - 0.54 z^{-1},\\ \bar{u}_k &= c_1 u_k + c_2 u_k^2 + c_3 u_k^3 = 0.52 u_k + 0.54 u_k^2 + 0.82 u_k^3,\\ \theta &= [a_1, a_2, b_1, b_2, c_1, c_2, c_3, d_1]^\mathsf{T} = [0.45, 0.56, 0.25, -0.35, 0.52, 0.54, 0.82, -0.54]^\mathsf{T}.\end{aligned}$$

In this example, the input $\{u_k\}$ is a persistently excited signal sequence and $\{v_k\}$ is a white noise sequence with zero mean and variances $\sigma^2 = 0.80^2$. The data length is taken as $L = 4000$, where the first 3500 samples are assigned for system identification and the remaining 500 samples are assigned for prediction and validation. The details are as follows.

1. Firstly, applying the AM-MISG algorithm and the AM-MIFSG algorithm with $\alpha = 0.94$ to estimate the parameters of considered system. Tables 1 and 2 show the parameter estimates and their errors with $p = 1, 2, 4$ and 6. Figures 2 and 3 indicate the parameter estimation errors $\delta := \|\hat{\theta}_k - \theta\|/\|\theta\|$ versus k.

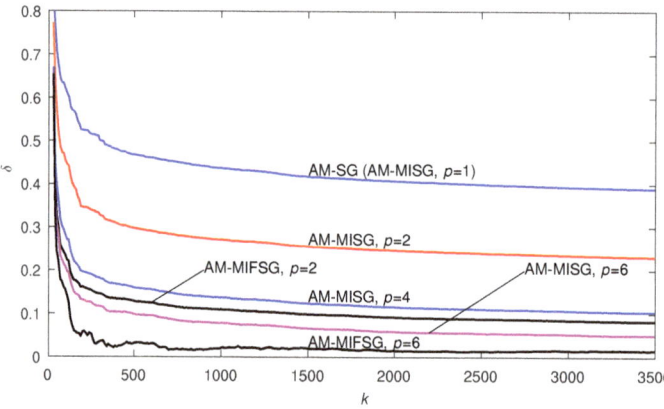

Figure 2. The AM-MISG estimation error δ versus k with $p = 1, 2, 4$ and 6 and the AM-MIFSG estimation error δ versus k with $p = 2$ and 6.

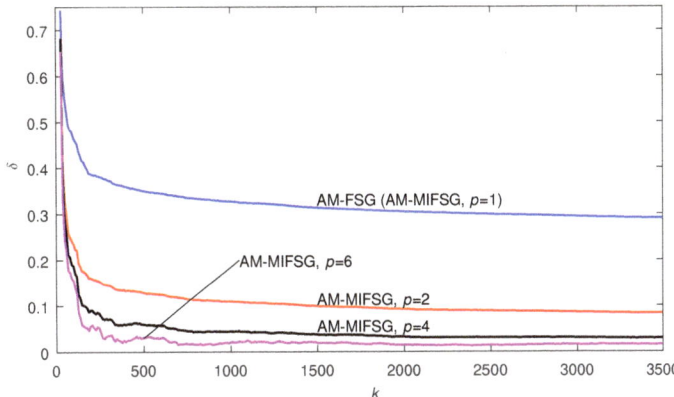

Figure 3. The AM-MIFSG estimation error δ versus k with $p = 1, 2, 4$ and 6.

Table 1. The AM-MISG estimates and errors $p = 1, 2, 4$ and 6.

p	k	a_1	a_2	b_1	b_2	c_1	c_2	c_3	d_1	$\delta(\%)$
1	100	0.03695	0.23553	−0.03327	−0.20875	0.27338	0.03195	0.49278	−0.31825	61.82507
	200	0.07113	0.34613	−0.03653	−0.29482	0.32261	0.07100	0.59507	−0.37276	52.41770
	500	0.10951	0.42150	−0.04098	−0.34040	0.35083	0.10812	0.64790	−0.39006	46.76773
	1000	0.13632	0.46020	−0.03759	−0.36082	0.36386	0.12852	0.67688	−0.39765	43.71611
	2000	0.16601	0.49396	−0.03164	−0.38052	0.37629	0.14894	0.70178	−0.40765	40.77833
	3000	0.18127	0.50703	−0.02857	−0.38630	0.38212	0.15952	0.71245	−0.41028	39.42179
2	100	0.08718	0.45055	−0.02857	−0.34926	0.38250	0.12367	0.64113	−0.44165	45.20563
	200	0.21252	0.53700	−0.00743	−0.40687	0.43140	0.19230	0.74170	−0.46339	34.63613
	500	0.27602	0.53381	0.01178	−0.39497	0.44676	0.23408	0.76740	−0.46163	29.78555
	1000	0.29539	0.53658	0.03287	−0.38875	0.45350	0.26006	0.78181	−0.45922	27.12588
	2000	0.31601	0.54589	0.05191	−0.39039	0.45935	0.28327	0.79322	−0.46058	24.67852
	3000	0.32385	0.54810	0.06104	−0.38751	0.46189	0.29484	0.79696	−0.46005	23.55543
4	100	0.27486	0.60968	0.03978	−0.39097	0.50049	0.23079	0.79164	−0.50729	28.18166
	200	0.37933	0.57741	0.09918	−0.37644	0.52086	0.30714	0.83586	−0.48517	19.67676
	500	0.38912	0.54069	0.13209	−0.34877	0.51837	0.35074	0.82239	−0.48274	16.01070
	1000	0.38933	0.54894	0.15511	−0.34580	0.51921	0.37894	0.82425	−0.48325	13.73325
	2000	0.40081	0.55745	0.17376	−0.34949	0.51943	0.40309	0.82417	−0.48612	11.58755
	3000	0.40202	0.55950	0.18102	−0.34717	0.51845	0.41433	0.82086	−0.48675	10.74226
6	100	0.35169	0.61755	0.08824	−0.35474	0.53581	0.30223	0.83886	−0.52654	20.81443
	200	0.40866	0.59002	0.16182	−0.34923	0.53654	0.37953	0.84693	−0.50426	13.13328
	500	0.42035	0.54952	0.18298	−0.32811	0.52821	0.42093	0.81930	−0.50377	9.82957
	1000	0.41823	0.56425	0.20204	−0.33346	0.53010	0.44667	0.82367	−0.50609	7.80786
	2000	0.42981	0.56613	0.21766	−0.34005	0.52962	0.46847	0.82213	−0.50962	5.89007
	3000	0.42678	0.56730	0.22265	−0.33911	0.52764	0.47733	0.81677	−0.51053	5.32836
True values		0.45000	0.56000	0.25000	−0.35000	0.52000	0.54000	0.82000	−0.54000	

Table 2. The AM-MIFSG estimates and errors with $p = 1, 2, 4$ and 6.

p	k	a_1	a_2	b_1	b_2	c_1	c_2	c_3	d_1	$\delta(\%)$
1	100	0.15708	0.59506	−0.10818	−0.36395	0.38627	0.08191	0.73420	−0.36577	46.47154
	200	0.27641	0.58549	−0.07338	−0.37468	0.42517	0.13842	0.81526	−0.37310	38.73219
	500	0.29432	0.55031	−0.03625	−0.35604	0.43184	0.17816	0.82178	−0.37342	34.99604
	1000	0.29755	0.54549	−0.01005	−0.35097	0.43470	0.20499	0.82736	−0.37378	32.70875
	2000	0.30878	0.55280	0.01095	−0.35501	0.43788	0.22916	0.83306	−0.37827	30.47638
	3000	0.31387	0.55453	0.02109	−0.35366	0.43918	0.24158	0.83382	−0.37952	29.40027
2	100	0.35829	0.62744	0.15273	−0.37663	0.55109	0.23357	0.78715	−0.59971	23.46638
	200	0.42746	0.58770	0.19005	−0.36598	0.56024	0.31456	0.81037	−0.55565	16.12428
	500	0.43121	0.55375	0.20459	−0.34652	0.55703	0.35970	0.79703	−0.54665	12.87485
	1000	0.42642	0.56004	0.21858	−0.34550	0.55880	0.38789	0.80138	−0.54408	10.92266
	2000	0.43305	0.56476	0.22928	−0.34967	0.55890	0.41208	0.80204	−0.54336	9.22414
	3000	0.43194	0.56535	0.23216	−0.34777	0.55770	0.42315	0.79911	−0.54236	8.52773
4	100	0.37614	0.62528	0.15435	−0.35937	0.56108	0.31419	0.83373	−0.62653	18.87062
	200	0.43032	0.60035	0.22302	−0.36272	0.55406	0.42250	0.82836	−0.56139	9.08948
	500	0.44845	0.55152	0.22616	−0.33881	0.54491	0.46286	0.79775	−0.55411	6.00591
	1000	0.44218	0.56900	0.23948	−0.34615	0.55133	0.48607	0.81308	−0.55361	4.43982
	2000	0.44975	0.56313	0.24931	−0.35057	0.55048	0.50632	0.81111	−0.55405	3.24848
	3000	0.44202	0.56473	0.25018	−0.34923	0.54786	0.51266	0.80477	−0.55341	3.01322
6	100	0.37154	0.63684	0.14344	−0.33325	0.54734	0.38404	0.84946	−0.62029	15.86854
	200	0.43128	0.61835	0.23888	−0.36493	0.53142	0.48850	0.83328	−0.55888	5.77068
	500	0.45536	0.55467	0.23160	−0.34012	0.52315	0.50947	0.79826	−0.55434	3.08071
	1000	0.44922	0.57123	0.24453	−0.35088	0.53625	0.52484	0.82763	−0.55500	2.04837
	2000	0.45296	0.55918	0.25378	−0.35400	0.53417	0.53985	0.82120	−0.55601	1.49579
	3000	0.44123	0.56369	0.25237	−0.35328	0.53073	0.54151	0.81199	−0.55532	1.53252
True values		0.45000	0.56000	0.25000	−0.35000	0.52000	0.54000	0.82000	−0.54000	

2. Secondly, to validate the influence of the fraction order α, in the AM-MIFSG algorithm, we take $p = 5$ and 6, and α =0.80, 0.90 and 0.92, respectively, the simulation results are shown in Tables 3 and 4, and Figures 4 and 5.

3. In the end, a different data set ($L_e = 500$ samples from $k = 3501$ to 4000) and the estimated model obtained by the AM-MIFSG algorithm with $p = 6$ and $\alpha = 0.92$ are used for model validation. The predicted output and true output are plotted in Figure 6 from $k = 3501$ to 3700 and Figure 7 from $k = 3501$ to 4000, where the average predicted output error is

$$\delta_e = \frac{1}{L_e}\left[\sum_{k=3501}^{4000}[\hat{y}_k - y_k]^2\right]^{1/2} = 0.0658,$$

and the dots line is the output $\hat{y}_k$ of the estimated model and the solid line is the true output y_k.

From Tables 1–4 and Figures 2–7, we can draw the following conclusions: (1) with the innovation length p increases, both the AM-MISG and the AM-MIFSG algorithm can give higher parameter estimation accuracy; (2) in general, the AM-MIFSG algorithm has a faster convergence rate than the AM-MISG algorithm in the same situation, and the introduction of the fractional-order can improve the parameter estimation accuracy; (3) the convergence rate of the AM-MIFSG increases as the fractional-order α increases, the α within the range of [0.90, 0.95] seems to be an appropriate choice which can give better estimation results for the Hammerstein output-error systems; (4) the estimated model obtained by the AM-MIFSG algorithm can well capture system dynamics.

Table 3. The AM-MIFSG estimates and errors with $\alpha = 0.80, 0.90$ and 0.92 ($p = 5$).

α	k	a_1	a_2	b_1	b_2	c_1	c_2	c_3	d_1	$\delta(\%)$
0.80	100	0.23581	0.59985	0.09673	−0.40126	0.35249	0.17777	0.82055	−0.48512	32.54342
	200	0.43614	0.59824	0.19561	−0.40423	0.39862	0.34638	0.91657	−0.45204	18.57503
	500	0.44156	0.53689	0.21507	−0.35179	0.39149	0.42275	0.86423	−0.46396	13.37312
	1000	0.43930	0.56210	0.23408	−0.35731	0.40462	0.46695	0.88603	−0.47323	11.19137
	2000	0.44866	0.55624	0.24854	−0.35742	0.40774	0.49743	0.88189	−0.48368	9.82294
	3000	0.43832	0.56284	0.25029	−0.35452	0.40630	0.50670	0.87138	−0.48769	9.37582
0.90	100	0.25965	0.58762	0.16306	−0.43464	0.41011	0.32565	0.75782	−0.62289	23.25864
	200	0.43757	0.58825	0.25570	−0.43466	0.45124	0.46639	0.85201	−0.55254	9.35233
	500	0.46161	0.53207	0.24689	−0.37626	0.45338	0.49568	0.83323	−0.55019	6.10290
	1000	0.45443	0.55626	0.25376	−0.37420	0.46646	0.51604	0.85868	−0.55042	5.04977
	2000	0.45629	0.55114	0.25927	−0.36918	0.46731	0.53246	0.85369	−0.55163	4.58180
	3000	0.44425	0.55646	0.25785	−0.36480	0.46517	0.53551	0.84496	−0.55137	4.29344
0.92	100	0.36347	0.63518	0.16075	−0.38255	0.49377	0.31352	0.84459	−0.60486	18.82927
	200	0.44017	0.60438	0.24641	−0.38609	0.49205	0.43973	0.85042	−0.53959	8.24557
	500	0.45824	0.54625	0.24047	−0.35090	0.48640	0.47823	0.81880	−0.53739	4.87871
	1000	0.45085	0.56619	0.25046	−0.35726	0.49783	0.50209	0.84203	−0.53918	3.35483
	2000	0.45508	0.55783	0.25789	−0.35824	0.49792	0.52152	0.83812	−0.54161	2.43473
	3000	0.44406	0.56191	0.25685	−0.35603	0.49533	0.52624	0.82969	−0.54185	2.13756
True values		0.45000	0.56000	0.25000	−0.35000	0.52000	0.54000	0.82000	−0.54000	

Table 4. The AM-MIFSG estimates and errors with $\alpha = 0.80, 0.90$ and 0.92 ($p = 6$).

α	k	a_1	a_2	b_1	b_2	c_1	c_2	c_3	d_1	$\delta(\%)$
0.80	100	0.26215	0.64436	0.06610	−0.35700	0.41270	0.26270	0.84456	−0.55007	27.25073
	200	0.42892	0.61731	0.19706	−0.36999	0.43708	0.42526	0.90034	−0.50010	12.53946
	500	0.44308	0.55154	0.21290	−0.33527	0.42696	0.47865	0.84239	−0.50720	8.39876
	1000	0.44126	0.56985	0.23379	−0.34728	0.44327	0.50887	0.87299	−0.51449	6.95024
	2000	0.44893	0.55908	0.24812	−0.35102	0.44420	0.53026	0.86479	−0.52163	6.06418
	3000	0.43731	0.56558	0.24875	−0.35051	0.44213	0.53393	0.85338	−0.52364	5.87107
0.90	100	0.32475	0.62263	0.16067	−0.40949	0.45832	0.35530	0.80222	−0.63261	18.70725
	200	0.44220	0.60427	0.26639	−0.41415	0.47360	0.48866	0.84739	−0.55477	7.38987
	500	0.46480	0.54124	0.24811	−0.36447	0.47263	0.50941	0.82077	−0.55289	4.30671
	1000	0.45622	0.56250	0.25474	−0.36674	0.48814	0.52662	0.85226	−0.55350	3.52333
	2000	0.45641	0.55356	0.25988	−0.36364	0.48753	0.54148	0.84452	−0.55464	3.17023
	3000	0.44318	0.55972	0.25723	−0.36073	0.48473	0.54283	0.83450	−0.55411	2.90188
0.92	100	0.38062	0.64556	0.16445	−0.36053	0.51634	0.33828	0.85954	−0.61558	17.41472
	200	0.44313	0.61605	0.25735	−0.38049	0.50254	0.46548	0.84460	−0.54615	6.92261
	500	0.46199	0.55073	0.24347	−0.34802	0.49615	0.49609	0.80967	−0.54393	3.60595
	1000	0.45377	0.56875	0.25282	−0.35650	0.51054	0.51672	0.83983	−0.54582	2.32078
	2000	0.45586	0.55746	0.25944	−0.35758	0.50944	0.53419	0.83334	−0.54795	1.60509
	3000	0.44334	0.56272	0.25702	−0.35604	0.50628	0.53692	0.82365	−0.54786	1.35745
True values		0.45000	0.56000	0.25000	−0.35000	0.52000	0.54000	0.82000	−0.54000	

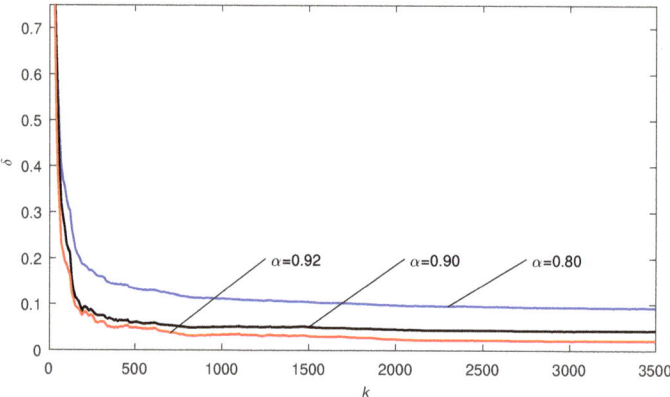

Figure 4. The AM-MIFSG estimation error δ versus k with $\alpha = 0.80, 0.90$ and 0.92 ($p = 5$).

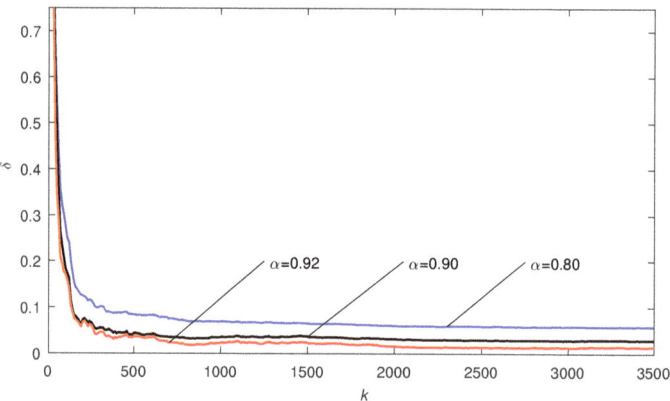

Figure 5. The AM-MIFSG estimation error δ versus k with $\alpha = 0.80, 0.90$ and 0.92 ($p = 6$).

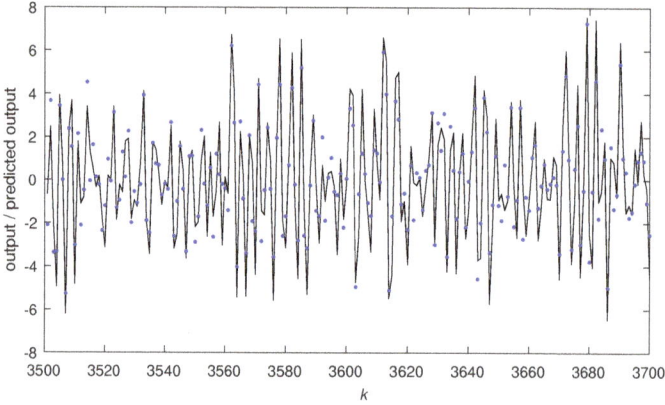

Solid line: the true output y_k, dots: the predicted output $\hat{y}_k$.

Figure 6. The predicted output $\hat{y}_k$ and true output y_k from $k = 3501$ to 3700.

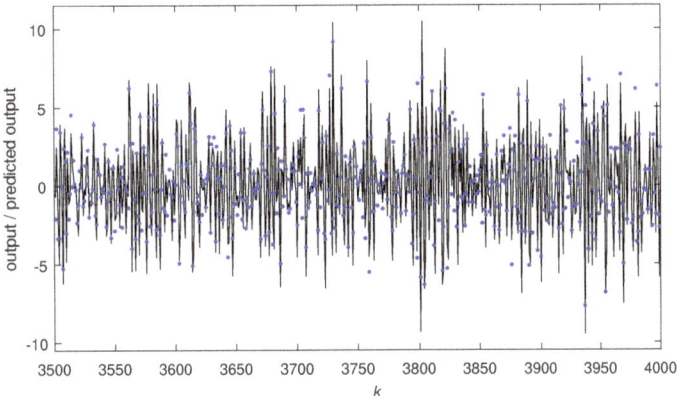

Solid line: the true output y_k, dots: the predicted output $\hat{y}_k$.

Figure 7. The predicted output $\hat{y}_k$ and true output y_k from $k = 3501$ to 4000.

7. Conclusions

This paper derives an AM-MIFSG estimation algorithm for Hammerstein output-error systems based on the key-term separation principle and auxiliary model identification idea. By means of the key-term separation principle, all the parameters in the linear and nonlinear blocks are separated, and the unknown variables in the identification model are replaced by the outputs of the auxiliary models. The analysis of the simulation results shows that the proposed algorithm obtains better parameter estimation performance than the AM-MISG algorithm. However, there also exist many topics that need to be further discussed. For example, is this algorithm still effective for systems with missing data? And is the performance of the algorithm can be improved by introducing a time-varying differential order α? These topics remain as open problems for future studies.

Author Contributions: Conceptualization, C.X.; methodology, Y.M.; software, C.X.; validation, C.X.; formal analysis, C.X.; investigation, C.X.; resources, Y.M.; data curation, Y.M.; writing-original draft preparation, Y.M.; writing—review and editing, C.X.; visualization, C.X.; supervision, Y.M.; project administration, Y.M.; funding acquisition, C.X. All authors have read and agreed to the published version of the manuscript.

Funding: This work was supported in part by the National Natural Science Foundation of China (No. 62103167), and in part by the Natural Science Foundation of Jiangsu Province (No. BK20210451), and the research project of Jiangnan University (Nos. JUSRP12028 and JUSRP12040).

Institutional Review Board Statement: Not applicable.

Informed Consent Statement: Not applicable.

Data Availability Statement: Not applicable.

Conflicts of Interest: The authors declare no conflict of interest.

References

1. Wang, L.J.; Guo, J.; Xu, C.; Wu, T.Z.; Lin, H.P. Hybrid model predictive control strategy of supercapacitor energy storage system based on double active bridge. *Energies* **2019**, *12*, 2134. [CrossRef]
2. Zhang, Y.; Yan, Z.; Zhou, C.C.; Wu, T.Z.; Wang, Y.Y. Capacity allocation of HESS in micro-grid based on ABC algorithm. *Int. J. Low-Carbon Technol.* **2020**, *15*, 496-505. [CrossRef]
3. Chen, H.T.; Jiang, B.; Chen, W.; Yi, H. Data-driven detection and diagnosis of incipient faults in electrical drives of high-speed trains. *IEEE Trans. Ind. Electron.* **2019**, *66*, 4716-4725. [CrossRef]
4. Chen, H.T.; Jiang, B.; Ding, S.X.; Huang, B. Data-driven fault diagnosis for traction systems in high-speed trains: A survey, challenges, and perspectives. *IEEE Trans. Intell. Transp. Syst.* **2020**. [CrossRef]

5. Ding, F.; Zhang, X.; Xu, L. The innovation algorithms for multivariable state-space models. *Int. J. Adapt. Control Signal Process.* **2019**, *33*, 1601–1608. [CrossRef]
6. Ding, F.; Liu, Y.J.; Bao, B. Gradient based and least squares based iterative estimation algorithms for multi-input multi-output systems. *Proc. Inst. Mech. Eng. Part I J. Syst. Control Eng.* **2012**, *226*, 43–55. [CrossRef]
7. Xu, L.; Song, G.L. A recursive parameter estimation algorithm for modeling signals with multi-frequencies. *Circuits Syst. Signal Process.* **2020**, *39*, 4198–4224. [CrossRef]
8. Xu, L.; Xiong, W.L.; Alsaedi, A.; Hayat, T. Hierarchical parameter estimation for the frequency response based on the dynamical window data. *Int. J. Control Autom. Syst.* **2018**, *16*, 1756–1764. [CrossRef]
9. Zhang, X.; Yang, E.F. Highly computationally efficient state filter based on the delta operator. *Int. J. Adapt. Control Signal Process.* **2019**, *33*, 875–889. [CrossRef]
10. Cuevas, E.; Díaz, P.; Avalos, O.; Zaldivar, D.; Pérez-Cisneros, M. Nonlinear system identification based on ANFIS-Hammerstein model using Gravitational search algorithm. *Appl. Intell.* **2018**, *48*, 182–203. [CrossRef]
11. Mukhopadhyay, S.; Mukherjee, A. ImdLMS: An imputation based LMS algorithm for linear system identification with missing input data. *IEEE Trans. Signal Process.* **2020**, *68*, 2370–2385. [CrossRef]
12. Sepulveda, N. E.; Sinha, J. Mathematical validation of experimentally optimised parameters used in a vibration-based machine-learning model for fault diagnosis in rotating machines. *Machines* **2021**, *9*, 155. [CrossRef]
13. Zhao, S.Y.; Yuriy, D.; Ahn, C.; Zhao, C.H. Probabilistic monitoring of correlated sensors for nonlinear processes in state space. *IEEE Trans. Ind. Electron.* **2020**, *67*, 2294–2303. [CrossRef]
14. Du, J.; Zhang, L.; Chen, J.; Li, J.; Jiang, X.; Zhu, C. Self-adjusted decomposition for multi-model predictive control of Hammerstein systems based on included angle. *ISA Trans.* **2020**, *103*, 19–27. [CrossRef] [PubMed]
15. Chen, H.T.; Jiang, B. A review of fault detection and diagnosis for the traction system in high-speed trains. *IEEE Trans. Intell. Transp. Syst.* **2020**, *21*, 450–465. [CrossRef]
16. Yang, M.; Wang, J.; Zhang, Y. Fault detection and diagnosis for plasticizing process of single-base gun propellant using mutual information weighted MPCA under limited batch samples modelling. *Machines* **2021**, *9*, 166. [CrossRef]
17. Chandra, S.; Hayashibe, M.; Thondiyath, A. Muscle fatigue induced hand tremor clustering in dynamic laparoscopic manipulation. *IEEE Trans. Syst. Man Cybern. -Syst.* **2020**, *50*, 5420–5431. [CrossRef]
18. Jalaleddini, K.; Kearney, R. E. Subspace identification of SISO Hammerstein systems: application to stretch reflex identification. *IEEE Trans. Biomed. Eng.* **2013**, *60*, 2725–2734. [CrossRef]
19. Ding, F.; Chen, H.; Xu, L.; Dai, J.; Li, Q.; Hayat, T. A hierarchical least squares identification algorithm for Hammerstein nonlinear systems using the key term separation. *J. Frankl. Inst.* **2018**, *355*, 3737–3752. [CrossRef]
20. Ding, J.; Cao, Z.; Chen, J.; Jiang, G. Weighted parameter estimation for Hammerstein nonlinear ARX systems. *Circuits Syst. Signal Process.* **2020**, *39*, 2178–2192. [CrossRef]
21. Kazemi, M.; Arefi, M. M. A fast iterative recursive least squares algorithm for Wiener model identification of highly nonlinear systems. *ISA Trans.* **2017**, *67*, 382–388. [CrossRef]
22. Hou, J.; Liu, T.; Wang, Q. G. Subspace identification of Hammerstein-type nonlinear systems subject to unknown periodic disturbance. *Int. J. Control* **2021**, *94*, 849–859. [CrossRef]
23. Wang, L.; Ji, Y.; Wan, L.; Bu, N. Hierarchical recursive generalized extended least squares estimation algorithms for a class of nonlinear stochastic systems with colored noise. *J. Frankl. Inst.* **2019**, *356*, 10102–10122. [CrossRef]
24. Wan, L.; Ding, F. Decomposition-and gradient-based iterative identification algorithms for multivariable systems using the multi-innovation theory. *Circuits Syst. Signal Process.* **2019**, *38*, 2971–2991. [CrossRef]
25. Loizou, N.; Richtárik, P. Momentum and stochastic momentum for stochastic gradient, newton, proximal point and subspace descent methods. *Comput. Optim. Appl.* **2020**, *77*, 653–710. [CrossRef]
26. Pan, J.; Jiang, X.; Wan, X.; Ding, W. A filtering based multi-innovation extended stochastic gradient algorithm for multivariable control systems. *Int. J. Control Autom. Syst.* **2017**, *15*, 1189–1197. [CrossRef]
27. Ma, H.; Pan, J.; Lv, L... Recursive algorithms for multivariable output-error-like ARMA systems. *Mathematics* **2019**, *7*, 558. [CrossRef]
28. Pan, J.; Ma, H.; Zhang, X. Recursive coupled projection algorithms for multivariable output-error-like systems with coloured noises. *IET Signal Process.* **2020**, *14*, 455–466. [CrossRef]
29. Ma, H.; Zhang, X.; Liu, Q.Y. Hayat, T. Partially-coupled gradient-based iterative algorithms for multivariable output-error-like systems with autoregressive moving average noises. *IET Contr. Theory Appl.* **2020**, *14*, 2613–2627. [CrossRef]
30. Khan, S.; Ahmad, J.; Naseem, I.; Moinuddin, M. A novel fractional gradient-based learning algorithm for recurrent neural networks. *Circuits Syst. Signal Process.* **2018**, *37*, 593–612. [CrossRef]
31. Ding, F.; Chen, T. Performance analysis of multi-innovation gradient type identification methods. *Automatica* **2007**, *43*, 1–14. [CrossRef]
32. Chaudhary, N. I.; Raja, M. A. Z. Identification of Hammerstein nonlinear ARMAX systems using nonlinear adaptive algorithms. *Nonlinear Dyn.* **2015**, *79*, 1385–1397. [CrossRef]
33. Chaudhary, N. I.; Raja, M. A. Z. Design of fractional adaptive strategy for input nonlinear Box-Jenkins systems. *Signal Process.* **2015**, *116*, 141–151. [CrossRef]
34. Wang, Y.; Li, M.; Chen, Z. Experimental study of fractional-order models for lithium-ion battery and ultra-capacitor: Modeling, system identification, and validation. *Appl. Energy* **2020**, *278*, 115736. [CrossRef]

35. Aslam, M. S.; Chaudhary, N. I.; Raja, M. A. Z. A sliding-window approximation-based fractional adaptive strategy for Hammerstein nonlinear ARMAX systems. *Nonlinear Dyn.* **2017**, *87*, 519–533. [CrossRef]
36. Chaudhary, N. I.; Raja, M. A. Z.; Khan, A. U. R. Design of modified fractional adaptive strategies for Hammerstein nonlinear control autoregressive systems. *Nonlinear Dyn.* **2015**, *82*, 1811–1830. [CrossRef]
37. Cheng, S.; Wei, Y.; Sheng, D.; Chen, Y.; Wang, Y. Identification for Hammerstein nonlinear ARMAX systems based on multi-innovation fractional order stochastic gradient. *Signal Process.* **2018**, *142*, 1–10. [CrossRef]
38. Chaudhary, N. I.; Raja, M. A. Z.; He, Y.; Khan, Z. A.; Machado, J. T. Design of multi innovation fractional LMS algorithm for parameter estimation of input nonlinear control autoregressive systems. *Appl. Math. Model.* **2021**, *93*, 412–425. [CrossRef]
39. Ding, F.; Shi, Y.; Chen, T. Auxiliary model-based least-squares identification methods for Hammerstein output-error systems. *Syst. Control Lett.* **2007**, *56*, 373–380. [CrossRef]
40. Zhang, Q. Nonlinear system identification with output error model through stabilized simulation. *IFAC Proc. Vol.* **2004**, *37*, 501–506. [CrossRef]
41. Vörös, J. Parameter identification of discontinuous Hammerstein systems. *Automatica* **1997**, *33*, 1141–1146. [CrossRef]
42. Vörös, J. Recursive identification of Hammerstein systems with discontinuous nonlinearities containing dead-zones. *IEEE Trans. Autom. Control* **2003**, *48*, 2203–2206. [CrossRef]
43. Vörös, J. Identification of nonlinear cascade systems with output hysteresis based on the key term separation principle. *Appl. Math. Model.* **2015**, *39*, 5531–5539. [CrossRef]
44. Mao, Y.; Ding, F.; Yang, E. Adaptive filtering based multi-innovation gradient algorithm for input nonlinear systems with autoregressive noise. *Int. J. Adapt. Control Signal Process.* **2017**, *31*, 1388–1400. [CrossRef]
45. Liu, Y.; Yu, L.; Ding, F. Multi-innovation extended stochastic gradient algorithm and its performance analysis. *Circuits Syst. Signal Process.* **2010**, *29*, 649–667. [CrossRef]
46. Shah, S. M. Riemann-Liouville operator-based fractional normalised least mean square algorithm with application to decision feedback equalisation of multipath channels. *IET Signal Process.* **2016**, *10*, 575–582. [CrossRef]
47. Li, C.; Qian, D.; Chen, Y. On Riemann-Liouville and caputo derivatives. In *Discrete Dynamics in Nature and Society*; Hindawi Limited: London, UK, 2011.
48. Li, M.H.; Liu, X.M. Maximum likelihood least squares based iterative estimation for a class of bilinear systems using the data filtering technique. *Int. J. Control Autom. Syst.* **2020**, *18*, 1581–1592. [CrossRef]
49. Ding, F.; Xu, L.; Meng, D.D. Gradient estimation algorithms for the parameter identification of bilinear systems using the auxiliary model. *J. Comput. Appl. Math.* **2020**, *369*, 112575. [CrossRef]
50. Li, M.H.; Liu, X.M. Maximum likelihood hierarchical least squares-based iterative identification for dual-rate stochastic systems. *Int. J. Adapt. Control Signal Process.* **2021**, *35*, 240–261. [CrossRef]
51. Xu, L.; Chen, F.Y.; Hayat, T. Hierarchical recursive signal modeling for multi-frequency signals based on discrete measured data. *Int. J. Adapt. Control Signal Process.* **2021**, *35*, 676–693. [CrossRef]
52. Ji, Y.; Kang, Z.; Zhang, C. Two-stage gradient-based recursive estimation for nonlinear models by using the data filtering. *Int. J. Control Autom. Syst.* **2021**, *19*, 2706–2715. [CrossRef]
53. Xu, L.; Yang, E.F. Auxiliary model multiinnovation stochastic gradient parameter estimation methods for nonlinear sandwich systems. *Int. J. Robust Nonlinear Control* **2021**, *31*, 148–165. [CrossRef]
54. Wang JW, Ji Y, Zhang C. Iterative parameter and order identification for fractional-order nonlinear finite impulse response systems using the key term separation. *Int. J. Adapt. Control Signal Process.* **2021**, *35*, 1562–1577. [CrossRef]
55. Zhang, X. Adaptive parameter estimation for a general dynamical system with unknown states. *Int. J. Robust Nonlinear Control* **2020**, *30*, 1351–1372. [CrossRef]
56. Zhang, X. Recursive parameter estimation methods and convergence analysis for a special class of nonlinear systems. *Int. J. Robust Nonlinear Control* **2020**, *30*, 1373–1393. [CrossRef]
57. Mao, Y.W.; Liu, S.; Liu, J.F. Robust economic model predictive control of nonlinear networked control systems with communication delays. *Int. J. Adapt. Control Signal Process.* **2020**, *34*, 614–637. [CrossRef]
58. Ding, F. State filtering and parameter estimation for state space systems with scarce measurements. *Signal Process.* **2014**, *104*, 369–380. [CrossRef]
59. Ding, F. Combined state and least squares parameter estimation algorithms for dynamic systems. *Appl. Math. Modell.* **2014**, *38*, 403–412. [CrossRef]
60. Li, M.H.; Liu, X.M. Iterative identification methods for a class of bilinear systems by using the particle filtering technique. *Int. J. Adapt. Control Signal Process.* **2021**, *35*, 2056–2074. [CrossRef]
61. Zhao, S.Y.; Huang, B.; Liu F. Linear optimal unbiased filter for time-variant systems without apriori information on initial conditions. *IEEE Trans. Autom. Control* **2017**, *62*, 882–887. [CrossRef]
62. Liu, Y.J.; Shi, Y. An efficient hierarchical identification method for general dual-rate sampled-data systems. *Automatica* **2014**, *50*, 962–970. [CrossRef]
63. Ding, F.; Qiu, L.; Chen, T. Reconstruction of continuous-time systems from their non-uniformly sampled discrete-time systems. *Automatica* **2009**, *45*, 324–332. [CrossRef]
64. Zhao, S.Y.; Yuriy, S.; Ahn, C.; Liu F. Adaptive-horizon iterative UFIR filtering algorithm with applications. *IEEE Trans. Ind. Electron.* **2018**, *65*, 6393–6402. [CrossRef]

65. Zhao, S.Y.; Huang, B. Trial-and-error or avoiding a guess? Initialization of the Kalman filter. *Automatica* **2020**, *121*, 109184. [CrossRef]
66. Ding, F.; Liu, G.; Liu, X.P. Parameter estimation with scarce measurements. *Automatica* **2011**, *47*, 1646–1655. [CrossRef]

Article

A Process Monitoring Method Based on Dynamic Autoregressive Latent Variable Model and Its Application in the Sintering Process of Ternary Cathode Materials

Ning Chen [1], Fuhai Hu [1], Jiayao Chen [1,*], Zhiwen Chen [1,2,*], Weihua Gui [1] and Xu Li [3]

[1] School of Automation, Central South University, Changsha 410083, China; ningchen@csu.edu.cn (N.C.); FuhaiHu@csu.edu.cn (F.H.); gwh@csu.edu.cn (W.G.)
[2] State Key Laboratory of High Performance Complex Manufacturing, Central South University, Changsha 410083, China
[3] Hunan Shanshan Energy Technology Co., Ltd., Changsha 410205, China; li.bo@shanshanenergy.com
* Correspondence: jychen@csu.edu.cn (J.C.); zhiwen.chen@csu.edu.cn (Z.C.)

Abstract: Due to the ubiquitous dynamics of industrial processes, the variable time lag raises great challenge to the high-precision industrial process monitoring. To this end, a process monitoring method based on the dynamic autoregressive latent variable model is proposed in this paper. First, from the perspective of process data, a dynamic autoregressive latent variable model (DALM) with process variables as input and quality variables as output is constructed to adapt to the variable time lag characteristic. In addition, a fusion Bayesian filtering, smoothing and expectation maximization algorithm is used to identify model parameters. Then, the process monitoring method based on DALM is constructed, in which the process data are filtered online to obtain the latent space distribution of the current state, and T^2 statistics are constructed. Finally, by comparing with an existing method, the feasibility and effectiveness of the proposed method is tested on the sintering process of ternary cathode materials. Detailed comparisons show the superiority of the proposed method.

Keywords: process monitoring; dynamics; variable time lag; dynamic autoregressive latent variables model; sintering process

1. Introduction

To ensure production safety and product quality, process monitoring technology has become an indispensable ingredient for industrial processes in recent years. It is commonly divided into model-based methods and data-driven methods. Compared with the former ones, the later ones can take advantage of the routine measurement and do not rely on process prior knowledge and precise mechanism models, which are unavailable or cost-intensive to obtained at times [1,2]. Therefore, they are widely used in modern industrial process.

During the past decades, many data-driven process monitoring methods have been published [3–6]. Kim et al. [7] proposed a probabilistic PCA to monitoring industrial processes, which firstly extracts redundant information from the variables and constructs feature distribution for monitoring, but it only extracts features of the input space. Zhao et al. [8] proposed the probabilistic PLSR process monitoring method to monitor quality-related faults, which can simultaneously consider the fault characteristics of the input and output spaces for monitoring. Furthermore, Chen et al. [9] proposed a probability-related PCA method for detecting incipient faults, which can greatly improve the detection ability of minor faults. Probabilistic framework modeling can overcome process noise [10]. However, the process monitoring methods currently proposed are all static methods, and the actual production processes are dynamical featured with variable time lag [11,12].

Process dynamics could refer to the mutual influence before and after the current sampling [13]. To deal with the dynamics of process, Ku et al. [14] built an augmented matrix and extended the static PCA model to the dynamic PCA (DPCA) for process monitoring. However, the introduction of augmented matrix increases the parameter dimensions called the curse of dimensionality [15]. Motivated by DPCA, Li et al. [16] proposed a dynamic latent variable model for monitoring the Tennessee Eastman process. In this model, the autoregressive model is used to extract data dynamic information, and PCA is performed to reduce redundancy between variables. It divides variable order reduction and dynamic information extraction into two stages, which makes the system complex and not easy to tune. In addition, compared with process variables, quality variables also contain useful fault information [17]. For this reason, Ge et al. [18] proposed a supervised linear dynamic system model process monitoring method. This method uses a first-order autoregressive equation to simulate the first-order dynamic [19] but does not take the variable time lag into account.

Variable time lag refers to the delay between the effects of variables [20]. The existing monitoring methods considering time lag are usually divided into two categories [21]. One is to find the time lag between variables and translate the data to eliminate the time lag and then establish a static process monitoring model for the processed data. For example, Wang et al. [22] proposed a spatial reconstruction method to identify system time lag, then aligned the data and established a monitoring model, but the alignment operation will destroy the data structure and cause data loss. The other idea is to use time lag as an unknown parameter of the process monitoring model and identify the parameters through a data-driven method. For example, Huber et al. [23] proposed to take the time lag as a parameter of a high-order state space system model and then solve it uniformly with the model parameters, but this method relies on the setting of the time lag parameter and the parameter identification method.

From the above discussions, it can be observed that the variable time lag characteristic of a process makes the previous work unfavorable. However, this characteristic is common in industrial processes [24,25]. To deal with this problem, this paper proposes a process monitoring method based on a dynamic autoregressive latent variable model. Firstly, from the data point of view, a linear dynamic model is constructed between process variables and quality variables, and the dynamic information of process input and process output is compressed to latent variables, and then a dynamic autoregressive latent variable model (DALM) is constructed for latent variables to extract variable time lag information. In addition, a fusion Bayesian filtering, smoothing and expectation maximization algorithm is used to identify model parameters. Then, the DALM is applied to the industrial monitoring process. The process variables are filtered through improved Bayesian filtering technology to obtain the latent space distribution of the current state, and the T^2 statistics of the latent space are constructed and monitored [26] to realize the process monitoring task. The main contribution can be concluded as (1) a process monitoring method based on dynamic autoregressive latent variable model is proposed in this paper; (2) a dynamic autoregressive latent variable model (DALM) is developed to extract variable time lag information; (3) a fusion Bayesian filtering, smoothing and expectation maximization algorithm is improved to identify model parameters; (4) based on the DALM, the T^2 statistics of the latent space are constructed to realize the process monitoring task.

The main structure of the paper is arranged as follows. In the second section, a dynamic autoregressive latent variable model is proposed, and the parameter identification algorithm of the model is derived in detail. A process monitoring method based on DALM is proposed in the third chapter. The fourth section uses the monitoring method to monitor the sintering process of the ternary cathode material to verify the monitoring performance of the proposed method. Finally, the last section concludes.

2. Modeling Method Based on Dynamic Autoregressive Latent Variable Model

This section proposes a dynamic autoregressive latent variable modeling method for the accurate modeling of dynamical industrial processes. Bayesian filtering and smoothing inference were used to obtain the spatial distribution of latent variables, and the parameters of the model were identified by combining with the EM algorithm.

2.1. Dynamic Autoregressive Latent Variable Model Structure

In order to consider the dynamic characteristics of the process, the traditional probability latent variable model [27] establishes the relationship between the current moment and the previous moment data, as shown in (1).

$$\begin{aligned} \mathbf{z}_t &= \mathbf{A}\mathbf{z}_{t-1} + \boldsymbol{\eta}_t^z, \\ \mathbf{x}_t &= \mathbf{B}_\mathbf{x}\mathbf{z}_t + \boldsymbol{\eta}_t^x, \\ \mathbf{y}_t &= \mathbf{B}_\mathbf{y}\mathbf{z}_t + \boldsymbol{\eta}_t^y, \end{aligned} \quad (1)$$

where the structure consists of a linear Gaussian dynamic equation and two linear Gaussian observation equations, $\mathbf{z}_t$ is the latent variable of the process state at time t; $\mathbf{x}_t$ and $\mathbf{y}_t$ are the process variable and quality variable at time t, respectively; $\mathbf{A}, \mathbf{B}_\mathbf{x}$ and $\mathbf{B}_\mathbf{y}$ are their own load matrix. The Gaussian dynamic equation is used to describe the dynamic relationship of the process data. The observation equation compresses the information of the process data into low-dimensional latent variables. Therefore, an accurate mathematical model can be established for the dynamic process, but the structure does not consider the time lag characteristics of the process.

In order to further consider the characteristics of process time lag, on the basis of dynamic probabilistic latent variable model (DPLVM) [13], the trend similarity analysis algorithm [22] was first used to obtain the time lag coefficient L of the current process, and then an autoregressive equation was constructed for the latent variables to describe the variable time lag information. Among them, the autoregressive equation models the process dynamics and time lag characteristics, and the linear observation equation models the cross-correlation of data. The probability graph model of the model is shown in Figure 1, and the mathematical expression is shown in (2).

$$\begin{aligned} \mathbf{z}_t &= \mathbf{A}\mathbf{h}_{t-1} + \boldsymbol{\eta}_t^z, \\ \mathbf{x}_t &= \mathbf{B}_\mathbf{x}\mathbf{z}_t + \boldsymbol{\eta}_t^x, \\ \mathbf{y}_t &= \mathbf{B}_\mathbf{y}\mathbf{z}_t + \boldsymbol{\eta}_t^y, \end{aligned} \quad (2)$$

where $\mathbf{z}_t \in R^d$ represents the latent variable of the process state at time t, $\mathbf{h}_{t-1} = \begin{bmatrix} \mathbf{z}_{t-1}^T & \mathbf{z}_{t-2}^T & \cdots & \mathbf{z}_{t-L}^T \end{bmatrix}^T \in R^{dL}$ is the augmented state variable containing the latent variables at time L in the past, $\mathbf{x}_t \in R^v$ is the observed value of the process variable at time t, $\mathbf{y}_t \in R^k$ is the observed value of the quality variable at time t and d, v, k, respectively, correspond to the dimensions of latent variables, process variables and quality variables. $\mathbf{A} \in R^{d \times dL}$ is the state transition matrix, L is the time lag value of the process, and $\mathbf{B}_\mathbf{x} \in R^{v \times d}$, $\mathbf{B}_\mathbf{y} \in R^{k \times d}$ are the state divergence matrices. $\boldsymbol{\eta}_t^z \in R^d$, $\boldsymbol{\eta}_t^x \in R^v$, and $\boldsymbol{\eta}_t^y \in R^k$ are Gaussian noise terms of latent variables, process variables and quality variables, respectively. Assuming that the noises are independent of each other, the distributions obeyed are $\boldsymbol{\eta}_t^z \sim N(0, \Sigma_\mathbf{z})$, $\boldsymbol{\eta}_t^x \sim N(0, \Sigma_\mathbf{x})$ and $\boldsymbol{\eta}_t^y \sim N(0, \Sigma_\mathbf{y})$, respectively. Latent variables represent the current state of the process, and this model is an extension of the traditional DPLVM.

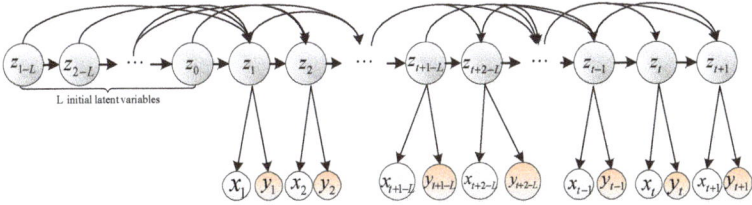

Figure 1. Probability graph model of dynamic autoregressive latent variable model.

2.2. Parameter Identification Based on EM Algorithm

Since only $\mathbf{x}_t$ and $\mathbf{y}_t$ can be observed in the process, and latent variables are abstracted to describe the state of the process and are unobservable, the EM algorithm was used to identify the parameters of the model [28]. Each iteration of the EM algorithm consisted of two steps: E step, seeking expectation (exception); M step, seeking maximization (maximization). This section uses Bayesian filtering and smoothing to infer the spatial distribution of latent variables, so as to solve the difficult problem of calculating latent variable statistics.

Under the framework of probability, the model assumed that the latent variables at the initial moment obeyed a Gaussian distribution with mean $\mathbf{u}_0$ and variance $\mathbf{V}_0$, that is, $\mathbf{z}_0, \mathbf{z}_{-1}, \cdots, \mathbf{z}_{1-L} \sim N(\mathbf{u}_0, \mathbf{V}_0)$. From the knowledge of probability theory [16], it is easy to get that the distribution of the latent variable $\mathbf{z}_t$, the process variable $\mathbf{x}_t$ and the quality variable $\mathbf{y}_t$ obey the Gaussian distribution, as shown in (3).

$$\begin{aligned}
\mathbf{z}_t|\mathbf{z}_{t-1}, \mathbf{z}_{t-2}, \cdots, \mathbf{z}_{t-L} &\sim N(\mathbf{A}_1\mathbf{z}_{t-1} + \mathbf{A}_2\mathbf{z}_{t-2} + \cdots + \mathbf{A}_L\mathbf{z}_{t-L}, \Sigma_\mathbf{z}), \\
\mathbf{x}_t|\mathbf{z}_t &\sim N(\mathbf{B}_\mathbf{x}\mathbf{z}_t, \Sigma_\mathbf{x}), \\
\mathbf{y}_t|\mathbf{z}_t &\sim N(\mathbf{B}_\mathbf{y}\mathbf{z}_t, \Sigma_\mathbf{y}).
\end{aligned} \tag{3}$$

The parameters that needed to be identified were denoted as $\Theta = \{\mathbf{A}, \mathbf{B}_\mathbf{x}, \mathbf{B}_\mathbf{y}, \mathbf{u}_0, \mathbf{V}_0, \Sigma_\mathbf{z}, \Sigma_\mathbf{x}, \Sigma_\mathbf{y}\}$, of which $\mathbf{A} = [\mathbf{A}_1, \mathbf{A}_2, \cdots, \mathbf{A}_L]$. According to the naive what you see is what you get thought, the parameter identification problem was transformed into the maximum observation data $\mathbf{x}_{1:T}, \mathbf{y}_{1:T}$. The log-likelihood function on the parameter Θ is shown in (4), where $\mathbf{x}_{1:T}$ represents the observation sequence of the process variable $\mathbf{x}_1, \mathbf{x}_2, \cdots, \mathbf{x}_T, \mathbf{y}_{1:T}$ represents the observation sequence $\mathbf{y}_1, \mathbf{y}_2, \cdots, \mathbf{y}_T$ of the quality variable, where T is the total number of training samples, namely,

$$\Theta^{new} = \arg\max_{\Theta} \log P(\mathbf{x}_{1:T}, \mathbf{y}_{1:T}|\Theta). \tag{4}$$

The EM algorithm [29] was used to solve the optimization problem of Equation (4). In the E step of the EM algorithm, the log-likelihood function $\log P(\mathbf{x}_{1:T}, \mathbf{y}_{1:T}, \mathbf{z}_{1-L:T}|\Theta)$ of the complete data had to be calculated with respect to the conditional expectation of the latent variable $\mathbf{z}_{1-L:T}$ to obtain the objective cost function (Q function), as shown in (5).

$$Q\left(\Theta|\Theta^{old}\right) = E_{\mathbf{z}_{1-L:T}|(\mathbf{x}_{1:T}, \mathbf{y}_{1:T}, \Theta^{old})}\{\log P(\mathbf{x}_{1:T}, \mathbf{y}_{1:T}, \mathbf{z}_{1-L:T}|\Theta)\}. \tag{5}$$

Actually, the likelihood function can be formulated by the application of the product rule of probability. From the model structure, the log-likelihood function of the complete data was expanded, as expressed by (6).

$$\begin{aligned}
&\log P\left(\mathbf{x}_{1:T}, \mathbf{y}_{1:T}, \mathbf{z}_{(-L+1):T}|\Theta\right) \\
&= \log\left\{P(\mathbf{z}_0, \mathbf{z}_{-1}, \cdots, \mathbf{z}_{-L+1})\prod_{t=1}^T P(\mathbf{z}_t|\mathbf{z}_{t-1}, \mathbf{z}_{t-2}, \cdots, \mathbf{z}_{t-L})P(\mathbf{x}_t|\mathbf{z}_t)P(\mathbf{y}_t|\mathbf{z}_t)\right\} \\
&= \log P(\mathbf{z}_0, \mathbf{z}_{-1}, \cdots, \mathbf{z}_{-L+1}) + \sum_{t=1}^T \log P(\mathbf{z}_t|\mathbf{z}_{t-1}, \mathbf{z}_{t-2}, \cdots, \mathbf{z}_{t-L}) + \sum_{t=1}^T \log P(\mathbf{x}_t|\mathbf{z}_t) \\
&\quad + \sum_{t=1}^T \log P(\mathbf{y}_t|\mathbf{z}_t).
\end{aligned} \tag{6}$$

For clear writing, we denote $E_{\mathbf{z}_T}(f(\mathbf{x}_t, \mathbf{y}_t, \mathbf{z}_t)) = E_{\mathbf{z}_{1-L:T}|(\mathbf{x}_{1:T}, \mathbf{y}_{1:T}, \Theta^{old})}(f(\mathbf{x}_t, \mathbf{y}_t, \mathbf{z}_t))$, then the expectation of the complete data likelihood function $\log P(\mathbf{x}_{1:T}, \mathbf{y}_{1:T}, \mathbf{z}_{(-L+1):T}|\Theta)$ with respect to the latent variable distribution $P(\mathbf{z}_{(-L+1):T}|\mathbf{x}_{1:T}, \mathbf{y}_{1:T}, \Theta^{old})$ is shown in (7).

$$
\begin{aligned}
Q\left(\Theta|\Theta^{old}\right) &= E_{\mathbf{z}_T}\{\log p(\mathbf{x}_{1:T}, \mathbf{y}_{1:T}, \mathbf{z}_{1-L:T}|\Theta)\} \\
&= -\frac{1}{2}\left\{\log|\mathbf{V}_0| + E_{\mathbf{z}_T}\left(\begin{bmatrix}\mathbf{z}_0\\\vdots\\\mathbf{z}_{-L+1}\end{bmatrix}^T \mathbf{V}_0^{-1}\begin{bmatrix}\mathbf{z}_0\\\vdots\\\mathbf{z}_{-L+1}\end{bmatrix}\right) - 2E_{\mathbf{z}_T}\left(\begin{bmatrix}\mathbf{z}_0\\\vdots\\\mathbf{z}_{-L+1}\end{bmatrix}^T\right)\mathbf{V}_0^{-1}\mathbf{u}_0 + \mathbf{u}_0^T\mathbf{V}_0^{-1}\mathbf{u}_0\right\} \\
&\quad -\frac{1}{2}\left\{T\log|\Sigma_\mathbf{z}| + \sum_{t=1}^{T}\left\{E_{\mathbf{z}_T}(\mathbf{z}_t^T\Sigma_\mathbf{z}^{-1}\mathbf{z}_t) - 2E_{\mathbf{z}_T}\left(\begin{bmatrix}\mathbf{z}_{t-1}\\\vdots\\\mathbf{z}_{t-L}\end{bmatrix}^T \mathbf{A}^T\Sigma_\mathbf{z}^{-1}\mathbf{z}_t\right) + E_{\mathbf{z}_T}\left(\begin{bmatrix}\mathbf{z}_{t-1}\\\vdots\\\mathbf{z}_{t-L}\end{bmatrix}^T \mathbf{A}^T\Sigma_\mathbf{z}^{-1}\mathbf{A}\begin{bmatrix}\mathbf{z}_{t-1}\\\vdots\\\mathbf{z}_{t-L}\end{bmatrix}\right)\right\}\right\} \\
&\quad -\frac{1}{2}\left\{T\log|\Sigma_\mathbf{x}| + \sum_{t=1}^{T}\left\{\mathbf{x}_t^T\Sigma_\mathbf{x}^{-1}\mathbf{x}_t - 2E_{\mathbf{z}_T}(\mathbf{z}_t^T)\mathbf{B}_\mathbf{x}^T\Sigma_\mathbf{x}^{-1}\mathbf{x}_t + E_{\mathbf{z}_T}(\mathbf{z}_t^T\mathbf{B}_\mathbf{x}^T\Sigma_\mathbf{x}^{-1}\mathbf{B}_\mathbf{x}\mathbf{z}_t)\right\}\right\} \\
&\quad -\frac{1}{2}\left\{T\log|\Sigma_\mathbf{y}| + \sum_{t=1}^{T}\left\{\mathbf{y}_t^T\Sigma_\mathbf{y}^{-1}\mathbf{y}_t - 2E_{\mathbf{z}_T}(\mathbf{z}_t^T)\mathbf{B}_\mathbf{y}^T\Sigma_\mathbf{y}^{-1}\mathbf{y}_t + E_{\mathbf{z}_T}(\mathbf{z}_t^T\mathbf{B}_\mathbf{y}^T\Sigma_\mathbf{y}^{-1}\mathbf{B}_\mathbf{y}\mathbf{z}_t)\right\}\right\} + \text{cons}\tan t.
\end{aligned}
\tag{7}
$$

Appendix A provides a detailed update of all parameters at step M. From (7), the related statistics of latent variables in the Q function include $E_{\mathbf{z}_T}(\mathbf{z}_t)$, $E_{\mathbf{z}_T}(\mathbf{z}_t\mathbf{z}_t^T)$ and $E_{\mathbf{z}_T}(\mathbf{z}_t\mathbf{z}_{t-i}^T)$, where $t = 0, 1, \cdots, T$, $i = 1, 2, \cdots, L$, in fact, these statistics can be passed. The posterior probability distribution of the latent variables obtained in the E step of the EM algorithm was obtained. The calculation results of these statistics are shown in (8). The detailed derivation process is shown in Appendix B.

$$
\begin{cases}
E_{\mathbf{z}_T}(\mathbf{z}_t) = E\left(\mathbf{z}_t|\mathbf{x}_{1:T},\mathbf{y}_{1:T},\Theta^{old}\right) = \mathbf{m}_t^1 \\
E_{\mathbf{z}_T}(\mathbf{z}_t\mathbf{z}_t^T) = E\left(\mathbf{z}_t\mathbf{z}_t^T|\mathbf{x}_{1:T},\mathbf{y}_{1:T},\Theta^{old}\right) = \mathbf{M}_t^{11} + \mathbf{m}_t^1(\mathbf{m}_t^1)^T \\
E_{\mathbf{z}_T}(\mathbf{z}_t\mathbf{z}_{t-i}^T) = E\left(\mathbf{z}_t\mathbf{z}_{t-i}^T|\mathbf{x}_{1:T},\mathbf{y}_{1:T},\Theta^{old}\right) = \mathbf{M}_t^{1(i+1)} + \mathbf{m}_t^1\left(\mathbf{m}_t^{(i+1)}\right)^T \\
E_{\mathbf{z}_T}(\mathbf{z}_t\mathbf{z}_{t-L}^T) = E\left(\mathbf{z}_t\mathbf{z}_{t-L}^T|\mathbf{x}_{1:T},\mathbf{y}_{1:T},\Theta^{old}\right) = \sum_{i=1}^{L}\mathbf{A}_i\left(\mathbf{M}_{t-1}^{iL} + \mathbf{m}_{t-1}^i(\mathbf{m}_{t-1}^L)^T\right)
\end{cases}
\tag{8}
$$

where $\mathbf{m}_t$ and $\mathbf{M}_t$ are the mean and covariance of the posterior probability distribution of the latent variable. Therefore, through E step and M step iterative update until the parameters converge, the optimized parameter set $\Theta^{opt} = \{\mathbf{A}, \mathbf{B}_\mathbf{x}, \mathbf{B}_\mathbf{y}, \mathbf{u}_0, \mathbf{V}_0, \Sigma_\mathbf{z}, \Sigma_\mathbf{x}, \Sigma_\mathbf{y}\}$ can be obtained.

3. Process Monitoring Method Based on Dynamic Autoregressive Latent Variable Model

In this section, the established dynamic autoregressive latent variable model is used for industrial process monitoring. At first, DALM was used to model the process data so that the current state information was reflected in the latent variables, and then the latent space at the current time was obtained by filtering the process data distribution, constructing statistics and monitoring them. Let us introduce the monitoring process in detail below.

Although the latent space was unobservable, the establishment of a data-driven DALM model based on the characteristics of the process data extracted the information of the process variables to the spatial distribution of the latent variables. The process input $\mathbf{X} = [\mathbf{x}_1, \mathbf{x}_2, \cdots, \mathbf{x}_N]$ and output $\mathbf{Y} = [\mathbf{y}_1, \mathbf{y}_2, \cdots, \mathbf{y}_N]$ needed to be pre-processed by the normalization method, as shown in (9).

$$
\begin{aligned}
\mathbf{X}^q &= (\mathbf{x} - \mathbf{u_x}) \cdot \mathbf{std}_\mathbf{x}^{-1}, \\
\mathbf{Y}^q &= (\mathbf{y} - \mathbf{u_y}) \cdot \mathbf{std}_\mathbf{y}^{-1},
\end{aligned}
\tag{9}
$$

where $\mathbf{u_x}$ and $\mathbf{u_y}$ are the means of the variables $\mathbf{X}$ and $\mathbf{Y}$, $\mathbf{std_x}$ and $\mathbf{std_y}$ are the variances of the variables $\mathbf{X}$ and $\mathbf{Y}$. Preprocessed data were filtered through the filtering algorithm to obtain the spatial distribution of the latent variables, as shown in (10).

$$
\mathbf{z}_t^q, \mathbf{z}_{t-1}^q, \cdots, \mathbf{z}_{t-L+1}^q | \mathbf{x}_{1:t}^q, \mathbf{y}_{1:t}^q \sim N(\mathbf{u}_t^q, \mathbf{V}_t^q) = N\left(\begin{bmatrix}\mathbf{u}_t^{1q}\\\vdots\\\mathbf{u}_t^{Lq}\end{bmatrix}, \begin{bmatrix}\mathbf{V}_t^{11q} & \cdots & \mathbf{V}_t^{1Lq}\\\vdots & \ddots & \vdots\\\mathbf{V}_t^{(L)1q} & \cdots & \mathbf{V}_t^{LLq}\end{bmatrix}\right).
\tag{10}
$$

Among them, $\mathbf{u}_t^{1q}$ and $\mathbf{V}_t^{11q}$ are the mean and variance of the latent space distribution, respectively, which were obtained from (11). The detailed derivation process is shown in (A12)–(A16).

$$\mathbf{u}_t^{1q} = \mathbf{g}_t^1 + \begin{bmatrix} \mathbf{G}_t^{1(L+1)} & \mathbf{G}_t^{1(L+2)} \end{bmatrix} \begin{bmatrix} \mathbf{G}_t^{(L+1)(L+1)} & \mathbf{G}_t^{(L+1)(L+2)} \\ \mathbf{G}_t^{(L+2)(L+1)} & \mathbf{G}_t^{(L+2)(L+2)} \end{bmatrix} \begin{bmatrix} \mathbf{x}_t - \mathbf{g}_t^{L+1} \\ \mathbf{y}_t - \mathbf{g}_t^{L+2} \end{bmatrix},$$

$$\mathbf{V}_t^{11q} = \begin{bmatrix} \mathbf{G}_t^{11} & \cdots & \mathbf{G}_t^{1L} \end{bmatrix} - \begin{bmatrix} \mathbf{G}_t^{1(L+1)} & \mathbf{G}_t^{1(L+2)} \end{bmatrix} \begin{bmatrix} \mathbf{G}_t^{(L+1)(L+1)} & \mathbf{G}_t^{(L+1)(L+2)} \\ \mathbf{G}_t^{(L+2)(L+1)} & \mathbf{G}_t^{(L+2)(L+2)} \end{bmatrix}^{-1} \begin{bmatrix} \mathbf{G}_t^{1(L+1)} & \mathbf{G}_t^{1(L+2)} \\ \vdots & \vdots \\ \mathbf{G}_t^{L(L+1)} & \mathbf{G}_t^{L(L+2)} \end{bmatrix}^T. \quad (11)$$

It can be seen from (A24) that the information of the data $\mathbf{X}_t^q = [\mathbf{x}_1^q, \mathbf{x}_2^q, \cdots, \mathbf{x}_t^q]$ and $\mathbf{Y}_t^q = [\mathbf{y}_1^q, \mathbf{y}_2^q, \cdots, \mathbf{y}_t^q]$ at the current and previous moments was filtered into the current latent variable, and the latent variable distribution at the current moment is shown in (12).

$$\mathbf{z}_t^q | \mathbf{x}_{1:t}^q, \mathbf{y}_{1:t}^q \sim N(\mathbf{u}_t^{1q}, \mathbf{V}_t^{11q}). \quad (12)$$

Because the latent space contains the current state of the process dynamics and variable time lag information, the process statistic T^2 was constructed for the current latent variable at time t, as shown in (13).

$$T_{t,q}^2 = E(\mathbf{z}_t^q | \mathbf{x}_{1:t}^q, \mathbf{y}_{1:t}^q)^T covariance(\mathbf{z}_t^q | \mathbf{x}_{1:t}^q, \mathbf{y}_{1:t}^q)^{-1} E(\mathbf{z}_t^q | \mathbf{x}_{1:t}^q, \mathbf{y}_{1:t}^q). \quad (13)$$

Among them, the mathematical expectation and variance of the latent variables on the observation data at the current moment are shown in (14).

$$E(\mathbf{z}_t^q | \mathbf{x}_{1:t}^q, \mathbf{y}_{1:t}^q) = \mathbf{u}_t^{1q},$$
$$covariance(\mathbf{z}_t^q | \mathbf{x}_{1:t}^q, \mathbf{y}_{1:t}^q) = \mathbf{V}_t^{11q}. \quad (14)$$

The probability of the latent variable obeyed the Gaussian distribution. Therefore, according to the definition of chi-square distribution, this statistic obeyed the chi-square distribution $\chi_\alpha^2(d)$ after data preprocessing. Then, combining to the latent variable dimension d of the model and the significance level α required by the industry, the control threshold $T_{\lim}^2$ of the process monitoring method was obtained, and then the statistics of each time data were calculated online and compared with the control threshold, to determine whether the process deviated from the normal state. The process monitoring logic is determined by (15).

$$T_{t,q}^2 < T_{\lim}^2 = \chi_\alpha^2(d). \quad (15)$$

Too large an α value will lead to a high false alarm rate, and too low an α will lead to a high false alarm rate; therefore, in practice, it is a balance between false alarms and missed alarms. This paper chose α as 0.01, which means that the false positive rate of normal data was 0.01. If $T_{t,q}^2 < T_{\lim}^2$, the system was in a normal state. Otherwise, the process located in a fault state, and further diagnosis and identification of the fault was required for process maintenance. The process of DALM modeling and online process monitoring is shown in Figure 2.

The main steps of the process monitoring method based on the DALM model were as follows:
Step 1: Collect process data, divide the training and test data sets and standardize them.
Step 2: Use the training data set to learn the parameters of the DALM model.
Step 3: Build the model and determine the control threshold.
Step 4: Filter the process data online to get the latent space distribution at the current moment.
Step 5: Calculate statistics and compare with the control threshold to determine whether the process is abnormal.

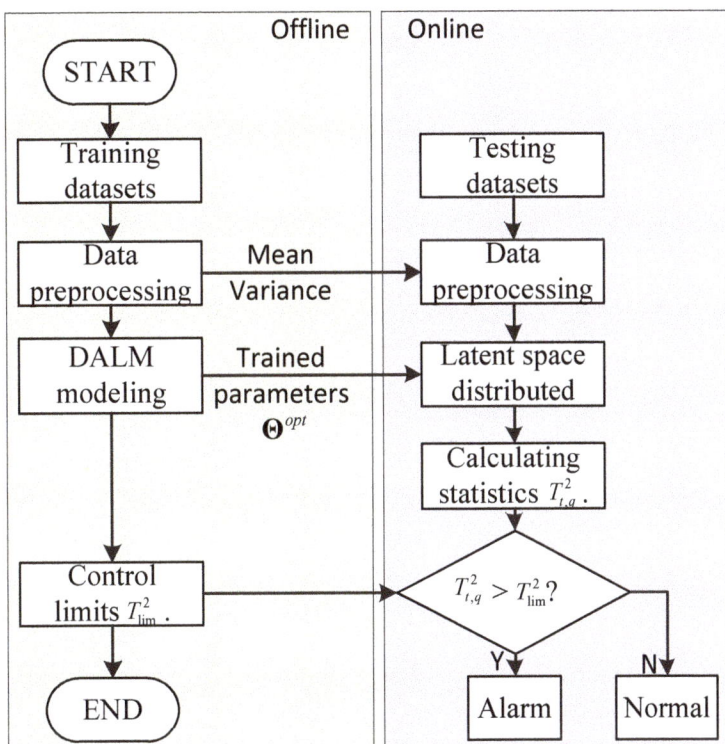

Figure 2. Flowchart based on DALM process monitor.

4. Case Study on the Sintering Process of Ternary Cathode Materials

In this section, the proposed process monitoring method based on the dynamic autoregressive latent variable model is used to monitor the sintering process of ternary cathode materials to verify the effectiveness of the method. First the sintering process technology of the ternary cathode material was introduced, then the model structure and parameter determination were introduced in detail, and finally the performance of the model was evaluated.

4.1. Introduction to the Sintering Process of Ternary Cathode Materials

The rapid development of the new energy industry has led to an extremely urgent demand for high-quality ternary cathode materials, and the sintering process of battery materials is the core and key process of battery preparation. This process consists of a series connection of a heating section, a constant temperature section and a cooling section, as shown in Figure 3. The optimal production state of a single temperature section cannot guarantee that the product performance indicators of the entire sintering process are within the optimal range; at the same time, changes in the sintering process, such as environmental humidity or temperature, also affect the stability of product performance indicators. In order to ensure the stability of product performance indicators as much as possible, while reducing energy consumption and material consumption, it is necessary to adjust the sintering parameters of the kiln according to the sintering state in real time, which leads to many variables in each temperature zone and series coupling, which makes the process data present complex process characteristics [29].

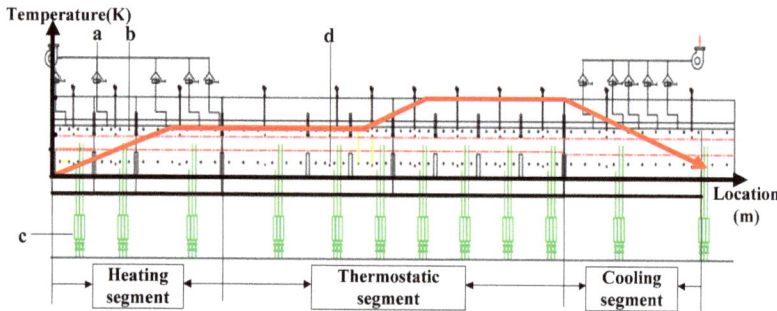

Figure 3. Structure diagram of sintering furnace for battery sintering.

The temperature field in the sintering process has a significant effect on the material properties. Over-firing will cause changes in the material morphology and internal structure, and under firing will not provide sufficient activation energy for chemical reactions. However, the decomposition reaction that occurs in the heating section is an endothermic process and requires sufficient heat supply, otherwise a reverse reaction will occur, resulting in inefficient water removal, which will affect the subsequent oxidation reaction. Therefore, the state of the heating section is very important to the sintering process. At the same time, the residual lithium content can directly reflect the quality of the product. In order to monitor the process status in real time, a monitoring model is established for the temperature and residual lithium content of the heating section.

Huang et al. [30] established a temperature field monitoring model based on the PBF equipment equation to monitor the dynamic sintering process of parts, but this method requires precise grinding tool structure parameters and can only monitor uniformly distributed temperature fields. Egorova et al. [31] tried to combine neural networks and PCA diagnosis method monitor and diagnose the sintering process. This method can locate the fault and diagnose the cause of the fault. However, the introduction of neural networks increases the time and space complexity of the system and ignores the system dynamic and time lag problems.

Due to the severe temperature interval coupling, the process variables exhibit complex characteristics, making the traditional static monitoring methods unable to achieve accurate monitoring results. The dynamic autoregressive latent variable model proposed in this section considers the dynamic and time lag information of the process at the same time, so it is more in line with the sintering process.

4.2. Determination of Model Parameters

This section establishes a monitoring model for the temperature and product quality in the heating section of the sintering process. The heating section contained seven temperature zones, and each temperature zone had two upper and lower temperature measuring points, but the temperature changes in the 4th to 7th temperature zones were not obvious. The temperature of the first three temperature zones was selected as the process variable x_t of the model. At the same time, the residual lithium content of the product reflects the quality of the battery, as does the quality variable y_t of the model, Table 1 lists the physical meaning of these variables.

Table 1. Selected variables in the sintering process.

No.	Measured Variables
1	Below temperature of 1st zone
2	Upper temperature of 1st zone
3	Below temperature of 2nd zone
4	Upper temperature of 2nd zone
5	Below temperature of 3rd zone
6	Upper temperature of 3rd zone
7	Lithium loss coefficient

To test the monitoring effect of the model under different faults, a total of 2200 continuous time data samples were collected on site with a sampling period of five minutes. The process included a total of three types of faults such as over-temperature, under-temperature and shutdown. For detailed status information, see Table 2.

Table 2. Process data description.

Date Types	Data Description	Time Durations
Normal	Normal data	1st–1000th
Fault 1	Normal samples in 1001st–1200th and abnormal samples of 3rd zone temperature rise in 1201st–1400th	1001st–1400th
Fault 2	Normal samples in 1401st–1600th and abnormal samples of 3rd zone temperature drop in 1601st–1800th	1401st–1800th
Fault 3	Normal samples in 1801st–2000th and abnormal samples of downtime fault in 2001st–2200th	1801st–2200th

First, analyze the dynamics of the data and the time lag characteristics of the variables from the data point of view. Figure 4 shows the autocorrelation and cross-correlation diagrams of process data.

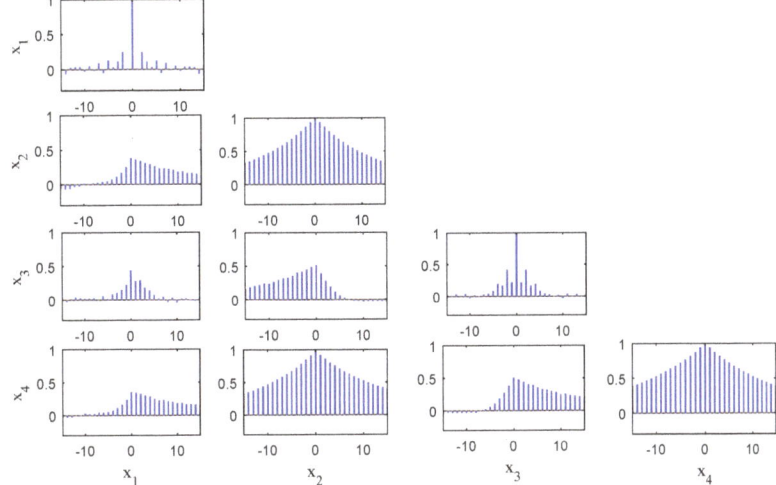

Figure 4. Correlation plots for the first four process variables.

Figure 4 shows the correlation and cross-correlation between the first four process variables. The value at time 0 in each figure represents the cross-correlation between variables; the value at non-zero time shows the autocorrelation between variables under different time lags. It is worth mentioning that the cross-correlation index can measure the redundancy of variable information, and the autocorrelation index can indirectly measure the dynamic and time delay information between variables. It can be seen that the cross-correlation performance between the variables was above 0.5, indicating that there was strong redundant information between the variables. At the same time, even if there was a difference of 10 sampling times, the autocorrelation between the variables was still very high, indicating that there were time lags and dynamic characteristics between the variables. Therefore, the establishment of a DALM model for the process can be considered. The emission equation of the model extracts the redundant information of the data, and the autoregressive equation of the model extracts the dynamic and time lag information of the variables. This paper uses the trend similarity algorithm, which constructs the trend similarity function according to the time lag feature and solves it, to determine the time lag coefficient, that is, L = 3.

To verify the rationality of the time lag coefficient, under different time lag coefficients, a dynamic autoregressive latent variable monitoring model was established respectively. Note: In order to avoid the latent variable dimension from interfering with the selection of the time lag coefficient the latent variable dimension selected by Akaike information criterion (AIC) was temporarily used [32].

The false alarm rate (false alarm rate, FAR) and fault detection rate (fault detection rate, FDR) were defined to evaluate and monitor performance indicators, as defined in (16).

$$FAR = \frac{N_{FAR}}{N_n} \quad FDR = \frac{N_{FDR}}{N_f}. \tag{16}$$

N_{FAR} represents the number of normal samples that were mistakenly detected as abnormal by the monitoring method, and N_n is the number of all normal samples. N_{FDR} represents the number of fault samples correctly monitored by the monitoring method, N_f is the number of all abnormal samples. Therefore, the closer the FAR is to the significance level, the better, and the closer the FDR is to 1, the better. The significance level of this work was set to 0.01.

The first 1000 normal samples were selected to train the model, and the data type fault 1 was used to test the monitoring effect of the model. Table 3 shows the indicators of the monitoring results of the new method under different time lag coefficients.

Table 3. FAR and FDR under different time lag.

Time Lag	2	3	4	5
FAR	0.165	0.050	0.270	\
FDR	0.665	1.000	0.905	\

The model did not converge when the time lag coefficient was 5, and when the model time lag coefficient was 3, the error and false alarm rate of the model were the best. Therefore, when the time lag coefficient was 3, the model gave the best performance. In order to visually see the monitoring results of the model, Figure 5 shows the monitoring T^2 diagram when the model's time lag coefficient was 2, 3 and 4.

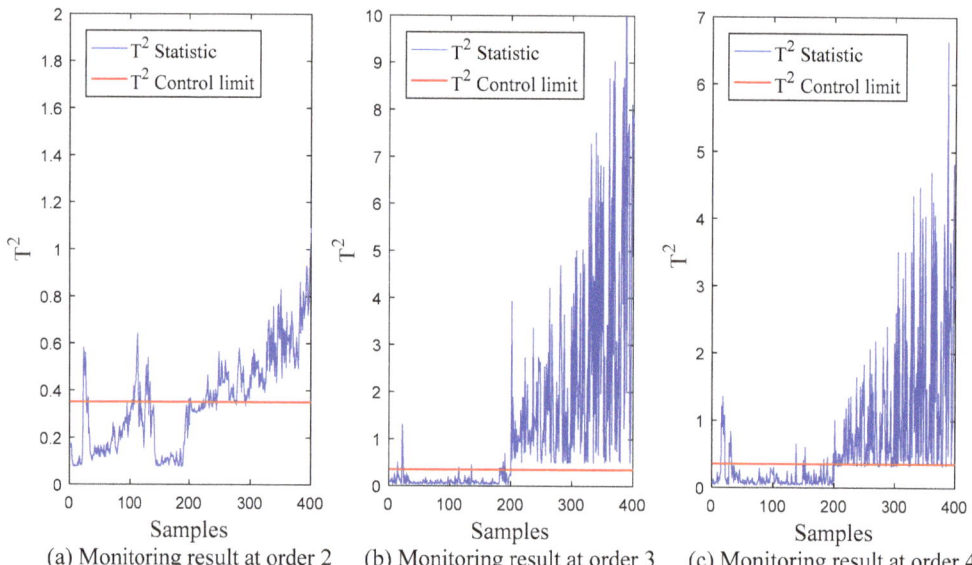

(a) Monitoring result at order 2 (b) Monitoring result at order 3 (c) Monitoring result at order 4

Figure 5. Monitoring performance under different time lag. (**a**) Monitoring result at order 2; (**b**) monitoring result at order 3; (**c**) monitoring result at order 4.

It can be seen from Figure 5 that when the model time lag coefficient was 2 and 4, it was easy to misclassify the sample. Especially in the fault interval of 201st–400th: the divided normal samples and abnormal samples were close to the monitoring threshold, which shows that the robustness of the model with this time lag is low; when the model had a time lag coefficient of 3, it is insensitive to the noise and the false alarms are the smallest. Hence, its N_{FAR} and N_{FDR} were the best. Therefore, the time lag coefficient obtained by the trend similarity identification algorithm enabled the model to obtain a better monitoring effect.

Next, the latent variable dimension was determined. The latent variable dimension is the result of comprehensively considering the complexity and accuracy of the model. The root mean square error (RMSE) is an indicator to measure the accuracy of the model. The expression is shown in (17).

$$RMSE = \sqrt{\sum_{i=1}^{N} \frac{|y_i - \hat{y}_i|}{N}}, \qquad (17)$$

where N is the number of test samples, $\hat{y}_i$ is the prediction of the true value y_i and $\bar{y}_i$ is the mean value of the true value of the test sample. Samples from the 1st to the 600th were used to train the model, and samples from the 601st to the 1000th were used as the test set. Table 4 shows the root mean square error of model prediction under different latent variable dimensions.

Table 4. RMSE under different latent variable dimensions.

Number of Latent Variables	1	2	3	4	5
RMSE	0.123	0.08	0.045	0.042	0.042

Table 4 shows that the prediction performance of the model tends to be stable after the latent variable dimension increased to 3, which was the balance point between model complexity and accuracy. It is worth mentioning that under the time lag coefficient, the latent variable dimension selected by the AIC algorithm was also 3, so the latent variable dimension was determined to be 3.

4.3. Model Performance Test

This section verifies the effect of the proposed monitoring method, and constructs a first-order dynamic process monitoring method: DPLVM [18] and static process monitoring method: PPLSR [33], which were used to compare with the proposed method. The latent variable dimensions of the model were adjusted to 3.

The first 1000 normal samples were used to train the parameters of the model, and the trained model was monitored for three types of different fault samples. In order to distinguish between normal and abnormal samples, the first 200 samples of each type of failure test set were normal samples, and the last 200 samples were their respective failure samples. Table 5 shows the FAR and FDR of different monitoring methods under different failure test sets, and the last line calculates the average value of different indicators.

Table 5. FAR and FDR of the three methods under different fault cases.

Faults	PPLSR		DPLVM		DALM	
	FAR	FDR	FAR	FDR	FAR	FDR
Fault 1	0.240	0.120	0.210	0.795	0.050	1.000
Fault 2	0.100	0.350	0.155	0.810	0.045	1.000
Fault 3	0.315	0.980	0.080	0.770	0.045	1.000
Average	0.218	0.483	0.148	0.792	0.047	1.000

It can be seen from Table 5 that the monitoring performance of the proposed method was better than that of the static model PPLSR and the first-order dynamic model DPLVM. Therefore, the detection performance was greatly improved after the autoregressive equation was added to the model to extract the dynamic and time lag information. Compared with the basic first-order dynamic DPLVM fault detection method, DALM considered the time lag characteristics, so the model performance was further improved. The detailed monitoring results of the three methods for the three types of faults are shown in Figures 6–8.

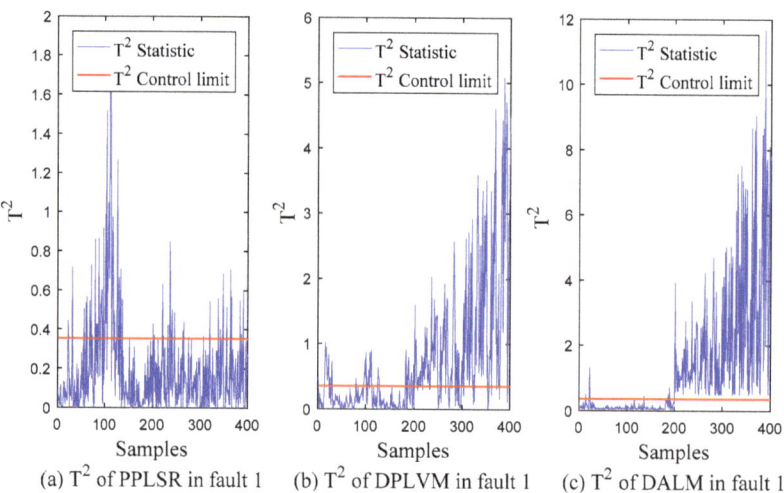

Figure 6. Monitoring results of fault 1. (**a**) T² of PPLSR in fault 1; (**b**) T² of DPLVM in fault 1; (**c**) T² of DALM in fault 1.

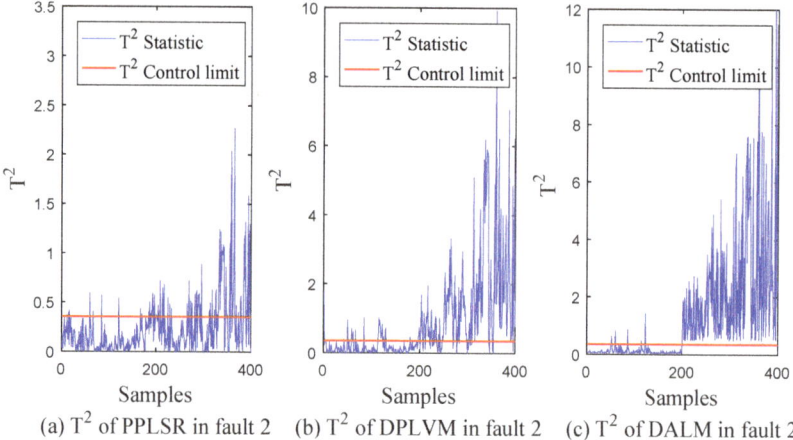

Figure 7. Monitoring results of fault 2. (**a**) T² of PPLSR in fault 2; (**b**) T² of DPLVM in fault 2; (**c**) T² of DALM in fault 2.

For each type of fault test set, the first 200 samples were in a normal state, and the last 200 samples were fault samples. It can be seen from Figure 8 that the static model PPLSR easily mistakenly classified normal samples into faulty samples, and it also easily classified faulty samples into normal samples. The error rate of the first-order dynamic model DPLVM was reduced a lot. Furthermore, the FAR based on the DALM fault detection method proposed in this paper was close to the significance level and the FDR was close to 1, verifying that its monitoring performance was greatly improved.

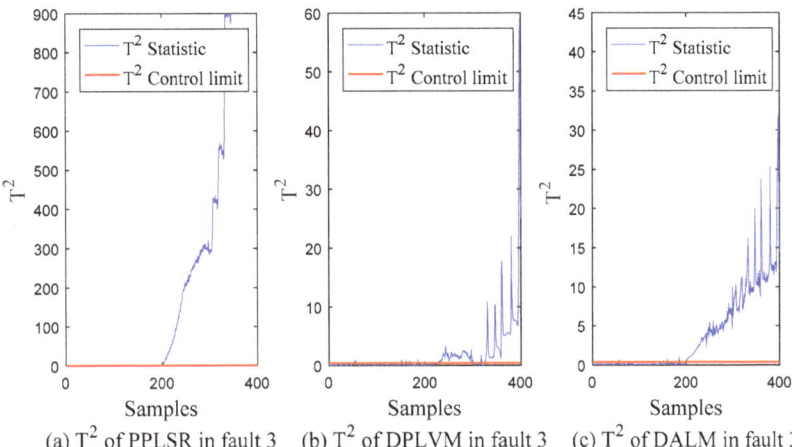

Figure 8. Monitoring results of fault 3. (**a**) T^2 of PPLSR in fault 3; (**b**) T^2 of DPLVM in fault 3; (**c**) T^2 of DALM in fault 3.

5. Conclusions

A process monitoring method based on the dynamic autoregressive latent variable model was proposed in this paper. Compared with the traditional DPLVM monitoring method, this method not only considered the dynamic characteristics of the process but also considered the complex time lag characteristics, integrated the time lag information into the model, and greatly improved the monitoring performance of the model in the time lag process. First, from the point of data, this method established a dynamic autoregressive latent variable model to adopt the characteristics of dynamics and variable time lag. Then a fusion Bayesian filtering, smoothing and expectation maximization algorithm was used to identify model parameters. Then, on the basis of the identified model, the improved Bayesian filtering technique was used to infer the latent variable distribution of the process state, and the T^2 statistic was constructed for the latent space and online monitoring is performed. Finally, the proposed method was applied to the monitoring of the sintering process of ternary cathode materials. Through industrial case studies, the modeling and monitoring results of the proposed method show that the DALM model was better than the static and first-order dynamic modeling process monitoring methods.

An important issue for process monitoring application in industrial processes is the multi-sampling rate problem. The method proposed in this paper assumed that the input and output data had the same sampling rate. If the sampling rate was inconsistent, some data were deleted by down-sampling. However, a more worthwhile way to try would be to combine semi-supervised learning methods, which can train data on unbalanced input and output data, thereby improving data utilization. Another practical problem is the non-linear relationship between process data, which is very common in industrial processes. How to effectively deal with this problem is worthy of further research in the near future to make the monitoring method more applicable.

Author Contributions: Conceptualization, N.C. and W.G.; methodology, F.H. and J.C.; software, F.H. and Z.C.; validation, J.C.; writing—review and editing, N.C., F.H., J.C., Z.C. and X.L.; supervision, N.C., Z.C. and W.G.; project administration, W.G. and X.L. All authors have read and agreed to the published version of the manuscript.

Funding: This work was funded in part by the Key Program of the National Natural Science Foundation of China (62033014) and in part by the Application Projects of Integrated Standardization and New Paradigm for Intelligent Manufacturing from the Ministry of Industry and Information Technology of China in 2016 and in part by the Fundamental Research Funds for the Central Universities of Central South University(2021zzts0700), in part by the Project of State Key Laboratory of High Performance Complex Manufacturing (#ZZYJKT2020-14).

Institutional Review Board Statement: Not applicable.

Informed Consent Statement: Not applicable.

Data Availability Statement: The data presented in this study are available upon request from the first author. The data are not publicly available due to intellectual property protection.

Acknowledgments: Not applicable.

Conflicts of Interest: The authors declare no conflict of interest.

Appendix A. Detailed Derivation of the M-Step

According to the EM algorithm, all the parameters of the DALM model can be updated in M steps. By maximizing the cost function $Q(\Theta|\Theta^{old})$, the estimated value Θ^{new} of the next iteration parameter was determined, which is shown in (A1).

$$\Theta^{new} = \underset{\Theta}{\operatorname{argmax}} Q(\Theta|\Theta_{old}). \tag{A1}$$

The Q function was applied to the partial derivative of the model parameters and the derivative was set to zero.

$$\frac{\partial Q(\Theta|\Theta_{old})}{\partial \Theta} = 0. \tag{A2}$$

The updated value of the model parameter Θ^{new} was obtained, as shown in (A3)–(A10).

$$\mathbf{u}_0^{new} = E_{\mathbf{z}_T}\left(\begin{bmatrix} \mathbf{z}_0 \\ \vdots \\ \mathbf{z}_{-L+1} \end{bmatrix}\right), \tag{A3}$$

$$\mathbf{V}_0^{new} = E_{\mathbf{z}_T}\left(\begin{bmatrix} \mathbf{z}_0 \\ \vdots \\ \mathbf{z}_{-L+1} \end{bmatrix}\begin{bmatrix} \mathbf{z}_0 \\ \vdots \\ \mathbf{z}_{-L+1} \end{bmatrix}^T\right) - E_{\mathbf{z}_T}\left(\begin{bmatrix} \mathbf{z}_0 \\ \vdots \\ \mathbf{z}_{-L+1} \end{bmatrix}\right) E_{\mathbf{z}_T}\left(\begin{bmatrix} \mathbf{z}_0 \\ \vdots \\ \mathbf{z}_{-L+1} \end{bmatrix}\right)^T, \tag{A4}$$

$$\mathbf{A}^{new} = \sum_{t=1}^{T} E_{\mathbf{z}_T}\left(\mathbf{z}_t \begin{bmatrix} \mathbf{z}_{t-1} \\ \vdots \\ \mathbf{z}_{t-L} \end{bmatrix}^T\right) \left[\sum_{t=1}^{T} E_{\mathbf{z}_T}\left(\begin{bmatrix} \mathbf{z}_{t-1} \\ \vdots \\ \mathbf{z}_{t-L} \end{bmatrix}\begin{bmatrix} \mathbf{z}_{t-1} \\ \vdots \\ \mathbf{z}_{t-L} \end{bmatrix}^T\right)\right]^{-1}, \tag{A5}$$

$$\Sigma_{\mathbf{z}}^{new} = \frac{1}{T}\sum_{t=1}^{T}\left(E_{\mathbf{z}_T}\left(\mathbf{z}_t \mathbf{z}_t^T\right) - 2\mathbf{A}^{new} E_{\mathbf{z}_T}\left(\begin{bmatrix} \mathbf{z}_{t-1} \\ \vdots \\ \mathbf{z}_{t-L} \end{bmatrix}\mathbf{z}_t^T\right) + \mathbf{A}^{new} E_{\mathbf{z}_T}\left(\begin{bmatrix} \mathbf{z}_{t-1} \\ \vdots \\ \mathbf{z}_{t-L} \end{bmatrix}\begin{bmatrix} \mathbf{z}_{t-1} \\ \vdots \\ \mathbf{z}_{t-L} \end{bmatrix}^T\right)\mathbf{A}^{newT}\right), \tag{A6}$$

$$\mathbf{B}_{\mathbf{x}}^{new} = \sum_{t=1}^{T}\mathbf{x}_t E_{\mathbf{z}_T}\left(\mathbf{z}_t^T\right)\left[\sum_{t=1}^{T} E_{\mathbf{z}_T}\left(\mathbf{z}_t \mathbf{z}_t^T\right)\right]^{-1}, \tag{A7}$$

$$\mathbf{B}_{\mathbf{y}}^{new} = \sum_{t=1}^{T}\mathbf{y}_t E_{\mathbf{z}_T}\left(\mathbf{z}_t^T\right)\left[\sum_{t=1}^{T} E_{\mathbf{z}_T}\left(\mathbf{z}_t \mathbf{z}_t^T\right)\right]^{-1}, \tag{A8}$$

$$\Sigma_{\mathbf{z}}^{new} = \frac{1}{T}\sum_{t=1}^{T}\left(\mathbf{x}_t\mathbf{x}_t^T - 2\mathbf{B}_{\mathbf{x}}^{new} E_{\mathbf{z}_T}(\mathbf{z}_t)\mathbf{x}_t^T + \mathbf{B}_{\mathbf{x}}^{new} E_{\mathbf{z}_T}\left(\mathbf{z}_t \mathbf{z}_t^T\right)\mathbf{B}_{\mathbf{x}}^{newT}\right), \tag{A9}$$

$$\Sigma_{\mathbf{y}}^{new} = \frac{1}{T}\sum_{t=1}^{T}\left(\mathbf{y}_t\mathbf{y}_t^T - 2\mathbf{B}_{\mathbf{y}}^{new} E_{\mathbf{z}_T}(\mathbf{z}_t)\mathbf{y}_t^T + \mathbf{B}_{\mathbf{y}}^{new} E_{\mathbf{z}_T}\left(\mathbf{z}_t \mathbf{z}_t^T\right)\mathbf{B}_{\mathbf{y}}^{newT}\right). \tag{A10}$$

The updated parameter set $\Theta^{new} = \left\{\mathbf{A}^{new}, \mathbf{B}_{\mathbf{x}}^{new}, \mathbf{B}_{\mathbf{y}}^{new}, \mathbf{u}_0^{new}, \mathbf{V}_0^{new}, \Sigma_{\mathbf{z}}^{new}, \Sigma_{\mathbf{x}}^{new}, \Sigma_{\mathbf{y}}^{new}\right\}$, E steps and M steps were iterated until the parameter Θ matrix converged, that is, satisfied (A11), where ς is a sufficiently small constant, and the model parameter identification was completed.

$$\|\Theta^{new} - \Theta^{old}\| < \varsigma. \tag{A11}$$

Among them, Θ^{old} is the parameter of the last iteration, and Θ^{new} is the parameter after this round of iteration. Only when the parameters obtained by two adjacent identifications converged did the algorithm stop calculating. Therefore, the parameter convergence can be guaranteed by the EM algorithm itself.

Appendix B. Detailed Derivation of the E-Step

In order to determine the statistics $E_{\mathbf{z}_T}(\mathbf{z}_t)$, $E_{\mathbf{z}_T}(\mathbf{z}_t\mathbf{z}_t^T)$ and $E_{\mathbf{z}_T}(\mathbf{z}_t\mathbf{z}_{t-i}^T)$, the forward and backward algorithm were employed. This is an iterative calculation method, which includes the forward filtering and backward correction step.

In the Bayesian filtering stage, the goal was to calculate the posterior probability of the latent variable $[\mathbf{z}_t, \mathbf{z}_{t-1}, \cdots, \mathbf{z}_{t-L+1}]$ with respect to the variable $\mathbf{x}_{1:t}, \mathbf{y}_{1:t}$ at time t, given the posterior distribution $\mathbf{z}_{t-1}, \mathbf{z}_{t-2}, \cdots, \mathbf{z}_{t-L}|\mathbf{x}_{1:t-1}, \mathbf{y}_{1:t-1} \sim N(\mathbf{u}_{t-1}, \mathbf{V}_{t-1})$ of the latent variable $[\mathbf{z}_{t-1}, \mathbf{z}_{t-2}, \cdots, \mathbf{z}_{t-L}]$ at the previous time $t - 1$ on the variable $\mathbf{x}_{1:t-1}, \mathbf{y}_{1:t-1}$, as shown in (A12) where $1 \le t \le T$,

$$\mathbf{z}_t, \mathbf{z}_{t-1}, \cdots, \mathbf{z}_{t-L+1}|\mathbf{x}_{1:t}, \mathbf{y}_{1:t} \sim N(\mathbf{u}_t, \mathbf{V}_t) = N\left(\begin{bmatrix} \mathbf{u}_t^1 \\ \vdots \\ \mathbf{u}_t^L \end{bmatrix}, \begin{bmatrix} \mathbf{V}_t^{11} & \cdots & \mathbf{V}_t^{1L} \\ \vdots & \ddots & \vdots \\ \mathbf{V}_t^{L1} & \cdots & \mathbf{V}_t^{LL} \end{bmatrix}\right). \quad (A12)$$

The joint probability distribution of the latent variables $[\mathbf{z}_t, \mathbf{z}_{t-1}, \cdots, \mathbf{z}_{t-L+1}]$ and $\mathbf{x}_t, \mathbf{y}_t$ with respect to the variable $\mathbf{x}_{1:t-1}, \mathbf{y}_{1:t-1}$ is shown in (A13).

$$\begin{aligned} &\mathbf{z}_t, \mathbf{z}_{t-1}, \cdots, \mathbf{z}_{t-L+1}, \mathbf{x}_t, \mathbf{y}_t|\mathbf{x}_{1:t-1}, \mathbf{y}_{1:t-1} \sim N(\mathbf{g}_t, \mathbf{G}_t) \\ &= N\left(\begin{bmatrix} \mathbf{g}_t^1 \\ \vdots \\ \mathbf{g}_t^{L+2} \end{bmatrix}, \begin{bmatrix} \mathbf{G}_t^{11} & \cdots & \mathbf{G}_t^{1(L+2)} \\ \vdots & \ddots & \vdots \\ \mathbf{G}_t^{(L+2)1} & \cdots & \mathbf{G}_t^{(L+2)(L+2)} \end{bmatrix}\right). \end{aligned} \quad (A13)$$

The parameters of (A13) can be calculated by (A15)

$$\begin{cases} \mathbf{g}_t^1 = E(\mathbf{z}_t|\mathbf{x}_{1:t-1}, \mathbf{y}_{1:t-1}) = \mathbf{A}\mathbf{u}_{t-1} \\ \mathbf{g}_t^i = E(\mathbf{z}_{t-i+1}|\mathbf{x}_{1:t-1}, \mathbf{y}_{1:t-1}) = \mathbf{u}_{t-1}^{i-1} \\ \mathbf{g}_t^{L+1} = E(\mathbf{x}_t|\mathbf{x}_{1:t-1}, \mathbf{y}_{1:t-1}) = \mathbf{C}\mathbf{g}_t^1 \\ \mathbf{g}_t^{L+2} = E(\mathbf{y}_t|\mathbf{x}_{1:t-1}, \mathbf{y}_{1:t-1}) = \mathbf{P}\mathbf{g}_t^1 \end{cases}, \quad (A14)$$

$$\begin{cases} \mathbf{G}_t^{11} = \mathrm{cov}(\mathbf{z}_t, \mathbf{z}_t|\mathbf{x}_{1:t-1}, \mathbf{y}_{1:t-1}) = \mathbf{A}\mathbf{u}_{t-1}\mathbf{A}^T + \Sigma_{\mathbf{z}} \\ \mathbf{G}_t^{1i} = \mathrm{cov}(\mathbf{z}_t, \mathbf{z}_{t-i+1}|\mathbf{x}_{1:t-1}, \mathbf{y}_{1:t-1}) = \mathbf{A}\begin{bmatrix} \mathbf{V}_{t-1}^{1(i-1)} \\ \vdots \\ \mathbf{V}_{t-1}^{L(i-1)} \end{bmatrix} \\ \mathbf{G}_t^{i1} = \left(\mathbf{G}_t^{1i}\right)^T \\ \mathbf{G}_t^{ij} = \mathrm{cov}(\mathbf{z}_{t-i+1}, \mathbf{z}_{t-j+1}|\mathbf{x}_{1:t-1}, \mathbf{y}_{1:t-1}) = \mathbf{V}_{t-1}^{(i-1)(j-1)} \\ \mathbf{G}_t^{(L+1)k} = \mathrm{cov}(\mathbf{x}_t, \mathbf{z}_{t-k+1}|\mathbf{x}_{1:t-1}, \mathbf{y}_{1:t-1}) = \mathbf{B}_\mathbf{x}\mathbf{G}_t^{1k} \\ \mathbf{G}_t^{(L+1)k} = \left(\mathbf{G}_t^{k(L+1)}\right)^T \\ \mathbf{G}_t^{(L+2)k} = \mathrm{cov}(\mathbf{y}_t, \mathbf{z}_{t-k+1}|\mathbf{x}_{1:t-1}, \mathbf{y}_{1:t-1}) = \mathbf{B}_\mathbf{y}\mathbf{G}_t^{1k} \\ \mathbf{G}_t^{(L+2)k} = \left(\mathbf{G}_t^{k(L+2)}\right)^T \\ \mathbf{G}_t^{(L+1)(L+1)} = \mathrm{cov}(\mathbf{x}_t, \mathbf{x}_t|\mathbf{x}_{1:t-1}, \mathbf{y}_{1:t-1}) = \mathbf{B}_\mathbf{x}\mathbf{G}_t^{11}\mathbf{B}_\mathbf{x}^T + \Sigma_\mathbf{x} \\ \mathbf{G}_t^{(L+1)(L+2)} = \mathrm{cov}(\mathbf{x}_t, \mathbf{y}_t|\mathbf{x}_{1:t-1}, \mathbf{y}_{1:t-1}) = \mathbf{B}_\mathbf{x}\mathbf{G}_t^{11}\mathbf{B}_\mathbf{y}^T \\ \mathbf{G}_t^{(L+2)(L+1)} = \left(\mathbf{G}_t^{(L+2)(L+1)}\right)^T \\ \mathbf{G}_t^{(L+2)(L+2)} = \mathrm{cov}(\mathbf{y}_t, \mathbf{y}_t|\mathbf{x}_{1:t-1}, \mathbf{y}_{1:t-1}) = \mathbf{B}_\mathbf{y}\mathbf{G}_t^{11}\mathbf{B}_\mathbf{y}^T + \Sigma_\mathbf{y} \end{cases}, \quad (A15)$$

where $i = 2, 3, \cdots, L; j = 2, 3, \cdots, L; k = 1, 2, \cdots, L$, therefore, according to the knowledge of conditional probability [34] and Appendix C, the mean value and variance of the latent variable filter distribution $\mathbf{z}_t, \mathbf{z}_{t-1}, \cdots, \mathbf{z}_{t-L+1} | \mathbf{x}_{1:t}, \mathbf{y}_{1:t} \sim N(\mathbf{u}_t, \mathbf{V}_t)$ were calculated as shown in (A16).

$$\mathbf{u}_t = \begin{bmatrix} \mathbf{g}_t^1 \\ \vdots \\ \mathbf{g}_t^L \end{bmatrix} + \begin{bmatrix} \mathbf{G}_t^{1(L+1)} & \mathbf{G}_t^{1(L+2)} \\ \vdots & \vdots \\ \mathbf{G}_t^{L(L+1)} & \mathbf{G}_t^{L(L+2)} \end{bmatrix} \begin{bmatrix} \mathbf{G}_t^{(L+1)(L+1)} & \mathbf{G}_t^{(L+1)(L+2)} \\ \mathbf{G}_t^{(L+2)(L+1)} & \mathbf{G}_t^{(L+2)(L+2)} \end{bmatrix}^{-1} \begin{bmatrix} \mathbf{x}_t - \mathbf{g}_t^{L+1} \\ \mathbf{y}_t - \mathbf{g}_t^{L+2} \end{bmatrix},$$

$$\mathbf{u}_t = \begin{bmatrix} \mathbf{G}_t^{11} & \cdots & \mathbf{G}_t^{1L} \\ \vdots & \ddots & \vdots \\ \mathbf{G}_t^{L1} & \cdots & \mathbf{G}_t^{LL} \end{bmatrix} - \begin{bmatrix} \mathbf{G}_t^{1(L+1)} & \mathbf{G}_t^{1(L+2)} \\ \vdots & \vdots \\ \mathbf{G}_t^{L(L+1)} & \mathbf{G}_t^{L(L+2)} \end{bmatrix} \begin{bmatrix} \mathbf{G}_t^{(L+1)(L+1)} & \mathbf{G}_t^{(L+1)(L+2)} \\ \mathbf{G}_t^{(L+2)(L+1)} & \mathbf{G}_t^{(L+2)(L+2)} \end{bmatrix}^{-1} \begin{bmatrix} \mathbf{G}_t^{1(L+1)} & \mathbf{G}_t^{1(L+2)} \\ \vdots & \vdots \\ \mathbf{G}_t^{L(L+1)} & \mathbf{G}_t^{L(L+2)} \end{bmatrix}^T. \quad (A16)$$

In the Bayesian smoothing stage, the goal was to calculate the posterior probability $\mathbf{z}_{t+1}, \mathbf{z}_t, \cdots, \mathbf{z}_{t-L+2} | \mathbf{x}_{1:T}, \mathbf{y}_{1:T} \sim N(\mathbf{m}_{t+1}, \mathbf{M}_{t+1})$ of the latent variable $[\mathbf{z}_{t+1}, \mathbf{z}_t, \cdots, \mathbf{z}_{t-L+2}]$ with respect to the variable $\mathbf{x}_{1:T}, \mathbf{y}_{1:T}$ at time $t + 1$ to calculate the posterior probability of the latent variable $[\mathbf{z}_t, \mathbf{z}_{t-1}, \cdots, \mathbf{z}_{t-L+1}]$ with respect to the variable $\mathbf{x}_{1:T}, \mathbf{y}_{1:T}$ at time t, as shown in (A17).

$$\mathbf{z}_t, \mathbf{z}_{t-1}, \cdots, \mathbf{z}_{t-L+1} | \mathbf{x}_{1:T}, \mathbf{y}_{1:T} \sim N(\mathbf{m}_t, \mathbf{M}_t) = N\left(\begin{bmatrix} \mathbf{m}_t^1 \\ \vdots \\ \mathbf{m}_t^L \end{bmatrix}, \begin{bmatrix} \mathbf{M}_t^{11} & \cdots & \mathbf{M}_t^{1L} \\ \vdots & \ddots & \vdots \\ \mathbf{M}_t^{L1} & \cdots & \mathbf{M}_t^{LL} \end{bmatrix}\right), \quad (A17)$$

where $0 \leq t \leq T$, T, there is $\mathbf{m}_T = \mathbf{u}_T$, $\mathbf{M}_T = \mathbf{V}_T$. In order to calculate the distribution, first, the posterior distribution of the latent variable $[\mathbf{z}_{t+1}, \mathbf{z}_t, \cdots, \mathbf{z}_{t-L+1}]$ was calculated with respect to the variable $\mathbf{x}_{1:t}, \mathbf{y}_{1:t}$, as shown in (A18).

$$\mathbf{z}_{t+1}, \mathbf{z}_t, \cdots, \mathbf{z}_{t-L+1} | \mathbf{x}_{1:t}, \mathbf{y}_{1:t} \sim N(\mathbf{d}_t, \mathbf{D}_t) = N\left(\begin{bmatrix} \mathbf{d}_t^1 \\ \vdots \\ \mathbf{d}_t^{L+1} \end{bmatrix}, \begin{bmatrix} \mathbf{D}_t^{11} & \cdots & \mathbf{D}_t^{1(L+1)} \\ \vdots & \ddots & \vdots \\ \mathbf{D}_t^{(L+1)1} & \cdots & \mathbf{D}_t^{(L+1)(L+1)} \end{bmatrix}\right). \quad (A18)$$

The parameter calculation of (A18) is shown in (A19)–(A20).

$$\begin{cases} \mathbf{d}_t^1 = E(\mathbf{z}_{t+1} | \mathbf{x}_{1:t}, \mathbf{y}_{1:t}) = \mathbf{A}\mathbf{u}_t \\ \mathbf{d}_t^i = E(\mathbf{z}_{t-i+2} | \mathbf{x}_{1:t}, \mathbf{y}_{1:t}) = \mathbf{u}_t^{i-1} \\ \mathbf{D}_t^{11} = \text{cov}(\mathbf{z}_{t+1}, \mathbf{z}_{t+1} | \mathbf{x}_{1:t}, \mathbf{y}_{1:t}) = \mathbf{A}\mathbf{V}_t\mathbf{A}^T + \Sigma_\mathbf{Q} \\ \mathbf{D}_t^{1i} = \text{cov}(\mathbf{z}_{t+1}, \mathbf{z}_{t-i+2} | \mathbf{x}_{1:t}, \mathbf{y}_{1:t}) = \mathbf{A} \begin{bmatrix} \mathbf{V}_t^{1(i-1)} \\ \vdots \\ \mathbf{V}_t^{L(i-1)} \end{bmatrix}, \\ \mathbf{D}_t^{i1} = \left(\mathbf{D}_t^{1i}\right)^T \\ \mathbf{D}_t^{ij} = \text{cov}\left(\mathbf{z}_{t-i+2}, \mathbf{z}_{t-j+2} | \mathbf{x}_{1:t}, \mathbf{y}_{1:t}\right) = \mathbf{V}_t^{(i-1)(j-1)} \end{cases} \quad (A19)$$

where $i = 2, 3, \cdots, L+1; j = 2, 3, \cdots, L+1$, and then the following distribution was calculated.

$$P(\mathbf{z}_{t-L+1} | \mathbf{z}_{t+1}, \mathbf{z}_t, \cdots, \mathbf{z}_{t-L+2}, \mathbf{x}_{1:T}, \mathbf{y}_{1:T}) = P(\mathbf{z}_{t-L+1} | \mathbf{z}_{t+1}, \mathbf{z}_t, \cdots, \mathbf{z}_{t-L+2}, \mathbf{x}_{1:t}, \mathbf{y}_{1:t}) = N(\mathbf{r}_t, \mathbf{R}_t). \quad (A20)$$

According to the knowledge of conditional probability [34] and Appendix C. The calculation of its mean and variance is shown in (A21).

$$r_t = d_t^{L+1} + \begin{bmatrix} D_t^{1(L+1)} \\ \vdots \\ D_t^{L(L+1)} \end{bmatrix}^T \begin{bmatrix} D_t^{11} & \cdots & D_t^{11} \\ \vdots & \ddots & \vdots \\ D_t^{L1} & \vdots & D_t^{LL} \end{bmatrix}^{-1} \begin{bmatrix} z_{t+1} - d_t^1 \\ \vdots \\ z_{t-L+2} - d_t^L \end{bmatrix},$$

$$R_t = D_t^{(L+1)(L+1)} - \begin{bmatrix} D_t^{1(L+1)} \\ \vdots \\ D_t^{L(L+1)} \end{bmatrix}^T \begin{bmatrix} D_t^{11} & \cdots & D_t^{11} \\ \vdots & \ddots & \vdots \\ D_t^{L1} & \cdots & D_t^{LL} \end{bmatrix}^{-1} \begin{bmatrix} D_t^{1(L+1)} \\ \vdots \\ D_t^{L(L+1)} \end{bmatrix}.$$

(A21)

The posterior probability of the latent variable $[z_{t+1}, z_t, \cdots, z_{t-L+1}]$ was obtained with respect to the variable $x_{1:T}, y_{1:T}$ as shown in (A22).

$$P(z_{t+1}, z_t, \cdots, z_{t-L+1}|x_{1:T}, y_{1:T})$$
$$= P(z_{t+1}, z_t, \cdots, z_{t-L+2}|x_{1:T}, y_{1:T}) P(z_{t-1}|z_{t+1}, z_t, \cdots, z_{t-L+2}, x_{1:T}, y_{1:T})$$
$$= N(h_t, H_t) = N\left(\begin{bmatrix} h_t^1 \\ \vdots \\ h_t^{L+1} \end{bmatrix}, \begin{bmatrix} H_t^{11} & \cdots & H_t^{1(L+1)} \\ \vdots & \ddots & \vdots \\ H_t^{(L+1)1} & \cdots & H_t^{(L+1)(L+1)} \end{bmatrix} \right).$$

(A22)

The mean and variance of the distribution were calculated as shown in (A23).

$$h_t = \begin{bmatrix} m_{t+1} \\ K \begin{bmatrix} m_{t+1}^1 - d_t^1 \\ \vdots \\ m_{t+1}^L - d_t^L \end{bmatrix} + d_t^{L+1} \end{bmatrix},$$

$$H_t = \begin{bmatrix} M_{t+1} & M_{t+1} K^T \\ K M_{t+1} & K M_{t+1} K^T + R_t \end{bmatrix} \text{ where } K = \begin{bmatrix} D_t^{1(L+1)} \\ \vdots \\ D_t^{L(L+1)} \end{bmatrix}^T \begin{bmatrix} D_t^{11} & \cdots & D_t^{11} \\ \vdots & \ddots & \vdots \\ D_t^{L1} & \vdots & D_t^{LL} \end{bmatrix}^{-1}.$$

(A23)

According to the Bayesian smoothing rule [31], the smooth distribution of the latent variable was obtained, as shown in (A24).

$$z_t, z_{t-1}, \cdots, z_{t-L+1} | x_{1:T}, y_{1:T} \sim N(m_t, M_t).$$

(A24)

The calculation of its mean and variance is shown in (A25).

$$m_t = \begin{bmatrix} h_t^2 \\ \vdots \\ h_t^{L+1} \end{bmatrix} \quad M_t = \begin{bmatrix} H_t^{22} & \cdots & H_t^{2(L+1)} \\ \vdots & \ddots & \vdots \\ H_t^{(L+1)2} & \cdots & H_t^{(L+1)(L+1)} \end{bmatrix}.$$

(A25)

Appendix C. Properties of Gaussian Distribution

Definition A1. (*Gaussian distribution*) *A random variable* $x \in R^n$ *has a Gaussian distribution with mean* $m \in R^n$ *and covariance* $P \in R^{n \times n}$ *if its probability density has the form.*

$$N(x|m, P) = \frac{1}{(2\pi)^{n/2} |P|^{1/2}} \exp\left(-\frac{1}{2}(x-m)^T P^T (x-m)\right),$$

(A26)

where $|P|$ is the determinant of the matrix P.

Lemma A1. (*Joint distribution of Gaussian variables*) *If random variables* $x \in R^n$ *and* $y \in R^m$ *have the Gaussian probability distributions.*

$$x \sim N(m, P),$$
$$y|x \sim N(Hx + u, R).$$

(A27)

then the joint distribution of **x**,**y** and the marginal distribution of **y** are given as (A28).

$$\begin{pmatrix} \mathbf{x} \\ \mathbf{y} \end{pmatrix} \sim \mathbf{N}\left(\begin{pmatrix} \mathbf{m} \\ \mathbf{Hm} + \mathbf{u} \end{pmatrix}, \begin{pmatrix} \mathbf{P} & \mathbf{PH}^T \\ \mathbf{HP} & \mathbf{HPH}^T + \mathbf{R} \end{pmatrix}\right),$$
$$\mathbf{y} \sim \mathbf{N}(\mathbf{Hm} + \mathbf{u}, \mathbf{HPH}^T + \mathbf{R}).$$
(A28)

Lemma A2. *(Conditional distribution of Gaussian variables) If the random variables **x** and **y** have the joint Gaussian probability distribution.*

$$\begin{pmatrix} \mathbf{x} \\ \mathbf{y} \end{pmatrix} \sim \mathbf{N}\left(\begin{pmatrix} \mathbf{a} \\ \mathbf{b} \end{pmatrix}, \begin{pmatrix} \mathbf{A} & \mathbf{C} \\ \mathbf{C}^T & \mathbf{B} \end{pmatrix}\right).$$
(A29)

then the marginal and conditional distributions of **x** and **y** are given as follows:

$$\begin{aligned} \mathbf{x} &\sim \mathbf{N}(\mathbf{a}, \mathbf{A}), \\ \mathbf{y} &\sim \mathbf{N}(\mathbf{b}, \mathbf{B}), \\ \mathbf{x}|\mathbf{y} &\sim \mathbf{N}(\mathbf{a} + \mathbf{CB}^{-1}(\mathbf{y} - \mathbf{b}), \mathbf{A} - \mathbf{CB}^{-1}\mathbf{C}^T), \\ \mathbf{y}|\mathbf{x} &\sim \mathbf{N}(\mathbf{b} + \mathbf{C}^T\mathbf{A}^{-1}(\mathbf{x} - \mathbf{a}), \mathbf{B} - \mathbf{C}^T\mathbf{A}^{-1}\mathbf{C}). \end{aligned}$$
(A30)

References

1. Chen, Z.W.; Cao, Y.; Ding, S.X.; Zhang, K.; Koenings, T.; Peng, T.; Yang, C.H.; Gui, W.H. A Distributed Canonical Correlation Analysis-Based Fault Detection Method for Plant-Wide Process Monitoring. *IEEE Trans. Ind. Inf.* **2019**, *15*, 2710–2720. [CrossRef]
2. Qin, S.J. Survey on data-driven industrial process monitoring and diagnosis. *Annu. Rev. Control* **2012**, *36*, 220–234. [CrossRef]
3. Venkatsubramanian, V.; Rengaswamy, R.; Yin, K.; Kavuri, S.N. A review of process fault detection and diagnosis Part I: Quantitative model-based methods. *Comput. Chem. Eng.* **2003**, *27*, 293–311. [CrossRef]
4. Venkatasubramanian, V.; Rengaswamy, R.; Kavuri, S.N. A review of process fault detection and diagnosis Part II: Quantitative model and search strategies. *Comput. Chem. Eng.* **2003**, *27*, 313–326. [CrossRef]
5. Venkatasubramanian, V.; Rengaswamy, R.; Kavuri, S.N.; Yin, K. A review of process fault detection and diagnosis Part III: Process history based methods. *Comput. Chem. Eng.* **2003**, *27*, 327–346. [CrossRef]
6. Chen, H.; Jiang, B.; Ding, S.; Huang, B.S. Data-driven fault diagnosis for traction systems in high-speed trains: A survey, challenges, and perspectives. *IEEE Trans. Intell. Transp. Syst.* **2020**. aarly access. [CrossRef]
7. Kim, D.S.; Lee, I.B. Process monitoring based on probabilistic PCA. *Chemometr. Intell. Lab. Syst.* **2003**, *67*, 109–123. [CrossRef]
8. Kong, X.Y.; Cao, Z.H.; An, Q.S.; Gao, Y.B.; Du, B.Y. Quality-Related and Process-Related Fault Monitoring With Online Monitoring Dynamic Concurrent PLS. *IEEE Access* **2018**, *6*, 59074–59086. [CrossRef]
9. Chen, H.; Jiang, B.; Chen, W.; Yi, H. Data-driven Detection and Diagnosis of Incipient Faults in Electrical Drives of High-Speed Trains. *IEEE Trans. Ind. Electron.* **2019**, *66*, 4716–4725. [CrossRef]
10. Ge, Z.Q. Process Data Analytics via Probabilistic Latent Variable Models: A Tutorial Review. *Ind. Eng. Chem. Res.* **2018**, *57*, 12646–12661. [CrossRef]
11. Si, Y.B.; Wang, Y.Q.; Zhou, D.H. Key-Performance-Indicator-Related Process Monitoring Based on Improved Kernel Partial Least Squares. *IEEE Trans. Ind. Electron.* **2021**, *68*, 2626–2636. [CrossRef]
12. Jiang, Y.; Fan, J.L.; Chai, T.; Lewis, F.L.; Li, J.N. Tracking Control for Linear Discrete-Time Networked Control Systems With Unknown Dynamics and Dropout. *IEEE Trans. Neural Netw. Learn. Syst.* **2018**, *29*, 4607–4620. [CrossRef] [PubMed]
13. Ge, Z.Q.; Chen, X.R. Dynamic Probabilistic Latent Variable Model for Process Data Modeling and Regression Application. *IEEE Trans. Control Syst. Technol.* **2019**, *27*, 323–331. [CrossRef]
14. Kruger, U.; Zhou, Y.Q.; Irwin, G.W. Improved principal component monitoring of large-scale processes. *J. Process Control* **2004**, *14*, 879–888. [CrossRef]
15. Chen, J.H.; Liu, K.C. On-line batch process monitoring using dynamic PCA and dynamic PLS models. *Chem. Eng. Sci.* **2002**, *57*, 63–75. [CrossRef]
16. Li, G.; Qin, S.J.; Zhou, D.H. A New Method of Dynamic Latent-Variable Modeling for Process Monitoring. *IEEE Trans. Ind. Electron.* **2014**, *61*, 6438–6445. [CrossRef]
17. Ge, Z.Q.; Song, Z.H.; Gao, F.R. Review of Recent Research on Data-Based Process Monitoring. *Ind. Eng. Chem. Res.* **2013**, *52*, 3543–3562. [CrossRef]
18. Ge, Z.Q.; Chen, X.R. Supervised linear dynamic system model for quality related fault detection in dynamic processes. *J. Process Control* **2016**, *44*, 224–235. [CrossRef]
19. Zhurabok, A.N.; Shumsky, A.E.; Pavlov, S.V. Diagnosis of Linear Dynamic Systems by the Nonparametric Method. *Autom. Remote Control* **2017**, *78*, 1173–1188. [CrossRef]

20. Ren, X.M.; Rad, A.B.; Chan, P.T.; Lo, W.L. Online identification of continuous-time systems with unknown time delay. *IEEE Trans. Autom. Control* **2005**, *50*, 1418–1422. [CrossRef]
21. Yang, Z.J.; Iemura, H.; Kanae, S.; Wada, K. Identification of continuous-time systems with multiple unknown time delays by global nonlinear least-squares and instrumental variable methods. *Automatica* **2007**, *43*, 1257–1264. [CrossRef]
22. Yalin, W.; Haibing, X.; Xiaofeng, Y. Multi-delay identification by trend-similarity analysis and its application to hydrocracking process. *Chem. Eng. News* **2018**, *69*, 1149–1157.
23. Drakunov, S.V.; Perruquetti, W.; Richard, J.P.; Belkoura, L. Delay identification in time-delay systems using variable structure observers. *Annu. Rev. Control* **2006**, *30*, 143–158. [CrossRef]
24. Shen, B.B.; Ge, Z.Q. Supervised Nonlinear Dynamic System for Soft Sensor Application Aided by Variational Auto-Encoder. *IEEE Trans. Instrum. Meas.* **2020**, *69*, 6132–6142. [CrossRef]
25. Prosper, H.B. Deep Learning and Bayesian Methods. In Proceedings of the 12th Conference on Quark Confinement and the Hadron Spectrum, Thessaloniki, Greece, 29 August–3 September 2016.
26. Alcala, C.F.; Qin, S.J. Reconstruction-based contribution for process monitoring. *Automatica* **2009**, *45*, 1593–1600. [CrossRef]
27. Lataire, J.; Chen, T. Transfer function and transient estimation by Gaussian process regression in the frequency domain. *Automatica* **2016**, *72*, 217–229. [CrossRef]
28. Panic, B.; Klemenc, J.; Nagode, M. Improved Initialization of the EM Algorithm for Mixture Model Parameter Estimation. *Mathematics* **2020**, *8*, 373. [CrossRef]
29. Chen, J.Y.; Gui, W.H.; Dai, J.Y.; Jiang, Z.H.; Chen, N.; Li, X. A hybrid model combining mechanism with semi-supervised learning and its application for temperature prediction in roller hearth kiln. *J. Process Control* **2021**, *98*, 18–29. [CrossRef]
30. Huang, X.K.; Tian, X.Y.; Zhong, Q.; He, S.W.; Huo, C.B.; Cao, Y.; Tong, Z.Q.; Li, D.C. Real-time process control of powder bed fusion by monitoring dynamic temperature field. *Adv. Manuf.* **2020**, *8*, 380–391. [CrossRef]
31. Egorova, E.G.; Rudakova, I.V.; Rusinov, L.A.; Vorobjev, N.V. Diagnostics of sintering processes on the basis of PCA and two-level neural network model. *J. Chemom.* **2018**, *32*, 2. [CrossRef]
32. De Waele, S.; Broersen, P.M.T. Order selection for vector autoregressive models. *IEEE Trans. Signal Process.* **2003**, *51*, 427–433. [CrossRef]
33. Wang, Z.Y.; Liang, J. A JITL-Based Probabilistic Principal Component Analysis for Online Monitoring of Nonlinear Processes. *J. Chem. Eng. Jpn.* **2018**, *51*, 874–889. [CrossRef]
34. Särkkä, S. *Bayesian Filtering and Smoothing*; Cambridge University Press: London, UK, 2013.

MDPI
St. Alban-Anlage 66
4052 Basel
Switzerland
Tel. +41 61 683 77 34
Fax +41 61 302 89 18
www.mdpi.com

Machines Editorial Office
E-mail: machines@mdpi.com
www.mdpi.com/journal/machines